Intelligent Systems Reference Library

Volume 130

Series editors

Janusz Kacprzyk, Polish Academy of Sciences, Warsaw, Poland
e-mail: kacprzyk@ibspan.waw.pl

Lakhmi C. Jain, University of Canberra, Canberra, Australia;
Bournemouth University, UK;
KES International, UK
e-mail: jainlc2002@yahoo.co.uk; jainlakhmi@gmail.com
URL: http://www.kesinternational.org/organisation.php

About this Series

The aim of this series is to publish a Reference Library, including novel advances and developments in all aspects of Intelligent Systems in an easily accessible and well structured form. The series includes reference works, handbooks, compendia, textbooks, well-structured monographs, dictionaries, and encyclopedias. It contains well integrated knowledge and current information in the field of Intelligent Systems. The series covers the theory, applications, and design methods of Intelligent Systems. Virtually all disciplines such as engineering, computer science, avionics, business, e-commerce, environment, healthcare, physics and life science are included.

More information about this series at http://www.springer.com/series/8578

Grzegorz J. Nalepa

Modeling with Rules Using Semantic Knowledge Engineering

 Springer

Grzegorz J. Nalepa
AGH University of Science and Technology
Kraków
Poland

ISSN 1868-4394 ISSN 1868-4408 (electronic)
Intelligent Systems Reference Library
ISBN 978-3-319-88294-9 ISBN 978-3-319-66655-6 (eBook)
https://doi.org/10.1007/978-3-319-66655-6

Printed on acid-free paper

This Springer imprint is published by Springer Nature
The registered company is Springer International Publishing AG
The registered company address is: Gewerbestrasse 11, 6330 Cham, Switzerland

To DeXa, my truest companion
To my Father, the first engineer
To my Mother, the first researcher

Contents

Part II Formal Models for Rules

Abbreviations

AI	Artificial Intelligence
ALSV(FD)	Attributive Logic with Set Values over Finite Domains
AMVCBC	Adaptable Model, View, Context-Based Controller
ARD	Attribute Relationship Diagrams
BPM	Business Process Management
BPMN	Business Process Model and Notation
BPMS	Business Process Management System
BR	Business Rules
BRA	Business Rules Approach
BRMS	Business Rules Management System
CAS	Context-Aware Systems
CEP	Complex Event Processing
CKE	Collaborative Knowledge Engineering
CLIPS	C Language Integrated Production System
DAAL	Description And Attributive Logic
DL	Description Logics
DMN	Decision Model and Notation
DSS	Decision Support System
DSL	Domain Specific Language
DT	Decision Table
ES	Expert Systems
FOL	First Order Logic
HaDEs	HeKatE Design Environment
HalVA	HeKatE Verification and Analysis
HaThoR	HeKatE Translation framework for Rules
HeaRT	Hybrid Rule RunTime
HeaRTDroid	HeaRT for Android
HeKatE	Hybrid Knowledge Engineering
HMR	HeKatE Meta Representation
HQEd	HeKatE Qt Editor

xi

IS	Intelligent System
KB	Knowledge Base
KBS	Knowledge-Based Systems
KE	Knowledge Engineering
KR	Knowledge Representation
LHS	Left-Hand Side
Loki	Logic-based Wiki
MDA	Model-Driven Architecture
MDD	Model-Driven Development
MDE	Model-Driven Engineering
MOF	Meta Object Facility
MVC	Model View Controller
OCL	Object Constraint Language
OMG	Object Management Group
OPS5	Official Production System
OWL	Web Ontology Language
PlWiki	Prolog-based Wiki
PL	Predicate Logic
RDF	Resource Description Framework
RHS	Right-Hand Side
RBS	Rule-Based Systems
SE	Software Engineering
SKE	Semantic Knowledge Engineering
UML	Unified Modeling Language
VARDA	Visual ARD+ Rapid Development Alloy
XMI	XML Metadata Interchange
XTT2	eXtended Tabular Trees, Version 2

List of Figures

Introduction

Knowledge is a deadly friend
when no one sets the rules

—KING CRIMSON

Sorrow never comes too late
and happiness too swiftly flies (...)
where ignorance is bliss
it is folly to be wise

—THOMAS GRAY

Motivation

This book is dedicated to a systematic presentation of my research results in the area of intelligent rule-based systems. I conducted this work with my Team, GEIST[1] together with colleagues, students at AGH UST in Kraków, with collaboration with many colleagues in Europe, in the years 2008–2016. My original research background combines several areas. As I have graduated from automation and robotics, I have a basic knowledge in automatic control, dynamic systems, as well as different hardware platforms, including embedded systems. On the other hand, in my Ph.D. thesis and afterwards it, I worked mainly in computer science and artificial intelligence. I focused on symbolic knowledge representation, reasoning, and knowledge engineering. In the group that I worked in, there was also strong research in the fields of data bases, software engineering, and formal methods. After my Ph.D., with my team I explored several new areas, including Semantic Web, Business Process Management, and more recently context-aware

---

[1] See the webpage http://geist.re.

systems. These topics influenced my research, emphasizing aspects of modeling, design and analysis of intelligent systems.

The core of my research in these years was related to rule-based knowledge representation methods. This was the main research thread, that I explored and extended in the areas mentioned above. Therefore, the single most important focus of this book is on *rules*. As such, this book aims at providing an academic, yet practical perspective on knowledge engineering with rule-based systems. The focus of the book is on the *Semantic Knowledge Engineering* approach (SKE) I developed. In the next section I introduce the structure and contents of the book. As the book contains the synthesis of my most important works in last eight years, I discuss specific results and publications that were the foundation for the specific chapters. These results were often delivered within the scope of the research projects that I was involved in. After that, I summarize the original and specific perspective of this book and the targeted audience. The book is enriched with selected examples of designs of intelligent systems that I briefly describe afterwards. After 20 years in research and development, it is clear to me that one of the most important factors in doing it, are the right people to cooperate with. I discuss in more detail particular collaborators and contributors in the last section of this introduction. Without these persons, this book might not have come into existence.

Structure of the Book

Organizing this material was not a trivial task for me. When I worked on specific topics there has always been a "big picture" of how to possibly incorporate research threads in a diverse, yet single and coherent structure. However, not all research produced the results that I wished, and clearly the resulting structure was different to what I had originally envisioned. While working on this book I took my habilitation monograph [1] as a starting point. I selected and then reworked the presentation of the most important results that provided the foundation for my post habilitation work.

I decided to partition the contents of the book into three different parts. The first part presents selected aspects of the state of the art in the area of rules in AI and software engineering (in a broad sense). In Chap. 1, I aimed at emphasizing the roots of my work in classic AI and knowledge representation. Chapter 2 identifies challenges as well as methods relevant for the engineering practice. Then, in Chap. 3. I discuss these recent applications of rules which are interesting for me, and relevant for the rest of the book. These are related to software engineering, business systems, and web applications.

The core part of the book includes several related models presented in the second part. Clearly, the most important one is XTT2, based on ALSV(FD) presented in Chap. 4. This main model was augmented with several others. The inference control in the XTT2 decision network can be delegated to a BP engine, as discussed in Chap. 5. The design of the structure on the XTT2 knowledge base can be efficiently supported and simplified by the ARD+ method, as presented in Chap. 6. Two important extensions of the base XTT2 formalism were developed. The first is related to uncertain (or incomplete) knowledge, and is covered in Chap. 7. The second one is oriented at providing knowledge interoperability, and is discussed in Chap. 8.

The models from the second part were developed and used in several specific domains discussed in the third part. This part begins with Chap. 9 which presents the SKE approach, including the specific design process for intelligent systems, and tools supporting it. After it, a number of case studies where SKE was used, are presented. I begin with the discussion of knowledge interoperability techniques in Chap. 10. This work shows how XTT2 -based systems can be translated to the languages used in classic rule-based system shells. The translation uses the model from Chap. 8. Then, I move to software engineering, to demonstrate in Chap. 11, how XTT2 can be used with UML to design software. In this theme, in Chap. 12, I show the application of XTT2 to the automation of test cases in software testing. In Chap. 13 I remain in the area of business applications where I present practical integration of business processes with rules encoded with XTT2. To this end the model previously introduced in Chap. 5 was used. The two next chapters consider knowledge representation on the Web. In Chap. 14 I discuss the DAAL language that allows for an integration of ALSV(FD) with Description Logics, opening up opportunities for using XTT2 rules in the Semantic Web applications. In Chap. 15 the LOKI platform is introduced which provides a Collaborative Knowledge Engineering environment. It is a web-based engineering environment based on the concept of Semantic Wikis. The last two chapters describe case studies closer to specific hardware platforms. In Chap. 16 I discuss the use of the SKE design and application of XTT2 rules to the modeling of control logic for mobile robots. Then in Chap. 17 I demonstrate how XTT2 rules can be put in operation on mobile platforms to support reasoning in context-aware systems. I use the model previously introduced in Chap. 7 to handle uncertainty which is omnipresent in such systems.

The contents of the book form a synthesis of selected important results that were partially made available in some previous works described next.

Previous Related Research

In this section I mention the works conducted, and papers published after the habilitation monograph [1], that are related to this book.

The first part of the book gives an introduction to the domain, as well as a certain perspective on it. The starting point for Chaps. 1–3 was the Chap. 2 of [1]. Clearly, the discussion was much extended and focused. It was also enriched and illustrated by selected examples and analysis from [2–4].

The second part presents the formal models I developed for rule-based systems. Chapter 4 is a reworked and extended version of the Chap. 4 of [1]. Chapter 5 discusses the model I developed with Dr. Krzysztof Kluza in his Ph.D. thesis [3]. It is based on the discussion in [5]. The results of these works were also presented in our papers [6–8]. Chapter 6 presents the formalized version of the ARD+ method. This chapter is partially based on parts of Chap. 5 of [1], our report [9], Chap. 6 of [3], and paper [10]. Chapter 7 discusses the model I developed with Dr. Szymon Bobek in his Ph.D. thesis [4]. It is based on the results presented in [11], and in [12]. Chapter 8 discusses the model I built with Dr. Krzysztof Kaczor in his Ph.D. thesis [2]. This chapter uses and extends the results from papers [13–15].

The discussion of case studies in the third part begins with the presentation of the SKE approach in Chap. 9. That chapter extends the Chap. 5 of [1], and includes more recent work on the HaDEs + toolset. Chapter 10 contains the presentation of selected practical results of the thesis [2]. It is also partially based on the paper [13]. Then Chap. 11 is a reworked version of Chap. 7 of [1]. Chapter 12 is a synthesis of results regarding software, preliminary published in [16–18]. In Chap. 13 selected practical results of the thesis [3] are discussed, also partially presented in [5]. Chapter 14 is partially based on the results from Chap. 6 of [1]. Chapter 15 presents a synthesis of work on semantic wikis, also partially published in [19–22]. Chapter 16 is based on the thesis [23] and on an extended version of [24]. Finally, the work in Chap. 17 is related to research discussed in [25–28]. The Chapter presents selected original results discussed in the Ph.D. Thesis [4]. Moreover, it presents some results also discussed in [11, 12].

The results presented in the book were often achieved in the scope of research projects I was involved in together with my team and my collaborators.

Projects

The first project that gave early ideas for my work was Regulus (KBN 8 T11C 019 17), where tabular knowledge representation methods were considered. An important first project was Mirella (KBN 4 T11C 027 24), as it provided the very first version of the XTT method. The primary motivation for this work and initial results were first delivered in the HeKatE project (N516 024 32/2878). HeKatE (Hybrid Knowledge Engineering) resulted in number of methods, including ALSV (FD) and ARD+. The objectives were to investigate the possible bridging of knowledge engineering and software engineering. The second important project was BIMLOQ (N516 422338). The main aim of the project was to build a declarative model for business processes, including business rules specification, with an emphasis on the analysis and optimization of those processes. In that project preliminary work on BP integration with semantic wiki was carried out. The next project was Parnas (N516 481240) which was aimed at developing tools for inference control and quality analysis in modularized rule bases, and provided the final formalized version of the XTT2 method.

All of the above mentioned projects regarded fundamental research in the area of computer science. An important step in the application of selected results was the Prosecco project (NCBiR PBS1/B3/14/2012). In this project my team had the opportunity to apply methods and tools we developed in the previous projects. This included the work on business processes, business rules, and semantic web methods.

Finally, I supervised three Preludium grants from NCN supporting work related to the theses of my Ph.D. students. This included the SaMURAI project (UMO-2011/03/N/ST6/00886) regarding the work on semantic interoperability of rules, that supported the thesis [2]. Then the HiBuProBuRul Project (UMO-2011/03/N/ST6/00909) supported the work on the integration of business rules and processes considered in the thesis [3]. Finally, the KnowMe Project (UMO-2014/13/N/ST6/01786) supported the thesis [3].

More information on the mentioned projects can be found on the webpage of the GEIST team at http://geist.re. I wish to thank NCN and NCBR for financing these research projects.

Perspective and Intended Audience

The domain of rule-based systems is a rich research field with a long tradition. Clearly, there are now many textbooks in this area. There is also a lot of important original research conducted in last 40 years. I discuss a selection of these in the first part of the book.

The perspective of this book, is on developing intelligent systems based on rules. Today this area is on the intersection of original knowledge engineering and software engineering, we aim at combining these approaches by appropriate models and tools. I am interested in the practical design of rule bases, using visual representations. Moreover, my focus is on formal models knowledge for the design and analysis of rule-based systems. Finally, a very important aspect, is the integration of rule-based systems with selected current methods and tools. I consider mainly selected areas in software engineering and business systems.

From my perspective, rules can serve as a high level decision and control logic. Systems can incorporate such logic in a heterogeneous manner. It can be combined with business processes for inference control, and with ontologies for extending vocabulary or type systems. The rule-based core can be considered as a dynamic system, where rules describe the transition between system states, thus capturing the dynamics of the system.

The intended audience of this book are researchers and practitioners in the field of intelligent systems and software engineering. From a conceptual point of view, the second part of the book contains formal models, which are the results of fundamental research. The applied research results, including practical applications of these models are described in the third part of the book. The book could also prove to be useful for Ph.D. students in the areas related to intelligent systems, software engineering, and business systems.

Examples in the Book

In the book I use a number of examples to illustrate the main methods and concepts. Some of them are general enough to be used in several chapters, while others are more specific so are used to demonstrate only selected features. Most of the examples are related to three main system cases described below.

The BOOKSTORE system case is first introduced in Chap. 4, and then used in Chaps. 14, and 15. The BOOKSTORE system is a simple rule-based book recommender. It is used to demonstrate the features of the ALSV(FD) and XTT2 . Furthermore, I use it to discuss applications of the SKE approach to the Semantic Web and CKE . The case was developed together with Weronika T. Adrian and first used in [1].

The PLOC system case is the extension of the previous case PLI used in [1]. The PLI case was first modeled by Szymon Książek. It is a rule-based decision support system that calculates car insurance rate, based on historical data in Poland. PlOC was extended and formalized by Krzysztof Kaczor and Krzysztof Kluza in their Ph.D. theses. The case is used in several chapters, including Chaps. 5, 6, 8, and 10, 13, to demonstrate the ARD+ design method, then the integration of rules and processes, and finally rule base interoperability.

The Cashpoint system case was also previously modeled with XTT2 and used in [1]. It is a simple model of an ATM system for cash withdrawal. The specification considered by us is based on the system presented by Denvir et al. in [29] and by Poizat and Royer in [30]. In this book it is used to demonstrate HaDEs + tools in Chap. 9, and the integration of the SKE design process with UML in Chap. 11.

Moreover, in Chaps. 7, and 17 several small cases from [4], developed with Szymon Bobek, are used in the discussion of context-aware systems.

Finally, during the HeKatE project, several cases for the SKE approach were developed. These are available online at http://ai.ia.agh.edu.pl/wiki/hekate:cases:start.

Acknowledgments

This book contains a synthesis of results of my research in last eight years. After all these all these years of doing research, I very much realize, that besides research topics and results, what matters a lot are the people I work with. I was very lucky in these years to meet and cooperate with many unique and talented persons. Moreover, research is done by individuals. Even if they form teams, their individuality is what makes these teams successful and their research fruitful. Without a number of individuals my research and this book would not be possible.

This books summarizes almost a decade of research, started a few years after my Ph.D. [31] in 2004. First of all I wish to thank Prof. Antoni Ligeza, who was the supervisor of my Ph.D. thesis, who has been a good colleague of mine since than. His concept of formalized algebra for rules laid foundation for the Ψ-trees and the original XTT method. Our shared work then included the ALSV(FD) logic which allowed me to formalize XTT2 in [1] and in [32]. We worked together on many of the previously mentioned projects, including Regulus, HeKatE, and BIMLOQ.

These projects would not be possible without a great number of talented young people. My special thanks and gratitude go to my Ph.D. students. I would like also thank them for their permission to include some examples and studies from their theses in this book. Krzysztof Kaczor worked for many years on the HQEd editor for XTT2 and we formulated a formalized model for rule interoperability [2]. Krzysztof Kluza extended my early work on integration of business rules and processes [33]. We provided a formalized model for such an integration in [3]. With Szymon Bobek, together we boldly went where no one in our team had gone before. We successfully developed research in context-aware systems, and learned many completely new methods and tools from the area of machine learning and data mining. Building on our previous work on XTT2 and the HEaRT engine, we delivered KNOWME in [4] and several important research papers, Szymon together with Mateusz Ślażyński developed the HEARTDROID engine. Krzysztof Kutt is finishing his thesis on collaborative knowledge engineering. Szymon and Krzysztof also gave me many valuable remarks on this book.

So far, I have supervised over 40 master and bachelor theses. Most of them were closely related to my research. A number of my students developed valuable results and we wrote papers together. Weronika T. Adrian worked with me on DAAL and other topics related to the Semantic Web. Piotr Hołownia developed PlNXT middleware, and then Maciej Makowski and Błażej Biesiada developed it to support XTT2 for programming Mindstorms NXT robots. Krzysztof Kotra developed the prototype of LOKI/PlWIKI, then Janusz Kamieński and Mirosława Ozgowicz enhanced it towards semantic wikis. Marta Woźniak developed the SBVR plugin for Loki and Urszula Ciaputa provided the BPMN plugin. Finally, Olgierd Grodzki worked on context-aware systems, including in-door micro-localization.

I am in debt with many distinguished professors that supported me in the last decade, and I wish to thank them for their support. Prof. Ryszard Tadeusiewicz originally attracted me to AI with his lectures and research, and for all of my years at AGH UST offered me his guidance and continuous support. Prof. Tomasz Szmuc shared knowledge about from software engineering and supported we when I needed it, including the Prosecco project. Prof. Mariusz Flasiński gave me valuable remarks on the "big picture on AI". Prof. Stan Matwin has supported me strongly since the first days of the Polish AI Society (PSSI), and also gave many important ideas about machine learning and my research. Prof. Jerzy Stefanowski, not only supported me with PSSI, but also shared many valuable comments regarding my research and its development towards machine learning. Prof. Paweł Gryboś has been encouraging me to finish writing this book and related papers for a long time —I am very grateful for his support. Last, but very much not least I want to thank Prof. Janusz Kacprzyk, who motivated and supported me since my habilitation, and made this book possible.

I wish to thank my good colleagues who supported me on many levels and from whom I learned a lot.[2] I would like to thank Prof. Marcin Szpyrka. We have been working together for over 17 years. I have always had a great respect and admiration for his passion for research, and endless motivation for his hard work. I also wish to thank him for all his valuable remarks he gave me on this book and the Ph. D. theses of my students. My thanks also go to Prof. José Palma Mendez for his friendliness and openness during all my stays in Spain. This included my recent stay in Murcia where he created a friendly working atmosphere and shared many books that helped me in improving this book. I wish to thank Prof. Jerzy Pamin for all his support and friendship. I very much appreciate all of our meetings, talks, and email conversations. Finally, I very much wish to thank my good friend, Prof. Joachim Baumeister, PD. We have been colleagues and friends for over a decade now. What started as a small collaboration on a startup workshop, turned out to be a great collaboration on the topics of knowledge engineering and software engineering. We developed KESE (http://kese.ia.agh.edu.pl) into a recognized scientific event and built a vibrant scientific community. This has been a very important professional and personal experience for me.

Thank you all!

Kraków 2014—Murcia 2016—Kraków 2017

References

1. Nalepa, G.J.: Semantic Knowledge Engineering. A Rule-Based Approach. Wydawnictwa AGH, Kraków (2011)
2. Kaczor, K.: Knowledge formalization methods for semantic interoperability in rule bases. PhD thesis, AGH University of Science and Technology (2015) (Supervisor: Grzegorz J. Nalepa)
3. Kluza, K.: Methods for Modeling and Integration of Business Processes with Rules. Ph.D. thesis, AGH University of Science and Technology (March 2015) Supervisor: Grzegorz J. Nalepa
4. Nalepa, G.J., Bobek, S.: Rule-based solution for context-aware reasoning on mobile devices. Comput. Sci. Inf. Syst. 11(1), 171–193 (2014)
5. of Business Processes Integrated with Business Rules, F.M.: Krzysztof kluza and grzegorz j. nalepa. Information Systems Frontiers (2016) submitted.
6. Kluza, K., Nalepa, G.J., Lisiecki, J.: Square complexity metrics for business process models. In Mach-Król, M., Pełech-Pilichowski, T. (eds.): Advances in Business ICT. Advances in Intelligent Systems and Computing, vol. 257, pp. 89–107. Springer (2014)

---

[2] On a more private note, I want to warmly thank Dorota and Chris for their support, and their help in improving my English in this book.

7. OMG: Business Process Model and Notation (BPMN): Version 2.0 specification. Technical report formal/2011-01-03, Object Management Group (January 2011)
8. Kluza, K., Kaczor, K., Nalepa, G.J.: Enriching business processes with rules using the Oryx BPMN editor. In Rutkowski, L., et al. (eds.): Artificial Intelligence and Soft Computing: 11th International Conference, ICAISC 2012: Zakopane, Poland, April 29–May 3, 2012. Lecture Notes in Artificial Intelligence, vol. 7268, pp. 573–581. Springer (2012)
9. Nalepa, G.J., Wojnicki, I.: ARD+ a prototyping method for decision rules. method overview, tools, and the thermostat case study. Technical Report CSLTR 01/2009, AGH University of Science and Technology (2009)
10. for Generation, A.M., of Business Processes with Business Rules, D.: Krzysztof kluza and grzegorz j. nalepa. Information and Software Technology (2016) submitted.
11. Bobek, S., Nalepa, G.J.: Uncertain context datamanagement in dynamic mobile environments. Future Gener. Comput. Syst. 66, 110–124 (2017)
12. Bobek, S., Nalepa, G.J.: Uncertainty handling in rule-based mobile contextaware systems. Pervasive and Mobile Computing (2016)
13. Kaczor, K., Nalepa, G.J.: Enabling collaborative modeling of rule bases by semantic interoperability methods. Future Generation Computer Systems (2015) Submitted.
14. Kaczor, K., Nalepa, G.J.: Formalization of production rule representation model for multilevel approach to rule interoperability. Data and Knowledge Engineering (2015) submitted.
15. Kaczor, K., Nalepa, G.J.: Encapsulation-driven approach to interchange of knowledge base structure. Lect. Notes Softw. Eng. 4(1), 66–72 (2016)
16. Nalepa, G.J., Kaczor, K.: Proposal of a rule-based testing framework for the automation of the unit testing process. In: Proceedings of the 17th IEEE International Conference on Emerging Technologies and Factory Automation ETFA 2012, Kraków, Poland, 28 September 2012. (2012)
17. Kutt, K.: Proposal of a rule-based testing framework. Master's thesis, AGH University of Science and Technology (July 2013) Supervisor: G. J. Nalepa.
18. Nalepa, G.J., Kutt, K., Kaczor, K.: Can the generation of test cases for unit testing be automated with rules? In Rutkowski, L., [et al.], eds.: Arti cial Intelligence and Soft Computing: 13th International Conference, ICAISC 2014: Zakopane, Poland. Volume 8468 of Lecture Notes in Arti cial Intelligence., Springer (201 548–599
19. Nalepa, G.J., Kluza, K., Kaczor, K.: Sbvrwiki a web-based tool for authoring of business rules. In: Rutkowski, L., et al. (eds.) Artificial Intelligence and Soft Computing: 14th International Conference, ICAISC 2015. Lecture Notes in Artificial Intelligence, pp. 703–713. Springer, Zakopane, Poland (2015)
20. Nalepa, G.J., Kluza, K., Ciaputa, U.: Proposal of automation of the collaborative modeling and evaluation of business processes using a semantic wiki. In: Proceedings of the 17th IEEE International Conference on Emerging Technologies and Factory Automation ETFA 2012, Kraków, Poland, 28 Sept 2012. (2012)
21. Nalepa, G.J.: Collective knowledge engineering with semantic wikis. J. Univers. Comput. Sci. 16(7), 1006–1023 (2010)
22. Nalepa, G.J.: Loki – semantic wiki with logical knowledge representation. In Nguyen, N.T. (ed.) Transactions on Computational Collective Intelligence III. Lecture Notes in Computer Science, vol. 6560, pp. 96–114. Springer, Berlin (2011)
23. Biesiada, B.: lnxt enhancements proposal. Master's thesis, AGH University of Science and Technology (2011)
24. Nalepa, G.J., Biesiada, B.: Declarative design of control logic for mindstorms NXT with XTT2 method. In Jedrzejowicz, P., Nguyen, N.T., Hoang, K., eds.: Computational Collective Intelligence. Technologies and Applications - Third International Conference, ICCCI 2011, Gdynia, Poland, September 21-23, 2011, Proceedings, Part II. Volume 6923 of Lecture Notes in Computer Science., Springer (2011) 150 159
25. Bobek, S.: Methods for modeling self-adaptive mobile context-aware systems. Ph.D. thesis, AGH University of Science and Technology (April 2016) Supervisor: Grzegorz J. Nalepa

26. Bobek, S.,Nalepa,G.: Compact representation of conditional probability for rule-basedmobile context-aware systems. In: Bikakis, A., Fodor, P., Roman, D. (eds.) Rules on the Web. From Theory to Applications. Lecture Notes in Computer Science. Springer International Publishing (2015)

27. Bobek, S., Dziadzio, S., Jaciów, P., 'la»y«ski, M., Nalepa, G.J.: Understanding Context with ContextViewer Tool for Visualization and Initial Preprocessing of Mobile Sensors Data. In: Modeling and Using Context: 9th International and Interdisciplinary Conference, CONTEXT 2015, Lanarca, Cyprus, November 2- 6,2015. Proceedings. Springer International Publishing, Cham (2015) 77–90

28. Bobek, S., Nalepa, G.J.: Incomplete and uncertain data handling in context-aware rule-based systems with modified certainty factors algebra. In: Bikakis, A., Fodor, P., Roman, D. (eds.) Rules on the Web. From Theory to Applications. Lecture Notes in Computer Science, vol. 8620, pp. 157–167. Springer International Publishing (2014)

29. Denvir, T., Oliveira, J., Plat, N.: The cash-point (ATM) 'Problem'. Form. Asp. Comput. 12(4), 211–215 (2000)

30. Poizat, P., Royer, J.C.: Kadl specification of the cash point case study. Technical report, IBISC, FRE 2873 CNRS - Universite d'Evry Val d'Essonne, France, Genopole Tour Evry 2, 523 place des terrasses de l'Agora 91000 Evry Cedex (2007)

31. Nalepa, G.J.: Meta-Level Approach to Integrated Process of Design and Implementation of Rule-Based Systems. PhD thesis, AGH University of Science and Technology, AGH Institute of Automatics, Cracow, Poland (September 2004)

32. Nalepa, G.J., Lige̜za, A., Kaczor, K.: Formalization and modeling of rules using the XTT2 method. Int. J. Artif. Intell. Tools 20(6), 1107–1125 (2011)

33. Nalepa, G.J.: Proposal of business process and rules modeling with the XTT method. In Negru, V., et al. (eds.) Symbolic and numeric algorithms for scientific computing, 2007. SYNASC Ninth international symposium. September 26–29, Los Alamitos, California, Washington, Tokyo. IEEE Computer Society. IEEE, CPS Conference Publishing Service, pp. 500–506 September 2007

Part I
Domain of Rule-Based Systems

In this first introductory part of the book, we give an overview of selected topics related to practical and current knowledge engineering with rule-based systems. Clearly, we do not to provide a thorough textbook-like treatment of this domain. In fact, a number of very good books are available. A short and concise introduction to expert systems is given in general Artificial Intelligence (AI) books [1–3]. A more detailed treatment of these topics was given in [4–8]. Important books on Intelligent Systems also provide a nice introduction to rules [9, 10] while giving a broader perspective. More recently, the book [1] provided a detailed treatment of formalized approaches to rule-based systems in general. Moreover, the edited book [12] gave a broad overview of recent important challenges in the domain of rules as well as applications of rule-based systems. We use them as a reference and starting point for this part of book, which was partitioned into three chapters.

Chapter 1 starts from the classic perspective of symbolic AI, where rules are treated as one of the main knowledge representation methods [13]. Among simple rules found in the production rule systems a number of other important methods can be identified [14]. In fact, when considering a formalized description some of them (e.g. decision tables and decision trees) can be converted to rules [11]. They will be in our focus in the remaining part of the book.

Chapter 2 addresses the selected important topics in practical knowledge engineering with rules [6, 15]. This includes issues of knowledge acquisition and modeling [16]. Since we focus on formalized methods for rule representation, it becomes possible to formally analyze the contents of rule bases [17]. Another important issue introduced in this chapter is the interchange of rule base knowledge between different systems.

The objective of Chap. 3 is to present important and recent applications of rule-based technologies. A much more detailed review is given in the book [12]. Here, we focus on applications related to business systems in a very broad sense. This includes business rules and processes, as well semantic representations. Moreover, we indicate the importance of rule-based approach in software engineering and context-aware systems.

To summarize, our perspective in this book, including this part, will be on reasoning with rules in intelligent systems. We are interested in the practical design of rule bases with different visual representations. Moreover, we focus on a formal description of knowledge for design and analysis. Finally, we consider the integration of rule-based systems with selected current methods and tools considered in software engineering and business systems.

References

1. Nilsson, N.J.: Artificial Intelligence: A New Synthesis, 1st edn. Morgan Kaufmann Publishers Francisco, CA, USA (1998) Inc., San Francisco (1998)
2. Negnevitsky, M.: Artificial Intelligence. A Guide to Intelligent Systems. Addison-Wesley, Harlow (2002). ISBN 0-201-71159-1
3. Mariusz, F.: Wstęp do sztucznej inteligencji. WN PWN (2011)
4. Giarratano, J.C., Riley, G.D.: Expert Systems. Thomson, Toronto (2005)
5. Genesereth,M.R., Nilsson, N.J.: Logical Foundations for Artificial Intelligence.Morgan Kaufmann Publishers Inc., Los Altos (1987)
6. Gonzalez, A.J., Dankel, D.D.: The Engineering of Knowledge-Based Systems: Theory and Practice. Prentice-Hall Inc, Upper Saddle River (1993)
7. Waterman, D.A.: A Guide to Expert Systems. Addison-Wesley Longman Publishing Co. Inc., Boston (1985)
8. David, J.M., Krivine, J.P., Simmons, R. (eds.): Second Generation Expert Systems. Springer, Secaucus (1993)
9. Torsun, I.S.: Foundations of Intelligent Knowledge-Based Systems. Academic Press, London (1995)
10. Hopgood, A.A.: Intelligent Systems for Engineers and Scientists, 2nd edn. CRC Press, Boca Raton (2001)
11. Ligęza, A.: Logical Foundations for Rule-Based Systems. Springer, Berlin (2006)
12. Giurca, A., Gašević, D., Taveter, K. (eds.) Handbook of Research on Emerging Rule-Based Languages and Technologies: Open Solutions and Approaches. Information Science Reference. Hershey, New York (May (2009)
13. Hendler, J., van Harmelen, F.: The semantic web: webizing knowledge representation. Hand book of Knowledge Representation. Elsevier, New York (2008)
14. van Harmelen, F., Lifschitz, V., Porter, B. (eds.): Handbook of Knowledge Representation. Elsevier Science, Amsterdam (2007)
15. Buchanan, B.G., Shortliffe, E.H. (eds.): Rule-Based Expert Systems. Addison-Wesley Publishing Company, Reading (1985)
16. Scott, A.C.: A Practical Guide to Knowledge Acquisition, 1st edn. Addison-Wesley Longman Publishing Co., Inc, Boston (1991)
17. Vermesan, A.I., Coenen, F. (eds.): Validation and Verification of Knowledge Based Systems. Theory, Tools and Practice. Kluwer Academic Publisher, Boston (1999)

Chapter 1
Rules as a Knowledge Representation Paradigm

Rules are a commonly used and natural way to express knowledge. They have been used for decades in AI, Computer Science, Cognitive Science and other domains. We start by discussing the AI roots of rules in Sect. 1.1 where we elaborate on different kinds and types of rules. We then focus on a more careful treatment of rules in symbolic AI. There, they constitute an approach which allows for the representation of knowledge and basic automated reasoning, see Sect. 1.2. Originally, one of the most important areas for rule applications were expert systems. While this perspective is much broader, we will briefly discuss it in Sect. 1.3. What makes rule-based representation and reasoning particularly interesting is the opportunity for the formalization of rule languages. Selected logic-based formalizations are considered in Sect. 1.4. Then in Sect. 1.5 we present in more detail a family of so-called *Attributive Logics* which play an important role in the remaining part of the book. Based on these concepts we introduce important requirements for a formalized description of rule-based systems in Sect. 1.6. We summarize the chapter in Sect. 1.7.

1.1 Rules in AI

Artificial Intelligence is often considered to be a field that aims to study and design *Intelligent Systems* (IS) [1, 2]. Since the 1970s, a number of successful paradigms for the engineering of such systems have been developed. From the application point of view, some of the most important are those that have focused on decision support. *Expert Systems* (ES) [3, 4], especially those that are rule-based proved to be the most successful. This is mainly due to the fact that rules allow for a powerful and declarative specification of knowledge that is easy to understand and use. The application areas of rules have been expanding. In last decades, they have been used in the area of business where they allow us to define and constrain aspects of business operations;

© Springer International Publishing AG 2018
G.J. Nalepa, *Modeling with Rules Using Semantic Knowledge Engineering*,
Intelligent Systems Reference Library 130,
https://doi.org/10.1007/978-3-319-66655-6_1

these are known as the so-called *Business Rules* (BR) [5]. They are also used as a complementary method of knowledge representation in *Semantic Web*[1] (SW) [6, 7] or *Business Processes Management* (BPM) [8].

The variety of rule applications resulted in the emergence of distinct rule representations. Today many classifications of rules can be found in the literature. Considering the logical aspects of inference with rules a basic distinction could be made between *deductive* and *abductive (derivation)* rules (used in forward and backwards chaining respectively, see Sect. 2.2). Moreover, there exist the concepts of *facts* (rules with no condition) and *constraint* rules (defining certain conditions that must hold) [9]. In [10] an interesting classification is introduced. It is oriented on rule exchange and follows OMG MDA [11]. On the "computation independent"[2] level three general types of rules are identified: integrity, derivation, and reaction. An extended classification is provided by the RULEML organization and is considered in [13]. They identify five distinctive types of rules:

- integrity rules provide certain constraints in the system, they can be found in languages such as SQL, OCL, but also form a separate programming paradigm [14].
- derivation rules allow for the delivery, and inference of some results, mostly in a goal-driven manner, for example in Prolog [15].
- reaction rules are considered as a means of handling event-related actions; important dialects include ECA rules [16], and recently Reaction RuleML [17].
- production rules are one of the most common, they are related to forward chaining and form the core of expert systems.
- transformation rules are important for the rule-based transformation of knowledge, an example of a language that uses such an approach is XSLT [18].

Furthermore, in a business rules approach [5] a high level BR classification scheme is considered which would encompass: terms, facts, and rules. Then the following types of rules are identified: mandatory constraint, guideline, action enabler, computation, inference. The Business Rules Group (BRG) [19] distinguishes between four categories of business rules [19]: definitions of business terms, facts relating terms to each other, constraints, and derivations. Finally, in machine learning [20] and data mining [21] *association rules*, expressing a certain correlation between features (attributes) are considered, as well as classification rules. However, the latter can be simply considered derivations.

While these classifications might not be complete, they are a good starting point. Clearly, the current rich selection of rule representations, and systems using them makes their practical development and use challenging. What will be important for us in this book are the differences in how these representations can be understood and processed in intelligent systems.

---

[1]The idea for the project was originally proposed in [6]. See https://www.w3.org/standards/semanticweb for currents standards.

[2]MDA introduces three general abstraction levels in system modeling: computation independent (CIM), platform-independent (PIM) and platform-specific (PSM) [12].

1.2 Selected Knowledge Representation Methods

Rules are a special case of what is termed in classic AI as *knowledge representation (method)* [22]. In order to build *Knowledge-Based Systems* we need to have some basic understanding of the terms *knowledge* and *knowledge representation*. One of the classic, and now often challenged, understandings of knowledge in epistemology is that it is "justified true belief" (JTB) [23]. In AI, as an engineering field we seek some practical definitions, and many may be found in AI handbooks. We may define knowledge as a "theoretical or practical understanding of a subject or a domain" [24]. In a classic book [9] "knowledge" is understood as a set of propositional beliefs, and as such is declarative. After [9, 25] *Knowledge Representation* (KR) may be understood as "symbolic encoding of propositions believed (by some agent)". Knowledge can be possessed by people as well as stored in computer systems. If it is intended to be processed by machines it has to have a well-defined representation and semantics. In this book we follow the classic logic-based AI, where mathematical logic is the most suitable language for describing and formalizing knowledge.[3]

Many different knowledge representation techniques which are suitable for automated reasoning were developed. Here, we understand *reasoning* as the "manipulation of symbols encoding propositions to produce representations of new propositions" [9]. The difference between different KR methods lies in their expressiveness, representation, etc. See [27] for a state of the art discussion of KR. In this section we mention only those that are deemed to be classic and which are useful for the discussion in the following chapters, in that they focus on rules.

Frames-Based Representation

Frames, first proposed by Minsky in the 1970s, are used to capture and represent knowledge in a frame-based expert system [28]. A frame is a data structure with typical knowledge about a particular object or concept which is described by a collection of slots (see Fig. 1.1). Frame-based representation has many advantages. Frames are suitable for a visual representation, which makes them more transparent and intuitive. Therefore they offer a natural way of representing real world knowledge within a computer system by using objects. Many frame-based systems have been developed. Some of them extended the original idea, e.g. with scripts [9]. One of the weaknesses of the original method was a lack of a clear standard and logic based formalization. However, from today's point of view, frame-based knowledge modeling can be thought of as a prototype of object-oriented modeling.

Conceptual Modeling with Semantic Networks

Semantic Networks (SN) are a classic AI knowledge representation technique developed independently by many researchers in several domains. Their introduction to the fields of AI and computer science is often attributed to Quillian [29]. In the beginning,

---

[3]Clearly this is not the only possibility. One can find a comprehensive resume of discussion on the suitability of logic for knowledge representation in [26].

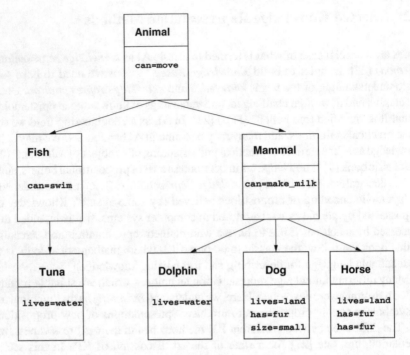

Fig. 1.1 Example of frames

this representation was developed as a way of representing human memory and language understanding. Since then, SN have been applied to many problems involving knowledge representation. A SN is usually represented as a labeled directed graph. Nodes commonly correspond to concepts, or sometimes physical objects or situations or concepts whereas edges (also called links) are used to express the relationships between them (see Fig. 1.2). Thanks to their visual form SN are an expressive knowledge representation. Moreover, the representation itself was not standardized. Originally, neither of them had a uniform logical representation. It is, however, easy to formalize it with logical calculi. However, the recent development of *Description Logics* [30], described later in this chapter is a successful attempt in this area. SN are most commonly used as definitional networks. However, a number of other types can also be identified [31].

Decision Rules

Besides frames and semantic networks, rules were another classic knowledge representation method. They have been in use in AI since 1970s and they have proved to be one of the most successful KR methods. They are so commonly used because they are transparent and easy to understand for humans. A simple rule can be written using an if... then... statement. It can be separated into two parts: the conditional part if..., and the conclusion (decision) part then.... The straightforward interpretation of such a rule is very intuitive: if the conditions are satisfied,

Fig. 1.2 Example of a
semantic network

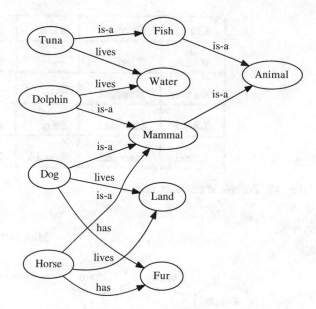

then the conclusions are drawn, or an action is executed. In so-called production
rule systems, described below, the execution (firing) of a rule would imply inserting
(asserting) new facts into the knowledge base. The wide use of rules resulted in many
specific rule representations. The differences between them lie in syntax, the way in
which they are processed, etc. Thus they offer a different expressiveness of a specific rule language. Sets of similar rules can also be represented in forms of decision
tables and trees. From our perspective, rules seem to be one of the most universal
and important KR methods. In fact, decision tables and decision trees can be automatically converted into a set of rules, and vice versa. Such an approach is especially
suitable for making the modeling process of rule bases easier (see Sect. 2.1).

Organizing Rules in Decision Tables

Decision Tables (DTs) [32] provide a tabular form for representing decision making
in forms of conditionals. In fact, they can be interpreted as a means for grouping
and expressing rules that have the same structure. Tables are compact and suitable
for visualization (see Fig. 1.3). In a decision table each row (or column, depending
on the notation, see [33]) can be interpreted as a single rule. Decision tables are
commonly used in different types of systems providing support for decision making
process. The decision tables representation technique was intensively developed by
Vanthienen [32, 34–37]. A very detailed discussion of different forms of decision
tables and their formalization was given by Ligęza in [33].

Inference with Decision Trees

Decision Trees constitute a very transparent method for the representation of decision
making [38] (see Fig. 1.4). The nodes of a tree contain conditional expressions while

Lives	Group	Size	Species
Water	Fish	---	Tuna
Water	Mammal	---	Dolphin
Land	Mammal	Big	Horse
Land	Mammal	Small	Dog

Fig. 1.3 Example of a decision table

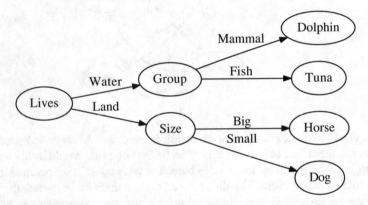

Fig. 1.4 Example of a decision tree

the edges correspond to the value of this expression. The final decision determined by the tree is given in leaves. The main advantage of decision trees is that they constitute a convenient method for the visual representation of a decision process. Trees facilitate the understanding of a decision process, tracing step-by-step or explaining an obtained decision. Some disadvantages of this representation are the redundancy of nodes. Moreover, there are important scalability issues, as there is the possibility of a combinatorial explosion of the nodes number.

Knowledge Bases

Specific KR methods are used to capture and store knowledge in a *Knowledge Base* (KB). In fact, a KBS is composed of two main components [39]:

1. A knowledge base that contains some domain-specific knowledge, but is encoded with the use of domain-independent method (selected KR method), and
2. An automated inference mechanism (engine) that infers new facts based on given facts and rules.

In the case of RBS these two components will in fact be a rule base and a rule-based inference engine.

Having a rule base, a number of automated reasoning techniques can be used by the inference engine. One of the two main inference strategies is *Forward Chaining* (Data Driven) which allows for the drawing of conclusions based upon the input knowledge. The second one is *Backward Chaining* (Goal Driven) which allows for the proving of statements in terms of the current knowledge [24], or demonstrate what facts are needed to satisfy given goals [9]. Both of these are commonly used in expert systems that are discussed below. To make the reasoning process a logical inference, the rule based knowledge needs to be formalized with the use of logic. As a result of this kind of formalization the knowledge base can have a well-defined semantics and expressive power. Moreover, it is possible to provide automated processing with verifiable theoretical characteristics. An in-depth discussion of formalization of rules was given in [33]. In Sect. 1.3 a short overview of important calculi will be given.

1.3 Expert Systems and Rules

Expert systems (ESs) [3, 40–42] are often considered to be one of the most successful applications of AI techniques that emerged in the 1970s. They were intended to help human experts (or substitute them) in solving problems that cannot be easily solved due to their complexity or size. They provided an efficient way of building DSS in a well defined domain (medicine, science, finance, etc.). Over the decades, several generations of expert systems were built [43], including first generation expert systems [40], and second generation [44]. Therefore, different rule-based tools in a large variety of domains were developed. They were applied in different areas including: Chemistry (e.g. CRYSALIS, TQMSTUNE), Electronics (e.g. ACE, IN-ATE, NDS), Medicine (e.g. PUFF, VM, ABEL), Engineering (e.g. REACTOR, DELTA), Geology (e.g. DIPMETER, LITHO), and Computer Science (e.g. PTRANS, BDS, XCON, XSEL) (see [3]). Most of them used similar architecture, which is in fact a special case and extension of the KBS architecture. The architecture of an ES consists of two main components: the knowledge base and the inference engine. The knowledge base allows for storing the domain knowledge with the help of a selected representation method. The inference engine makes use of knowledge as it allows for its processing and thus solving problems formulated by the user. A more general approach, based on this observation, for building such systems was developed. It provides a framework allowing for quick building of rule-based decision support systems which implements the generic (i.e. independent from knowledge) architecture elements. An application of this approach for production rule representation is known as production systems' shells.

Production Systems

Production System Shells [3] are frameworks that support a knowledge engineer in building an ES by providing a generic inference engine and a skeleton (empty) knowledge base. The only action that is performed by a knowledge engineer is to encode the knowledge by using provided rule language. After that, the system

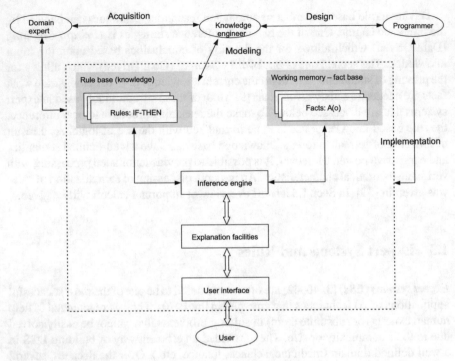

Fig. 1.5 Architecture of a rule-based expert system, after [24]

is ready to perform the inference process in order to solve the requested queries. This approach notably reduces the time needed for implementation of such systems. Their architecture (see Fig. 1.5) provides all the necessary mechanisms that work independently from knowledge concerning the specific domain. It consists of four main components:

Knowledge base – this constitutes a repository for storing knowledge in the form of rules and facts. Sometimes this component is partitioned into two, where one contains only facts, while the other contains only rules.

Inference engine – this provides algorithms that allow for the processing the encoded knowledge in terms of inference tasks and requested goals. Such algorithms must be generic and independent from the knowledge domain.

Explanation mechanism – this allows for the tracing of the reasoning and thus demonstrate why the inference algorithm produced a given conclusion. This is a very useful feature especially when the inference process involves a large number of facts and rules.

User interface – this provides means of user interaction. It allows for the use of above mentioned components, and defines the problems that must be solved by system.

Production system shells use rules as the knowledge representation.[4] The conclusion part of the production rule contains actions that are performed when the conditional part is satisfied. Usually, performing an action has impact on a knowledge base by producing (adding) or removing some information. In turn, any change done in the knowledge base must be taken into account by the inference process regardless of the inference mode. Production system shells often provide two modes of inference: forward chaining and backward chaining from which forward chaining is the basic strategy. The efficient implementation of the inference algorithm was considered to be a challenge in the early rule-based systems where the brute force algorithms were used. This challenge was overcome by the RETE [45] algorithm which was proposed by Forgy in the 1970s. There were many different implementations of such systems, with OPS5 [46] being the most important, while CLIPS [3] was the most commonly known.

OPS5

Official Production System (OPS5) [46] was the first computer program that implemented the idea of problem-solving processes using a set of condition-action rules as knowledge representation and a forward chaining inference engine based upon RETE [45]. Thanks to OPS5, rules became the dominant knowledge representation method in expert systems, and OPS5 became popular among expert system developers. Additionally, in some versions of the language, an invoked action could create a new rule, which made the system capable of "learning".

CLIPS

C Language Integrated Production System (CLIPS) [3, 47] was developed in 1984 at NASA's Johnson Space Center. The main goal was to facilitate the building of expert systems taking into the account portability, low cost, and easy integration into external systems. The inference and representation capabilities of CLIPS are similar but more powerful than those in OPS5. The syntax of the rule language, provided by CLIPS, is based upon LISP syntax which was the base syntax for expert system software tools at that time. Currently, the CLIPS rule language is a multi paradigm programming language that provides support for rule-based, object-oriented and procedural programming.

The wide spread and acceptance of CLIPS resulted in the development of its new incarnation, JESS [48]. JESS was entirely written in the JAVA language by Friedman-Hill at Sandia National Laboratories in Livermore, CA in the 1990s. Consequently, JESS can be used for building web-based software that provides reasoning capabilities.

Rule-based systems allow for representing knowledge in a declarative way and, moreover, they intuitively reflect a natural process of thinking and reasoning. The classic rule-based systems can be thought of as high-level programming tools. In

---

[4]The term "production" is understood technically. When a rule in such a system is fired (run) it can create a new fact which is added to the contents of the knowledge base. We can say that the rule "produced" the fact. Sometimes such rules are simply called "productions".

many areas there are more requirements when it comes to system design and implementation. It is often the case that some sort of formalized method of specification is expected. Moreover, using formal models for rules opens up opportunities for formalized analysis of rule-based systems. Therefore, in the next section a short overview of commonly used formalization methods is given.

1.4 Formal Calculi for Rules

Rules provide a powerful knowledge representation rooted in the concepts of logic [49]. Knowledge formulation within a rule base can be based upon various logical calculi (e.g. those that are propositional, attributive, first-order, or even higher order) or performed on the basis of engineering intuition. In fact, modern rule-based shells mostly follow the classical paradigm where the rule language is a programming solution with no formal definition.

Currently there exists different rule representations that are used in multiple systems. In such systems rules languages are mostly programming solutions with no clear logical formulation However, informal rule representations can cause many problems:

1. *Processing problem* – due to ambiguity, the encoded knowledge can be processed differently compared from that of the intention of the knowledge engineer.
2. *Validation problem* – there is no possibility of checking whether the developed knowledge base meets user requirements [50].
3. *Verification problem* – the lack of precise semantics prevents checking to see whether the knowledge base contains logical errors.
4. *Interoperability problem* – the knowledge translation involving such representation becomes hardly possible due to the ambiguous semantics.

As a matter of fact, it can be observed, that a *formalization* of the rule language is an emerging solution. It can bring a number of practical benefits such as:

- *a reliable design process* – formalized rule language allows for the possibility of partially formalizing the design process which can lead to better design error detection at early stages as well as simplifying the transitions between design stages,
- *clear definition of expressive power* – a strict definition of not only syntax but also the semantics of the rule language allows for the definition of the expressiveness of the formulae that it is based upon,
- *superior control of knowledge base quality* – formal methods can be used to identify logical errors in rule formulation, and
- *knowledge interoperability* – semi-formalized translation to other knowledge representation formats is possible.

In the next section the most common logical formulations of rules are given, following the presentation in [33]. A much more in-depth discussion can be found in [26].

Propositional Logic

The detailed discussion concerning expressiveness, syntax and semantics of PL can be found in [33]. A rule in *Propositional Logic* (PL) is commonly represented using a *Horn Clause* which constitutes an important form of knowledge representation in rule-based systems. A simple Horn clause ψ can be written as follows:

$$\psi: \neg p_1 \vee \neg p_2 \vee \cdots \vee \neg p_n \vee h \tag{1.1}$$

A Horn clause may contain one positive literal at most. According to the Formula (1.1), any Horn clause containing the positive literal h can be transformed into the form of rule r i.e.:

$$r: p_1 \wedge p_2 \wedge \cdots \wedge p_n \to h$$

where:

- $\psi = r$
- p_i and h are propositional symbols,
- $p_1 \wedge p_2 \wedge \cdots \wedge p_n = LHS(r)$ and is called the *Left Hand Side* of the rule r or the *Conditional Part*, and
- $h = RHS(r)$ and is called the *Right Hand Side* of the rule r or the *conclusion/decision part*.

In order to assign a meaning to the propositional symbols p_i and h the following notation is used:

$$p_i \stackrel{def}{=} \text{'definition'}$$

Let us consider the following example of a rule "*who is my boss*" which defines a rule that says "*Tom is my boss if he is a manager of the department in which I work*". First of all, we must define the meaning of the propositional symbols:

$$p_1 \stackrel{def}{=} \text{'Tom is the manager of the department of Computer Science'}$$
$$p_2 \stackrel{def}{=} \text{'I work in the department of Computer Science'}$$
$$h \stackrel{def}{=} \text{'Tom is my boss'}$$

Now the rule can be defined:

$$r: p_1 \wedge p_2 \to h$$

It is important to note that a propositional symbol can be assigned only a unique meaning. The simplicity and efficient reasoning capabilities of PL was the reason for its wide use in rule-based systems.

First-Order Predicate Calculus

First order logic (FOL) is one of the most important logical systems. In computer science FOL is mainly used for the formalization of programs and their components, and as a basic knowledge representation language in logic programming and AI. The detailed discussion concerning the expressiveness, syntax and semantics of FOL can be found in [33]. Thanks to the improved expressiveness of FOL, the rules can express more complex knowledge in a more precise way. In general FOL rule is represented as a Horn clause in the following way:

$$\psi: \neg p_1 \vee \neg p_2 \vee \cdots \vee \neg p_n \vee h \tag{1.2}$$

where: h is a literal (either a positive or negative one). Similarly as in the case of PL, considering the definition of implication and De Morgan's laws, such a clause can be written in the form of a rule:

$$p_1 \wedge p_2 \wedge \cdots \wedge p_n \rightarrow h \tag{1.3}$$

where: $p_1, p_2, \ldots, p_n$ and h are some literals.

Let us write the "*who is my boss*" example using FOL-based notation:

$$works\_in\_department(Y, D) \wedge is\_manager(X, D) \rightarrow boss(X, Y) \tag{1.4}$$

where:

- *works_in_department*, *is_manager* and *boss* are predicates which can be defined as follows:

$$works\_in\_department(X, D) \overset{def}{=} \text{'X works in department D'}$$
$$is\_manager(X, D) \overset{def}{=} \text{'X is a manager of department D'}$$
$$boss(X, Y) \overset{def}{=} \text{'X is a boss of Y'}$$

- X, Y, D are variables (or terms in general).

Using FOL, this example can be more precisely defined that in PL. It is worth noting that the above rule can be used for any facts that belong to the appropriate relations, unlike in PL, where the rule is defined for only three specific facts (*me*, *tom*, *computer Science*).

Among different programming languages and paradigms, the *Logic Programming* is particularly important for RBS. This paradigm consists of a direct application of subset of FOL for the declarative encoding of knowledge and the application of a specific strategy of resolution theorem proving for inference [33]. Currently, the

two most commonly used languages that allow logic programming are DATALOG
[51] and PROLOG [15, 52]. Both languages are declarative and use Horn rules as a
knowledge representation.

The PROLOG syntax for rules is as follows:

```
h :- q1, q2, ..., qn.
```

that corresponds to rule defined with the help of formula (1.3). The rule from the
example "*who is my boss*" can be easily modeled in PROLOG in the following way:

```
boss(X,Y) :- worksInDepartment(Y,D), isManager(X,D).
```

Additionally, the following set of facts can also be defined:

```
worksInDepartment(me, computerScience).
worksInDepartment(bob, computerScience).
worksInDepartment(baul, computerScience).
isManager(tom, computerScience).
isManager(john, automatics).
```

This rule and the given set of fact constitute a complete PROLOG program, which can
be subsequently queried. In order to ask "*who is my boss*", the following query can
be defined:

```
boss(Boss, me).
```

After the query is processed, the following response is displayed:

```
Boss = tom ;
```

More information concerning PROLOG programming can be found in [15].

DATALOG is a powerful query language dedicated for deductive databases. It is
a declarative logic language in which each formula is a function-free Horn clause.
It allows for building queries and database updates that are expressed in the logic
language. The basic syntax is similar to PROLOG, there are, however, many significant differences between these two languages. The most important of them is that
DATALOG statements can be expressed in any order and are guaranteed to terminate
if they work on finite sets. In comparison to PROLOG, DATALOG imposes many other
language syntax limitations.

First order logic is very often used as a starting point for the development of the
dedicated formalisms like Common Logic or Description Logics.

Common Logic Framework

Common Logic (CL) [53] is a framework for a family of logic languages that is
intended to facilitate the exchange of knowledge in computer-based systems. Therefore, it cannot be used as rule representation formalism for automated processing.

The framework allows for a variety of different syntactic forms, called dialects, all expressible within a common XML-based syntax and all sharing a single semantics. Common Logic possesses an ISO standard, and is published as *ISO/IEC 24707:2007 – Information technology – Common Logic (CL): a framework for a family of logic-based languages* [53]. That document includes the specifications of three CL dialects, the Common Logic Interchange Format (CLIF), the Conceptual Graph Interchange Format (CGIF), and an XML-based notation for Common Logic (XCL). The semantics of these dialects are defined by their translation to the abstract syntax and semantics of CL.

CL provides the full expressiveness of FOL with equality. CL syntax and semantics provide the full range of first-order syntactic forms, with their usual meanings. This is why, any conventional first-order syntax can be directly translated into CL without loss of information or alteration of meaning. On the other hand, in comparison to FOL, CL provides certain novel features. Many of them have been motivated directly by the ideas arising from new work on languages for the Semantic Web. Among the most important of these features, is a syntax which is *signature-free* and permits *higher-order* constructions such as quantification over classes or relations while preserving a first-order model theory [54], and a semantics which allows theories to describe intentional entities such as classes or properties e.g.:

$$\forall F (Symmetric(F) \rightarrow \forall x \forall y (F(x, y) \rightarrow F(y, x)))$$

The signature free feature allows predicates to take a variable number of arguments:

$$brother(tom, emma) \qquad \text{and} \qquad brother(tom, emma, john)$$

are the same predicates.

The detailed information concerning CL can be found in [53]. Common Logic is used as a standard logic for knowledge representation and automated reasoning. In this context, CL can be used directly as a representation which can be processed by different automated reasoning tools.

Description Logics

Rule support within Semantic Web is an active research area as the dynamic nature of the Web requires some sort of actions which may be defined by means of rules. Description Logics [30] (DL) are based upon the FOL and they constitutes the subset of it in which the syntax is restricted to formulae containing binary predicates. Perhaps the most prominent application of DLs is the formalization of ontology languages, OWL and OWL2. The DL knowledge base has two components: *TBox* and an *ABox*. The TBox contains intensional knowledge in the form of a *terminology* (hence the term TBox) and is built through declarations that describe the general properties of concepts. The ABox contains extensional knowledge, also called *assertional* knowledge (hence the term ABox), knowledge that is specific to the individuals of the domain of discourse. Intensional knowledge is usually thought not to change and extensional knowledge is usually thought to be contingent, or dependent upon a single set of circumstances, and therefore is subject to occasional or even constant change.

DL uses a different vocabulary for operationally-equivalent notions than the FOL e.g. FOL *classes* (the set of all structures satisfying a particular theory) are called *concepts* in DL, *properties* (*predicates*) known from FOL are called *roles* in DL, and *objects* correspond to DL *individuals*. There can be different approaches for expressing rules in DL. Below we give only simple formulations of rules. In fact, combining production rules is non trivial in DL systems. This issue will be discussed in much more depth in Chap. 14. The comprehensive study concerning DL can be found in [30, 55].

The following formula $\exists livesIn.\exists locatedIn.EUCountry \sqsubseteq EUCitizen$ is a simple DL rule. This rule corresponds to FOL rule having the following form: $livesIn(X, Y) \wedge locatedIn(Y, Z) \wedge EUCountry(Z) \rightarrow EUCitizen(X)$. Moreover, the chains of predicates can also be expressed as a rule-like DL statement: $hasParent \circ hasBrother \sqsubseteq hasUncle$. This statement corresponds to the following FOL rule: $hasParent(X, Y) \wedge hasBrother(Y, Z) \rightarrow hasUncle(X, Z)$.

In the case of the "*who is my boss*" example, the rule cannot be expressed in DL in the form corresponding to Formula (1.4). This is because, the *boss* role cannot be expressed by any chain of roles. However, this rule can be expressed by introducing an inverse role $department\_has\_worker \equiv works\_in\_department^-$ in the following way:

$$is\_manager \circ department\_has\ worker \sqsubseteq boss$$

F-Logic Representation

F-Logic [56] is a deductive, object-oriented database language which combines the declarative semantics and expressiveness of deductive database languages with the rich data modeling capabilities supported by the object oriented data model [57]. It supports all the typical aspects of OO paradigm i.e. it provides the concept of object having a complex internal structure, class hierarchies and inheritance, typing and encapsulation. It provides a complete interpretation of all possible propositions that precisely define their semantics. What is more, it has a sound and complete resolution-based proof theory. A small number of fundamental concepts that come from OO programming have direct representation in the F- LOGIC other secondary aspects of this paradigm are easily modeled as well. In a sense, F- LOGIC stands in the same relationship to the object-oriented paradigm as a classical predicate calculus to relational programming.

The F- LOGIC knowledge base consists of definitions of types, facts and rules. Only Horn rules in the form $head \leftarrow body$ are used. In such a rule $head$ is an F-molecule (the simplest kind of formula) and $body$ is a conjunction of F-molecules. The "*who is my boss*" use case can be modeled as following:

Definition of data types:

$$person\ [name \quad \Rightarrow string;]$$
$$department\ [name \quad \Rightarrow string;$$
$$manager \Rightarrow employee;]$$
$$employee\ [person \quad \Rightarrow person;$$
$$affiliation \Rightarrow department;$$
$$boss \quad \Rightarrow employee;]$$

Deductive rule:

$$E[boss \rightarrow M] \leftarrow E : employee \wedge D : department \wedge$$
$$\wedge E[affiliation \rightarrow D[manager \rightarrow M : employee]]$$

Apart from object-oriented databases, another important application of F- LOGIC is the area of frame-based languages. As these languages are built around the concepts of complex objects, inheritance, and deduction, the concept of Frame-Logic was derived from this [56].

Extended Representations in Modal Logics

Modal logics [58] extend the expressiveness of classic logic by introducing modal operators, e.g. *necessity* or *possibility*. Such logics are considered as logics which make statements about different worlds. The logical expression followed by a *necessity* operator must be satisfied in every world while expression followed by a *possibility* operator must be true in at least one world.

One of the most common applications of modal logics (aletheic and deontic) in the context of knowledge representation is SBVR (see Sect. 3.1). SBVR requires that each sentence in this language must be followed by one of the modal operators. It is assumed that in a case when the modal operator is not specified, then the aletheic operator of *necessity* is used. SBVR also allows for the definition of vocabulary used in the business. In the case of *who is my boss* example, the rule can be expressed in the following way:

```
(It is necceasry) Person1 is a boss of Person2
    if Person1 is a manager of some Department D
        where Person2 works.
```

Aletheic and deontic modal logics are not the only modal formalisms that exist. There are many different logics providing modalities. For more information see [59]. However, from a practical point of view, less expressive languages are often easier to handle. The use of ordinary logic like PL or FOL for building expert systems can be sometimes difficult because of problems concerning the application of such formalisms to real-world cases. Then, the use of *Attributive Logic* (AL) [33] can be a solution for such problems due to its intuitiveness and transparency. It will be introduced below, as it is the main logical foundation of the rule-based formalisms that are considered in this book.

1.5 Introduction to Attributive Logics

Knowledge representations based upon *attributes* are not only common, but also very intuitive, as it is related to technical ways of presentation. In such a case the behavior of a physical system is described by providing the values of system variables. This kind of logic is omnipresent in various applications e.g. relational database tables [60], attributive decision tables and trees [61, 62], and attributive RBS. It is also often applied to describe the state of dynamic systems and autonomous agents. While propositional and predicate logics have well-elaborated syntax and semantics, presented in numerous books covering logic for AI and knowledge engineering [41, 59, 63], or computer science [42, 49], discussion of attribute-based logical calculi is rare. Apparently, it is often assumed that attributive logic is some kind of a *technical language* equivalent (with respect to its expressive power) to propositional calculus, and as such it is not worth a detailed discussion. Such an in-depth study was provided by Ligęza in [33].

In order to define the characteristics of the system one selects some specific sets of attributes and assigns them some values. This way of describing an object and system properties is both simple and intuitive. Such languages provide a number of features, making them an efficient tool for the practical representation and manipulation of knowledge. As is stated in [33], these features can be as follows:

- *introduction of variables* — attributes play the role of *variables*; the same attribute can take different values and there is no need to introduce new propositional symbols;
- *specification of constraints* — since attributes play the role of variables, using relations between attribute values it is possible to specify constraints;
- *parametrization* — attributes may also play the role of parameters to be instantiated at some desired point of inference.

As a result of these advantages, the attributive logic is more expressive than the propositional logic, and furthermore, it stays intuitive and transparent.

The most typical way of thinking about *Attributive Logic* (AL) for knowledge specification may be presented as follows: (1) one has to define *facts*, typically of the form $A = d$ or $A(o) = d$ where A is a certain attribute, o is the object of interest and d is the attribute value, $d \in \mathbb{D}$, where $\mathbb{D}$ is a domain of the attribute A, i.e. a set of all possible values of an attribute; (2) facts are perceived as atomic formulae of propositional logic; (3) the syntax and semantics of propositional calculus are freely used. This basic approach is sometimes extended with the use of certain syntax modifications.

After Klösgen and Żytkow [61], rules in attributive logic may take the form:
$A_1 \in V_1 \wedge A_2 \in V_2 \wedge \ldots A_n \in V_n \longrightarrow A_{n+1} = d$. Following this line of extended
knowledge specification, various relational symbols can be introduced, e.g. $A_i > d$
(for ordered sets; this can be considered as a shorthand for $A_i \in (\mathbb{D}_i \setminus V_i)$, where V_i is
the set of all the values of A_i less than or equal to d) or $A_i \neq d_i$ i.e. $A_i \in (\mathbb{D}_i \setminus \{d_i\})$.
Modifying the syntax in such a way preserves the limitation that an attribute can
only take a single value at a time. Furthermore, without providing clearly defined
semantics for the language and some formal inference rules, it may lead to practical
problems. Therefore, a need for a more expressive language is apparent.

The simplest formula that can be considered as a knowledge item in attributive
language takes the form of a triple (o, A, V). Such a triple can be expressed in the
following way:

$$A(o) = V$$

where, o corresponds to an object, A is an attribute describing this object and V is
the value of this attribute for this object. In this way, a rule defined using attributive
language takes the following form:

$$r_i : (A_1(o) = d_{i1}) \wedge (A_2(o) = d_{i2}) \wedge \ldots \wedge (A_n(o) = d_{in}) \longrightarrow$$
$$\longrightarrow (H_1(o) = h_{i1}) \wedge (H_2(o) = h_{i2}) \wedge \ldots \wedge (H_m(o) = h_{im})$$

In case of the *who is my boss* example, the rule can be written as following:

$$works\_in\_department(o_1) = is\_manager\_of(o_2) \longrightarrow boss(o_1) = name(o_2)$$

In this rule, o_1 and o_2 play role of objects (that have complex internal structure)
while $works\_in\_department, is\_manager\_of, boss, name$ are attributes describing these objects.

Due to the intuitive way of knowledge representation, various forms of rule-based
systems, which are based upon attributive logic, are considered. Among the most
promising of them one can distinguish the following forms: Attributive Decision
Tables, Attributive Decision Trees, Tabular Trees, Attributive Rule-Based Systems.

In [33] a thorough discussion of attributive logics was given. It includes a formal
framework of SAL *(Set Attributive Logic)* that provides syntax, semantics and inference rules for calculus where attributes can take *set values*. In SAL it is assumed that
the *attribute A_i* is a function (or partial function) assigning certain properties from
a given domain to the object of interest. There we consider both *simple* attributes
taking a single value at any moment of time, and those that have a *generalized* form,
taking a set of values at a time. SAL is a step for the extension of attributive logics
towards practical applications. Nevertheless, it still has limited expressive power and
the semantics of the atomic formulae is limited.

SAL was an important step in the development of the *Semantic Knowledge Engineering* (SKE) approach [64] which is the foundation of the results presented in this

book. It provided the basis for development of the *Attributive Logic with Set Values over Finite Domains* (ALSV(FD)) [65]. Using it the XTT2 rule formalism was later formulated [66]. It will be discussed in detail in Chap. 4.

The formalization of systems description can be considered on many different levels and from different perspectives. These issues are discussed in the following section.

1.6 Rule Languages and Formalization

Objectives of Rule Languages

Considering the discussion in previous sections, from our perspective, the concept of *rule language* can have several important and distinct meanings. First of all, a rule language can be a certain well-defined notation for *encoding and storing* rules. In such a case only its syntax has to be defined. Such a language can be oriented on rule execution, thus being close (in terms of its goals) to general programming languages. The focus of such languages is on rule syntax and mostly data types, and data structure manipulation facilities. Examples of such languages are CLIPS, Jess, or Drools.

Another objective of rule language might be rule *interchange and translation*. In such a case the language is a notation that offers a richer syntax than the programming language, which allows for the expression of different types of rules (perhaps not all of them would be present in every rule base). Examples of such languages are RuleML and RIF.[5] In the case of these languages the semantics of rules is also considered, although not always fully defined.

The third case is a language which is a formalized knowledge representation for rules.[6] Both the syntax and semantics of such languages are formally defined. In most of the cases such languages serve not only to represent rules, but are more general knowledge representation languages [27]. Examples include F-Logic [56], or more recently Description Logics [30]. In fact, in our research we proposed a dedicated formalized language for rules based upon attributive logic [65] called XTT2 [66]. With formalized languages the inference is well-defined, and interchange much simplified. The design issue can also be better addressed. The limitation of formalized representation can be a lower flexibility and expressiveness when compared to solutions like CLIPS. This is due to the fact that "programming rule languages" have often vague (undefined) semantics.

---

[5]There also exist rule-based languages that are meant to express the transformations of data, such as XSLT. While it uses rules, it can be argued its main purpose is not to represent rules.

[6]In fact formalized rule languages are an important class of rule notations. They are often referred to as logic-based rule languages, as the formalization is mostly provided with means of logic.

The focus in this book is on the third case of rule languages. We aim to demonstrate how such a formalization can be provided to cover both the syntax and semantics of the language. Moreover, the possible benefits of the formalization for the design and implementation of RBS will be emphasized. Finally, we wish to explore the use of such a language in number of applications, as well as the integration of RBS into software engineering methods and tools.

The Benefits of Formalized Rule Languages

A number of benefits can be attained by using formalized rule languages. Probably the most apparent the model checking of RBS. Indeed even more practical benefits can be observed:

- *reliable design process* – a formalized rule language opens the possibility to partially formalize the design process which can lead to better design error detection at early stages as well as to simplify the transitions between design stages,
- *clear definition of expressive power* – a strict definition of not only syntax but also the semantics of the rule language allows for the definition of the expressiveness of the formulae it is based upon,
- *superior control of knowledge base quality* – formal methods can be used to analyze the properties of the rule base (e.g. redundancy) and identify logical errors in rule formulation, and
- *knowledge interoperability* – a semi-formalized translation to other knowledge representation formats is possible.

Besides these benefits, the limitations of formalized rule languages can also be identified. They mostly consider constrained expressiveness, usually in terms of data structure, manipulation or control of the inference flow. This also has impact on the rule execution environment, which has to be verified so the properties of the whole running systems are preserved. Furthermore, the integration of formalized systems can be challenging, and the interfaces have to be strictly defined.

1.7   Summary

This chapter has served as a short introduction to the topic of rules as discussed in various areas in AI. Clearly, the issues we covered are just a selection of topics from this very broad field. Reflecting the focus of this book, we gave a short introduction to knowledge representation methods. We also discussed classic expert systems that use rules as their core programming mechanism. Formalization of rule-based representations will be important in some of the following chapters, so its foundations were covered. Some emphasis was put on the attributive logic which will be developed in the following chapters. At the end of the chapter we provided different perspectives on rule languages and their formalization.

Following this introduction, next we will move on the practical issues related to construction of rule-based systems. Then, the first part of the book will be concluded with a presentation of selected recent applications of rules.

References

1. Nilsson, N.J.: Artificial Intelligence: A New Synthesis, 1st edn. Morgan Kaufmann Publishers Inc., San Francisco (1998)
2. Russell, S., Norvig, P.: Artificial Intelligence: A Modern Approach, 3rd edn. Prentice-Hall, Harlow (2009)
3. Giarratano, J., Riley, G.: Expert Systems. Principles and Programming, 4th edn. Thomson Course Technology, Boston (2005). ISBN 0-534-38447-1
4. Ligęza, A.: Expert systems approach to decision support. Eur. J. Oper. Res. **37**(1), 100–110 (1988)
5. von Halle, B.: Business Rules Applied: Building Better Systems Using the Business Rules Approach. Wiley, New York (2001)
6. Berners-Lee, T., Hendler, J., Lassila, O.: The Semantic Web. Scientific American (2001)
7. Hitzler, P., Krötzsch, M., Rudolph, S.: Foundations of Semantic Web Technologies. Chapman & Hall/CRC, Boca Raton (2009)
8. Dumas, M., La Rosa, M., Mendling, J., Reijers, H.A.: Fundamentals of Business Process Management. Springer, Berlin (2013)
9. Brachman, R., Levesque, H.: Knowledge Representation and Reasoning, 1st edn. Morgan Kaufmann (2004)
10. Wagner, G., Damásio, C.V., Antoniou, G.: Towards a general web rule language. Int. J. Web Eng. Technol. **2**(2/3), 181–206 (2005)
11. Miller, J., Mukerji, J.: MDA Guide Version 1.0.1. OMG (2003)
12. Kleppe, A., Warmer, J., Bast, W.: MDA Explained: The Model Driven Architecture: Practice and Promise. Addison Wesley, Boston (2003)
13. OMG: Production Rule Representation (OMG PRR) version 1.0 specification. Technical Report formal/2009-12-01, Object Management Group (2009). http://www.omg.org/spec/PRR/1.0
14. Dechter, R.: Constraint Processing. The Morgan Kaufmann Series in Artificial Intelligence. Morgan Kaufmann, San Francisco (2003)
15. Bratko, I.: Prolog Programming for Artificial Intelligence, 3rd edn. Addison Wesley, Harlow (2000)
16. Berndtsson, M., Mellin, J.: In: E.C.A Rules, pp. 959–960. Springer US, Boston, MA (2009)
17. Paschke, A., Kozlenkov, A., Boley, H., Athan, T.: Specification of reaction ruleml 1.0. Technical report, RuleML (2016). http://reaction.ruleml.org/spec
18. Clark, J.: XSL Transformations (XSLT) version 1.0 W3C recommendation 16 november 1999. Technical report, World Wide Web Consortium (W3C) (1999)
19. Hay, D., Kolber, A., Healy, K.A.: Defining Business Rules - what they really are. Final Report. Technical report, Business Rules Group (2000)
20. Flach, P.: Machine Learning: The Art and Science of Algorithms That Make Sense of Data. Cambridge University Press, New York (2012)
21. Han, J., Kamber, M.: Data Mining: Concepts and Techniques. Morgan Kaufmann Publisher, San Francisco (2000)
22. Nadel, L. (ed.): Knowledge representation. Encyclopedia of Cognitive Science, vol. 2, pp. 671–680. Macmillan Publishers Ltd., Basingstoke (2003)
23. Ichikawa, J.J., Steup, M.: The Analysis of Knowledge. In: Stanford Encyclopedia of Philosophy. Center for the Study of Language and Information (CSLI), Stanford University (2014)
24. Negnevitsky, M.: Artificial Intelligence. A Guide to Intelligent Systems. Addison-Wesley, Harlow (2002). ISBN 0-201-71159-1

25. Hendler, J., van Harmelen, F.: The semantic web: webizing knowledge representation. Handbook of Knowledge Representation. Elsevier, New York (2008)
26. van Harmelen, F., Lifschitz, V., Porter, B. (eds.): Knowledge representation and classical logic. Handbook of Knowledge Representation. Elsevier Science, Amsterdam (2007)
27. van Harmelen, F., Lifschitz, V., Porter, B. (eds.): Handbook of Knowledge Representation. Elsevier Science, Amsterdam (2007)
28. Minsky, M.: A framework for representing knowledge. The Psychology of Computer Vision. McGraw-Hill, New York (1975)
29. Quillian, M.R.: Semantic memory. Semantic Information Processing. MIT Press, Cambridge (1968)
30. Baader, F., Calvanese, D., McGuinness, D.L., Nardi, D., Patel-Schneider, P.F. (eds.): The Description Logic Handbook: Theory, Implementation, and Applications. Cambridge University Press, Cambridge (2003)
31. Sowa, J.: Semantic Networks. Encyclopedia of Artificial Intelligence. Wiley, New York (1987). revised and extended for the second edition (1992)
32. Vanthienen, J., Dries, E., Keppens, J.: Clustering knowledge in tabular knowledge bases. In: Proceedings Eighth IEEE International Conference on Tools with Artificial Intelligence, pp. 88–95 (1996)
33. Ligęza, A.: Logical Foundations for Rule-Based Systems. Springer, Berlin (2006)
34. Vanthienen, J., Mues, C., Wets, G.: Inter-tabular verification in an interactive environment. In: Vanthienen, J., van Harmelen, F. (eds.) EUROVAV, pp. 155–165. Katholieke Universiteit Leuven, Belgium (1997)
35. Vanthienen, J., Mues, C., Aerts, A., Wets, G.: A modularization approach to the verification of knowledge based systems. In: 14th International Joint Conference on Artificial Intelligence (IJCAI'95) - Workshop on Validation & Verification of Knowledge Based Systems, Montreal, Canada, 20–25 Aug 1995
36. Vanthienen, J., Robben, F.: Developing legal knowledge based systems using decision tables. In: ICAIL, pp. 282–291 (1993)
37. Vanthienen, J., Dries, E.: Illustration of a decision table tool for specifying and implementing knowledge based systems. In: ICTAI, pp. 198–205 (1993)
38. Quinlan, J.R.: Simplifying decision trees. Int. J. Man-Mach. Stud. **27**(3), 221–234 (1987)
39. Hopgood, A.A.: Intelligent Systems for Engineers and Scientists, 2nd edn. CRC Press, Boca Raton (2001)
40. Buchanan, B.G., Shortliffe, E.H.: Rule Based Expert Systems: The Mycin Experiments of the Stanford Heuristic Programming Project. The Addison-Wesley Series in Artificial Intelligence. Addison-Wesley Longman Publishing Co., Inc., Boston (1984)
41. Jackson, P.: Introduction to Expert Systems, 3rd edn. Addison–Wesley, Harlow (1999)
42. Liebowitz, J. (ed.): The Handbook of Applied Expert Systems. CRC Press, Boca Raton (1998)
43. Waterman, D.A.: A Guide to Expert Systems. Addison-Wesley Longman Publishing Co. Inc., Boston (1985)
44. David, J.M., Krivine, J.P., Simmons, R. (eds.): Second Generation Expert Systems. Springer, Secaucus (1993)
45. Forgy, C.: Rete: a fast algorithm for the many patterns/many objects match problem. Artif. Intell. **19**(1), 17–37 (1982)
46. Brownston, L., Farrell, R., Kant, E., Martin, N.: Programming Expert Systems in OPS5. Addison-Wesley, Boston (1985)
47. Riley, G.: CLIPS - A Tool for Building Expert Systems (2008). http://clipsrules.sourceforge. net
48. Friedman-Hill, E.: Jess in Action, Rule Based Systems in Java. Manning, Greenwich (2003)
49. Ben-Ari, M.: Mathematical Logic for Computer Science. Springer, London (2001)
50. Tepandi, J.: Verification, testing, and validation of rule-based expert systems. In: Proceedings of the 11-th IFAC World Congress, pp. 162–167 (1990)
51. Rosati, R.: DL+log: Tight integration of description logics and disjunctive Datalog. In: Proceedings of the Tenth International Conference on Principles of Knowledge Representation and Reasoning (KR 2006), pp. 68–78 (2006)

52. Nilsson, U., Małuszyński, J.: Logic, Programming and Prolog, 2nd edn. Wiley, Chichester (2000). http://www.ida.liu.se/~ulfni/lpp
53. Delugach, H.: ISO/IEC 24707 Information Technology–Common Logic (CL) – A Framework for a Family of Logic-Based Languages. The formally adopted ISO specification (2007)
54. Marcja, A., Toffalori, C.: A Guide to Classical and Modern Model Theory. Trends in Logic: Studia Logica Library. Kluwer Academic Publishers, Dordrecht (2003)
55. Baader, F., Horrocks, I., Sattler, U.: Description logics. Handbook of Knowledge Representation. Elsevier, New York (2008)
56. Kifer, M., Lausen, G., Wu, J.: Logical foundations of object-oriented and frame-based languages. J. ACM **42**(4), 741–843 (1995)
57. Lukichev, S.: Towards Rule Interchange and Rule Verification. Ph.D. thesis, Brandenburg University of Technology (2010). urn:nbn:de:kobv:co1-opus-20772; http://d-nb.info/1013547209
58. Blackburn, P.: Modal Logic. Cambridge University Press, Cambridge (2001)
59. Genesereth, M.R., Nilsson, N.J.: Logical Foundations for Artificial Intelligence. Morgan Kaufmann Publishers Inc., Los Altos (1987)
60. Connolly, T., Begg, C., Strechan, A.: Database Systems, A Practical Approach to Design, Implementation, and Management, 2nd edn. Addison-Wesley, Harlow (1999)
61. Klösgen, W., Żytkow, J.M. (eds.): Handbook of Data Mining and Knowledge Discovery. Oxford University Press, New York (2002)
62. Pawlak, Z.: Rough Sets. Theoretical Aspects of Reasoning about Data. Kluwer Academic Publishers, Dordrecht (1991)
63. Torsun, I.S.: Foundations of Intelligent Knowledge-Based Systems. Academic Press, London (1995)
64. Nalepa, G.J.: Semantic Knowledge Engineering A Rule-Based Approach. Wydawnictwa AGH, Kraków (2011)
65. Nalepa, G.J., Ligęza, A.: HeKatE methodology, hybrid engineering of intelligent systems. Int. J. Appl. Math. Comput. Sci. **20**(1), 35–53 (2010)
66. Nalepa, G.J., Ligęza, A., Kaczor, K.: Formalization and modeling of rules using the XTT2 method. Int. J. Artif. Intell. Tools **20**(6), 1107–1125 (2011)

Chapter 2
Knowledge Engineering with Rules

Intelligent systems that use knowledge representation and reasoning methods described in the previous chapter are commonly referred to as *Knowledge-Based Systems* (KBS). The domain that considers building KBS is most generally referred to as *Knowledge Engineering* (KE). It involves a number of development methods and processes. A general KE process can be found in several text-books in this field [1, 2]. It involves the identification of a specific problem area, the acquisition and encoding of the knowledge using the selected language, evaluation and mainte nance of the intelligent system. According to [3–5] this process can be divided into the following phases:

1. Problem identification and conceptualization,
2. Knowledge acquisition and categorization,
3. Knowledge modeling including knowledge base design,
4. Selection and application of inference mechanisms,
5. Knowledge evaluation, verification and validation,
6. System maintenance including, knowledge interoperability issues.

Each phase is dependent on the previous one. However, in practice they usually overlap considerably. Moreover, the process is highly iterative, where each iteration allows for a more precise refinement of the system model. It is common to visualize dependencies between these phases in a waterfall like manner, similar to the classic software life cycle model in *Software Engineering* (SE) [6].

In this book we are mostly concerned with selected tasks from phases 4 to 6. This chapter introduces some of these topics with respect to RBS in more detail. In Sect. 2.1 we discuss the modeling of acquired knowledge with the use of rules and other representation supporting the design. Two main groups of participants take part in these activities. The first one consists of *Domain Experts* who posses the requisite knowledge for problem solving in some specific domain. The second one consists of *knowledge engineers*, i.e. persons capable of designing, building and testing a RBS

© Springer International Publishing AG 2018
G.J. Nalepa, *Modeling with Rules Using Semantic Knowledge Engineering*,
Intelligent Systems Reference Library 130,
https://doi.org/10.1007/978-3-319-66655-6_2

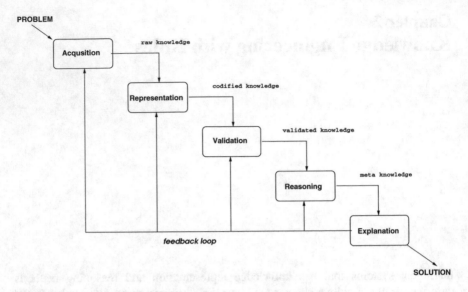

Fig. 2.1 Knowledge engineering process

using the KE methods. Once a rule set is built, an inference mechanism should be considered, see Sect. 2.2. As we discuss in Sect. 2.3, in the case of large rule sets their structure, including relations between rules have to be considered. The quality of the rule base should be analyzed during the design, or at least after it. Selected methods for this task are mentioned in Sect. 2.4. If possible, the rule base should be kept independent of the specific system implementation. To make this possible a specific rule interchange method can be used, see Sect. 2.5. Finally, as today RBS are not usually stand-alone systems, some specific architectures for their integration can be used, see Sect. 2.6. The chapter is summarized in Sect. 2.7 (Fig. 2.1).

2.1 Knowledge Acquisition and Rule Base Modeling

Challenges in Acquiring Knowledge

The knowledge engineering process includes the acquisition of new knowledge from different sources e.g. mainly from domain experts, but also different knowledge bases, available examples, etc. Knowledge acquisition is a tedious task. In fact, it is often referred to as the *Knowledge Aquisition Bottleneck*, described in several classic books e.g. one by Hayes-Roth [7, 8]. In a straightforward, if somewhat naive form, the process consists in extraction, refinement, transfer and input of the knowledge to the appropriate knowledge base. Knowledge acquisition usually starts with reviewing documents and reading books, papers and manuals related to the problem domain, as well as reviewing the old cases. Very often the knowledge is also retrieved from a domain expert, selected during the previous phase, who takes part in the guided interviews with predefined schemas [9]. During such interviews, an expert answers a series of questions that aim to elicit knowledge. Moreover, in some cases a certain domain specific methodology can also be applied. Subsequently the acquired knowledge is studied and analyzed. Then the entire process is repeated again until no new knowledge is collected. Clearly, this is an inherently iterative process.

There are a number of problems with practical knowledge acquisition:

- *syntactic mismatch* – human expert knowledge does not easily fit into any formal language; in most cases it goes beyond the expressive power of formal languages,
- *semantic mismatch* – there is no way to represent the semantics of the human expert knowledge within the only syntactically encoded knowledge base,
- *contextual and hidden knowledge* – most of inter-human communication rely on assuming some common ontology, contextual knowledge and implicit knowledge, mostly unavailable for computers, and often tacit,
- *common sense knowledge gap* – it is challenging to express all common sense human knowledge,
- *knowledge verbalization* – human experts have difficulties with smart verbalization of knowledge,
- *knowledge vagueness* – knowledge is often uncertain, imprecise, incomplete, etc.

There are many techniques supporting knowledge acquisition, including: guided interviews, i.e. following some predefined schemas, or observations, monitoring and storing expert decisions in an automated or semi-automated way. A number of classic books have been published in this area, for example [10] provides an in depth discussion. Van Harmelen provides a working recipe to start with an analysis of several representative case studies [11]. Besides the main KE approach, the use of machine learning is commonly advised for acquiring knowledge from data. This includes rule induction from databases of sets of case-bases, learning from examples, and rule generation. They allow knowledge to be generated according to the provided model of the system or according to the processes performed within the system. Today, the automated acquisition of knowledge by the machine learning approach is still an active area of research in AI [12, 13]. However, these methods are outside of the scope of this book. As today KBS are mainly software based, the knowledge acquisition phase is very often performed using SE techniques related to requirements analysis.

Knowledge Modeling

Modeling of knowledge is an evolutionary process. During the KE process new knowledge is collected and added to the knowledge base. Thus functionality of the KBS system is improved and it gradually evolves into a final version. The problems that may occur during this phase are mainly related to the vagueness of the knowledge, as it can also be imprecise or incomplete. Other problems, which can occur during this phase, include syntax errors that may appear when the knowledge is encoded into a specific language. It is also possible that some modeling errors may be caused by the complexity of the knowledge model. Due to the large amount of rules or complex dependencies, the modeled knowledge may not reflect the acquired knowledge in an appropriate way. Therefore, in order to make the modeling process more efficient a number of visual methods have been developed [14]. The visual (or semi-visual) languages facilitate the modeling phase making it more transparent for the knowledge engineer [15]. Therefore, the selection of an appropriate modeling technique can make this phase more efficient and improve the maintainability of the resulting KB.

As the number of rules identified in the system is increasing, it may be difficult to model and manage them. Thus, in complex systems having rule sets consisting of thousands of rules, various forms of rule set representations are used. Such forms as tables or trees are logically equivalent to a set of rules, but they are easier to understand and maintain. Therefore, below we discuss how these two knowledge representations can be used during the modeling process to represent the rule base.

Decision Tables for Rule Modeling

Decision tables are used to group sets of rules that are similar with respect to a set of formulas present in the preconditions and conclusions or actions [16]. Here, we start with the most basic logical form of a propositional rule (Horn clause [17]) which can be expressed as follows:

$$rule: p_1 \wedge p_2 \wedge \cdots \wedge p_n \rightarrow h$$

The canonical set of rules is one that satisfies the following assumptions [16]:

• all rules use the same propositional symbols in the same order, and
• the rules differ only with respect to using the negation symbol before the propositional symbol.

A complete canonical set of rules is a set containing all the possible combinations of using the negation sign, and it can be expressed as follows (# means either nothing or the negation symbol) [16]:

$$rule_1 : \#p_1 \wedge \#p_2 \wedge \cdots \wedge \#p_n \longrightarrow \#h_1$$
$$rule_2 : \#p_1 \wedge \#p_2 \wedge \cdots \wedge \#p_n \longrightarrow \#h_2$$
$$\vdots$$
$$rule_m : \#p_1 \wedge \#p_2 \wedge \cdots \wedge \#p_n \longrightarrow \#h_m$$

A set of rules which is not in the canonical form, can always be transformed into an equivalent canonical set. Typically, such canonical sets of rules are used for creating decision tables.

The binary decision table corresponding to the above schema is presented in Table 2.1, where vs and ws are values of conditional and decision attributes. In the presented table, each rule is specified in a single row, in which the first n columns specify the conditions under which a specific conclusion is fulfilled (e.g. specific actions can be executed or some conclusion statements can be inferred). As conditions are limited to binary logic, this is often too limited for knowledge encoding.

To enhance the expressive power and knowledge representation capabilities Attributive Logic can be used, see [18] for Attribute-Value Pair Table (AV-Pair Table)

Table 2.1 Decision table

p_1	p_2	$\ldots$	p_n	h
v_{11}	v_{12}	$\ldots$	v_{1n}	w_1
v_{21}	v_{22}	$\ldots$	v_{2n}	w_2
$\vdots$	$\vdots$	$\vdots$	$\ddots$	$\vdots$
v_{k1}	v_{k2}	$\ldots$	v_{kn}	w_k

Table 2.2 Attributive decision table

Rule	p_1	p_2	$\ldots$	p_j	$\ldots$	p_n	h_1	h_2	$\ldots$	h_m
r_1	v_{11}	v_{12}	$\ldots$	v_{1j}	$\ldots$	v_{1n}	w_{11}	w_{12}	$\ldots$	w_{1m}
r_2	v_{21}	v_{22}	$\ldots$	v_{2j}	$\ldots$	v_{2n}	w_{21}	w_{22}	$\ldots$	w_{2m}
$\vdots$	$\vdots$	$\vdots$		$\vdots$		$\vdots$	$\vdots$	$\vdots$		$\vdots$
r_i	v_{i1}	v_{i2}	$\ldots$	v_{ij}	$\ldots$	v_{in}	w_{i1}	w_{i2}	$\ldots$	w_{im}
$\vdots$	$\vdots$	$\vdots$		$\vdots$		$\vdots$	$\vdots$	$\vdots$		$\vdots$
r_k	v_{k1}	v_{k2}	$\ldots$	v_{kj}	$\ldots$	v_{kn}	w_{k1}	w_{k2}	$\ldots$	w_{km}

or [16] for Attributive Decision Table (AD-Table). An example of such a decision table is shown in Table 2.2. A row of an AD-Table represents a rule, expressed as follows:

$$r_i : (p_1 = v_{i1}) \land (p_2 = v_{i2}) \land \cdots \land (p_n = v_{in}) \rightarrow h_1 = w_{i1} \land h_2 = w_{i2} \land \cdots \land h_m = w_{im}$$

The conditions of such a rule can take several values from a specified domain. Moreover, this approach can be extended in order to allow for specifying an attribute value as an interval or a subset of the domain [16]. An example of such a table determining a rented car category[1] based upon the driver's age and driving license holding period is presented in Table 2.3. A second rule of this table corresponds to the following rule:

$$r_2 : (driver\_age < 21) \land (driving\_license\_hp < 2) \rightarrow rented\_car\_cat = A$$

More extensions to such decision tables, e.g. allowing for specifying an attribute value as an interval or a subset of the domain, can be found in [16]. Clearly the tabular knowledge representation has a very intuitive interpretation. Furthermore, through the examination of the table it is possible to obtain a better understanding of the contents of the rule base. It also allows for the identification of potential problems,

---

[1] The rented car category corresponds to Euro Car Segment classification, see: http://en.wikipedia.org/wiki/Euro_Car_Segment.

Table 2.3 Example of decision table

Driver age	Driving license holding period	Rented car category
<18	Any	None
<21	<2	A
<21	>=2	{A, B}
>=21	<2	{A, B, C}
>=21	>=2	{A, B, C, D }

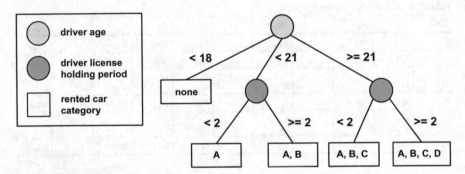

Fig. 2.2 Example of a decision tree corresponding to the decision table

such as missing values. In a way, decision tables allow for structuring rule bases. On the other hand, decision trees are useful for modeling how the knowledge is actually used in the inference process.[2]

Decision Trees for Rule Modeling

Decision trees allow for organizing rules in a hierarchical manner. As they show the dependencies between conditions and decisions, this clarifies the thinking about the consequences of certain decisions being made [19]. A decision tree has a flowchart-like structure in which a node represents an attribute and branches from such a node represent the attribute values. The end nodes (leaves) represent the final decision values. An example of a decision tree corresponding to the decision table from Table 2.3 is presented in Fig. 2.2.

Trees, especially Binary Decision Trees (BDTs), are very popular visual decision structures. They are used both in computer science and in other disciplines [16]. BDT consists of simple tests. Each test is an evaluation of a propositional statement which can be either true or false. Thus, a BDT leads through a series of tests to arrive eventually at some decision. Such a form of knowledge representation allows for the clear presentation of the decision process. Unfortunately, decision trees become much more complex if each attribute has a large number of different values because of the redundancy of nodes.

---

[2]Furthermore, they are useful to be integrated with user dialogs.

Both trees and tables are also useful from a practical point of view, as they provide a visual representation of the rule base. In this book, we assume after [16] that the transformation between rules, tables, and trees is always possible (having some syntactic restrictions). However, it is not always straightforward. These visual models help during the design process. After it they are translated to a set of rules that can then be processed using some of the common techniques described below. Similar to the acquisition phase, today a number of SE techniques are used in rule modeling. While we do not mention them here, they will be briefly described in Sect. 3.5.

2.2 Automated Inference with Rules

In rule-based systems the inference process requires two main elements. The first is the knowledge base, i.e. a rule base encoded in the format appropriate for the inference engine. The second is the inference engine itself. It uses specific inference algorithms to analyze the contents of the rules base, identify rules that can be fired, and fires them. It is generally assumed that the engine and algorithm are independent from the encoded knowledge and allow for the processing of knowledge from any domain. Important aspects that determine the operations of the inference engine include the inference mode and tasks.

The *inference mode* defines how the knowledge contained in the rule base is processed. In RBS two main inference modes are used [20]:

- *Forward chaining* is the *data-driven* (or bottom-up) reasoning. This mode of reasoning starts from the existing knowledge stored as facts and continues until no further conclusions can be drawn The engine checks the preconditions (LHS) of rules to see which ones can be fired. Then actions in RHS are executed. Therefore, the main drawback of this inference mode is that many rules may be executed that have nothing to do with the established goal.
- *Backward chaining* is a reverse process to forward chaining and is called *goal-driven* reasoning. In this mode the system has a goal (a hypothetical solution) and the inference engine attempts to find the evidence to prove it with the help of the facts stored within the fact base. Inference engine processes rules in a backward manner (in comparison to the previous mode) i.e. from RHS to LHS, and search for such rules that RHS matches to the desired goal. If LHS of a certain rule can be proven to be true in terms of the facts then the goal is also proven. If not, then the elements of LHS go to the list of goals and the entire process repeats.

In fact, sometimes a *mixed* mode allows for the combination of forward and backward inference modes.

Both of the inference modes can be applied to different kinds of problems. However, according to [21] forward chaining is a natural way to design expert systems for analysis and interpretation. For example, DENDRAL, an expert system for determining the molecular structure of unknown soil, used forward chaining [22]. In turn, the backward chaining expert systems are mostly used for diagnostic purposes. For instance, MYCIN, a medical expert system for diagnosing infectious blood diseases, used backward chaining [23].

An *inference task* is a scenario of using rules that is performed by an inference engine working in a given inference mode. Thus, it is important to distinguish between inference modes, like forward and backward chaining, and inference tasks. Therefore, a given inference task can be performed in different inference modes. One can introduce several inference tasks, such as:

- The *final consequence* inference task determines the evaluation of a given set of rules in order to infer all possible conclusions based upon the existing facts and the facts drawn during inference. Performing this task, the inference engine must take the changes of the fact base (state) into account and reevaluate rules that can be affected by these changes [21].
- The *single-pass consequence* inference task is similar to final consequence, but in contrast to it, the inference engine does not track changes of the system's state. This ensures that the rules get evaluated against existing facts only one time and then prevent them from being reevaluated.
- The *specific-pass consequence* inference task is, in turn, similar to a single-pass consequence, but additionally it explicitly defines the order of rules in which they must be executed.
- The *consequence reduction* inference task forces an inference engine to answer a question if a given hypothesis can be proved to be true according to existing facts. An inference engine tries to find such a sequence of rules that allows for the expression of the hypothesis by means of the existing facts.

Now let us consider the most important inference algorithms.

A typical forward chaining inference process performed in RBS is an iterative process consisting of the steps described below.

1. *Match* – search for all rules having satisfied their LHS.
2. *Conflict set resolution* – selection of rule for executing.
3. *Action* – invoking actions from the consequent part of the selected rule.
4. *Return* – performing the next iteration (return to the step 1).

The same process for backward chaining is analogous but it differs in direction of rule processing [21]. Among these four steps, the first step is the bottleneck of inference because it requires to match facts stored within the fact base to rules in order to check if a given rule has satisfied their LHS or not. The implementation of the naive algorithm involves iteration over all matching of the all combinations of facts to all rules. Nevertheless, such an approach is inefficient and thus other algorithms have been developed.

An important and now commonly used inference algorithm is RETE [24]. This algorithm allows the naive approach to be avoided and makes the *match* step much more efficient. It is based upon the two major ideas. The first one is related to knowledge compilation where each knowledge base is compiled and the set of all rules is transformed into so-called *discrimination network* that represents all the rules in the form of directed and acyclic graph. The second idea is to store the information concerning facts satisfying a certain condition within a corresponding node. As a result of this, operations performed during this step are limited to monitoring only changes (adding or removing) made in the fact base. When such a change is observed, it is passed through the network in order to identify rules having satisfied their LHS.

The idea of a discrimination network can also be found in other inference algorithms like TREAT [25] or GATOR [26]. In comparison to RETE, TREAT provides a more efficient memory management because it does not store any redundant information. In turn, GATOR is the most general one, as it uses several optimization methods for building a discrimination network. In particular cases they may take the forms of networks used by RETE or TREAT. Therefore, both of them are considered to be special cases of GATOR. More details on the comparisons of these algorithms can be found in [27–29].

2.3 Structure of Rule Bases

A basic discussion of rule-based systems is most commonly focused on building single rules, or constructing relatively small sets of rules. This is justified in simple cases, or studies of rule extraction algorithms. However, in engineering practice the size, structure, properties, and quality of such rule sets are very important. In fact, larger rule sets should be more commonly referred to as "rule bases" as rules in such a set are most often interrelated.

Structure on the Design Level

Relations of rules in a rule base can be expressed explicitly in rules. Examples include rules for decision control, where the execution of certain rules explicitly calls other rules. Moreover, there are cases of rewriting systems, where some rules can be modified by others. Furthermore, rule sets can be explicitly partitioned into groups operating in or in relation to given situations. There might be implicit relations between rules. Probably the most common and important cases are rules that share the same attributes. Even if such rules are not grouped together they could be related to similar situations, or objects. What makes such cases even more complicated are logical relations between such rules. This can result in contradicting or excluding rules which can lead to an unexpected operation of the system. There are different solutions to address these issues.

A simple solution, common in rule-based shells is the introduction of modularization of the rule base. CLIPS offers functionality for organizing rules into so-called modules. They allow for the restriction of access to their elements from other modules, and can be compared to global and local scoping in programming languages. In CLIPS each module has its own pattern-matching network for rules and its own agenda (rules to be fired). Jess provides a similar module mechanism that helps to manage large rule bases. Modules provide a control mechanism: rules in a module will fire only when this module has the focus, and only one module can have focus at a time. In general, although any Jess rule can be activated at any time, only rules in a module having focus will fire. Although this is a step towards introducing the structure of the rule base, still all rules are checked against the facts. In certain cases this mechanism can improve management of the rule base, as the large set of rules can be partitioned into smaller ones. It can also have a positive impact upon the performance of the inference process as not all of the rules need to be analyzed. A similar approach was employed in the Drools system. However, as Drools moved away from a CLIPS-like inference in large rule bases to a dedicated process engine, it will be described in the subsequent section on inference.

Another approach is to introduce structure into the model of the knowledge base during the design. Using visual representation methods such as decision tables can simplify the grouping of rules sharing the same attributes [30]. A decision tree can also be used to represent a group of rules but emphasizing the inference process. There exist hybrid representations such as XTT2 (eXtended Tabular Trees) that combine tables with trees [31]. While tables group rules with the same attributes, a high level inference network allows the inference process to be controlled. The tables can be connected during the design process to denote relations between groups of rules. These connections can be further used by the inference engine to optimize the inference process. For more details see [29]. In [32] a complete design and integration approach for formalized rule-based systems was introduced. It is called *Semantic Knowledge Engineering* (SKE) as it put emphasis on the proper interpretation of rule based knowledge as well as on its integration with other software engineering paradigms. In the subsequent parts of this chapter we will briefly discuss how inference control and integration are handled in SKE.

Improving Inference in Structured Rule Bases

The structuring of the rule base can be reflected during the inference process. CLIPS modules allow for the restriction of access to their elements from other modules, and can be compared to global and local scoping in other programming languages. The modularization of the knowledge base helps managing rules, and improves the efficiency of rule-based system execution. In CLIPS each module has its own pattern-matching network for its rules and its own agenda. When a *run* command is given, the agenda of the module which is the current focus is executed. Rule execution continues until another module becomes the current focus, no rules are left on the agenda, or the return function is used from the RHS of a rule. Whenever a module that was focused on runs out of rules on its agenda, the current focus is removed from the focus stack and the next module on the focus stack becomes the current focus. Before a rule executes, the current module is changed to the module in which the executing rule is defined (the current focus). The current focus can be dynamically switched in RHS of the rule with a *focus* command. A similar mechanism is present in Jess.

The Drools platform introduced RuleFlow tool. It is a workflow and process engine that allows for the advanced integration of processes and rules. It provides a graphical interface for processes and rules modeling. Drools have a built-in functionality to define the structure of the rule base which can determine the order of the rules evaluation and execution. Rules can be grouped into a ruleflow-groups which defines the subset of rules that are evaluated and executed. The ruleflow-groups have a graphical representation as the nodes on the *ruleflow* diagram. The ruleflow-groups are connected with the links what determines the order of its evaluation. A *ruleflow* diagram is a graphical description of a sequence of steps that the rule engine needs to take, where the order is important.

More recently, Drools 5 moved from a dedicated flow control engine into the integration of a rule-based reasoning system with a complete *Business Process Management Systems* (BPMS). In this case rule-based modules or subsystems can be called arbitrarily by a high-level flow control mechanism. In this case it is a Business Process engine jBPM. This approach to controlling the rule-based inference will be described in the section regarding integration of RBS with other systems.

2.4 Knowledge Base Analysis

Knowledge-based systems are widely used in areas where high performance and reliability are important. In some cases a failure of a such system may have serious consequences. Therefore, it is crucial to ensure that the system will work correctly in every possible situation.

Verification and validation have been discussed by many authors, including [33–42]. Verification concerns proving correctness of the set of rules in terms of some verifiable characteristics [43]. In fact features such as consistency, completeness, and various features of correctness may be efficiently checked only by using formal methods [15]. In turn, validation is related to checking if the system provides correct answers to specific inputs. In other words, validation consists in assuring that the system is sound and fits user requirements. Boehm [43] characterized the difference as follows: "Verification is building the system right. Validation is building the right system."

Here, it is assumed that:

- *Verification* is a process aimed at checking if the system meets its constraints and requirements (after Tepandi [41], Andert [33], and Nazareth [35]).
- *Testing* is a process which seeks to analyze system operation by comparing system responses to known responses for special input data (after Tepandi [41]).
- *Validation* is a case of testing, aimed at checking if the system meets user requirements (after Tepandi [41]).

A comprehensive overview of this field has been presented by Vermesan in [44]. Moreover, a summary result of analysis techniques and tools is presented in an online report by Wentworth et al. [45].

According to [46] verification and validation procedures of the system can be understood as one of the following: anomaly detection, formal verification, parallel use, rule base visualization to aid review, code review, testing. That study shows, that verification of the rule based systems is dominated by testing and code review. This approach highly depends on human skills, since an incorrectly written test may produce the wrong results. Formal verification and anomaly detection are not so widely used despite the fact that these methods usually have strong logical foundations and in most cases exceed the testing and debugging approach. These methods are able to check the following characteristics of the rule base [47]:

- *Redundancy* – is knowledge specified in an efficient way, so that redundancy is avoided, it includes identical rules, subsumed rules, equivalent rules, unusable rules (those that are never fired).
- *Inconsistency* – is the knowledge specified within a rule base consistent, where both internal, and external logical consistency, referring to the consistency of the real world (model) can be considered, it includes ambiguous rules, conflicting, ambivalent rules, logical inconsistency.
- *Minimal representation* – is specified knowledge in some minimal form, i.e. it cannot be replaced by another, more efficient and more concise representation: it includes reduction of rules, canonical reduction of rules, specific reduction of rules, the elimination of unnecessary attributes.
- *Completeness* – is knowledge complete, can the system always produce the output, it includes logical completeness, specific (physical) completeness, detection of incompleteness, identification of missing rules.
- *Determinism* – is provided output unique and repeatable under the same input conditions, it includes pairwise rule determinism, and complete system determinism,
- *Optimality* – is knowledge sufficient for producing the best available solutions. It includes optimal form of individual rules, locally optimal solutions, optimal knowledge covering, and optimal solutions.

According to a comparison of existing verification tools in [48] one can draw a conclusion, that the main reason why formal verification is not widely used among expert system developers is that it requires *formal knowledge representation*. In fact, most of these tools are usually based upon propositional or predicate logic. MELODIA [49] used propositional logic and a flat rule base. CLINT [50], COVER [51], PREPARE [52], SACCO [53] used predicate logic and a flat rule base. Moreover, IN- DEPTH [39] introduces a hierarchical representation, COVADIS [54] used simple production rules language with a flat rule base, whereas KRUST [55] used frames.

However, common expert system shells such as CLIPS, JESS or DROOLS do not provide formal knowledge representation, so it is not possible to apply formal methods to these tools. Although there are some analysis tools that are dedicated to aforementioned shells like CRSV-CLIPS [48, 56] for CLIPS, DROOLS Verifier, their aim is not to provide formal verification, but to offer a framework for writing tests.

Verification of knowledge in RBS is typically considered the last stage of the design procedure [5]. It is assumed that it will be performed on a complete, specified knowledge base; see the important works of Preece and Vermesan [44, 57, 58]. As such it is costly and difficult. In this book we advocate approaches that introduce a formalized description of rule base. As a result of these, formal verification of the rule base is possible during its design.

Once designed and analyzed the rule base needs to be maintained. An important and challenging aspect of maintenance includes the use of the same rule base in a different system. This involves the possible re-coding of rules in another rule language and it might be a challenging process. As it is one of the areas addressed in this book, we will analyze it in much more detail in the following section.

2.5 Rule Interchange and Interoperability

Having a complete, possibly verified and validated system, it is desirable to ensure that there is a method for sharing knowledge with other systems, representations and tools. Such methods bring many significant advantages especially for knowledge maintenance that becomes easier and more efficient. Some of them may be summarized in the following way:

- *Rapid development* – as a result of efficient interoperability methods, a new knowledge base may be supplemented by already existing knowledge that can even be modeled in different representation.
- *Cross-representation knowledge engineering* – knowledge can be modeled and verified by tools dedicated for different representation. This allows for the wider reuse of already existing tools and prevents from the unnecessary development of new ones.
- *Heterogeneous knowledge bases* – an intelligent system may use different sources/ pieces of knowledge that are translated into required representation on the fly. This creates a great opportunity for collaborative methods for knowledge engineering.

In the classic approach, a RBS was considered as standalone i.e. the implemented system was self-contained and was usually separated from other elements in its environment. Today, the rule-based knowledge representation found application in several areas that are oriented towards collaboration and integration with external technologies. Together with the increasing number of rules application areas, the number of different rule representations is also growing. The differences between these representations ensure that rule-based knowledge cannot be easily shared among different rule bases. Usually, the naive translation methods do not take rule base semantics into account which leads to a semantic mismatch before and after translation. This persistent problem is called *rule interoperability problem* and it has been known since classic expert systems [21].

In general, the methodology of interoperability must take two aspects into account: syntax that is used for knowledge encoding and semantics. On each of these two levels some problems can be identified, including ambiguous semantics, different expressiveness, and syntactic power. Over time, many different methods and approaches to the knowledge interoperability problem were developed. Some of them are general-purpose, i.e. aim to provide a framework for translation between many different representations. Others are dedicated for a certain set of representations that share similar assumptions and thus have similar semantics. The most important of these are described below.

KIF

Knowledge Interchange Framework (KIF) [59] constitutes one of the first implementations of the formal knowledge interoperability approach that uses unified intermediate representation providing declarative semantics. KIF was intended to be a formal language for the translation of knowledge among disparate computer programs providing the possibility of precise definition of semantics. It was not limited only to rules but also supports other representation techniques like frames, graphs, natural language, etc. It is important to note that KIF was not intended as a primary language for interaction with human users (though it can be used for it). Different programs could interact with their users in whatever forms that are most appropriate to their applications. The formal definition (specification) of KIF is complicated. It provides very complex meta-model consisting of a large number of classes. Moreover, its complexity led to very weak tool support and currently there are few tools that support KIF even partially. In the past, KIF was used in the context of classic expert systems. A new implementation of KIF appeared as one of the dialects of CL called *Common Logic Interchange Format* (CLIF) [60].

RIF

Rule Interchange Format (RIF) [61, 62] is a result of research conducted by Rule Interchange Format Working Group. This group was established by the World Wide Web Consortium (W3C) in 2005 to create a standard for exchanging rules among rule systems, in particular among web rule engines. Although originally envisioned by many as a *rule layer* for the Semantic Web, in reality the design of RIF is based upon the observation that there are many rule languages in existence, and what it is important to exchange rules between them. The approach taken by the group was to design a family of languages, called dialects with rigorously specified syntax and semantics of different rule systems. Currently RIF is part of the infrastructure for the Semantic Web, along with SPARQL, RDF and OWL.

Unlike CL, RIF dialects do not share the same expressiveness and semantics. Nevertheless, it is assumed that all of them include RIF-Core dialect [63] which defines an XML syntax for definite Horn rules without function symbols, i.e. DATALOG, with standard first-order semantics. Until now, the development of the RIF framework was focused on two kinds of dialects: (1) Logic-based dialects – *Basic Logic Dialect* (RIF-BLD) [64] and a subset of the RIF *Core Dialect*, they include languages that employ some kind of logic, such as FOL (often restricted to Horn logic) or non-FOL underlying the various logic programming languages; (2) Dialects for rules with actions – the *Production Rule Dialect* (RIF-PRD) [65]: it includes production rule systems, such as JESS, DROOLS and JRules, as well as reactive (or event-condition-action) rules, such as Reaction RULEML and XChange. Unfortunately, the RIF specification provided by the W3C working group has several limitations and deficiencies that makes RIF application difficult.[3]

---

[3]Lukichev shows that RIF does not specify guidelines regarding how to implement a transformation from a source rule language into the RIF and, what is more, how to verify the correctness of already

PRR

Production Rule Representation (PRR) [68] is an OMG standard for production rule representation that was developed as a response for Request For Proposals from 2002–2003 to address the need for a representation of production rules in UML models (i.e. business rule modeling as part of a modeling process). It adopts the rule classification scheme (supplied by the RULEML Initiative) and supports only production ones. It provides the MOF-based meta-model and profile that are composed of a core structure referred to as PRR Core and a non-normative abstract OCL-based syntax for the expressions, defined as an extended PRR Core meta-model referred to as PRR OCL [68]. PRR Core is a set of classes that enable the production of rules and rule sets to be defined in a purely platform independent way without having to specify OCL, in order to represent conditions and actions. As conditions and actions are "opaque" and simply strings, it is suitable to support rule expressions in both formal and informal way in order to be useful also for people without KE skills. This expressiveness was reached by very general definition of production rule semantics. This representation defines rule as the dynamic elements that allows for the chaining state of the rule-based systems and provides operational semantics for rules, the forward-chaining of production rules and rule sets.

RuleML

Rule Markup Language (RULEML) [69, 70] is defined by the RULEML Initiative.[4] This initiative aims to develop an open, vendor neutral XML/RDF-based rule language allowing for the exchange of rules between various systems including: distributed software components on the web, heterogeneous client-server systems found within large corporations, etc. RULEML is intended to be used in Semantic Web and this is why it offers XML-based language syntax for rules. In turn, the abstract syntax of this language is specified by means of a version of Extended BNF, very similar to EBNF notation used for XML. The foundation for the kernel of RULEML is the DATALOG (constructor-function-free) sub-language of Horn logic. Its expressiveness allows for the expression of both forward and backward-chaining rules in XML. It also supports different kinds of rules: derivation rules, transformation rules, reaction rules and production rules. The formal model of RULEML is comprehensively described in [71]. RULEML, as a general rule interoperability format, can be customized for various semantics of underlying rule languages that should be represented and translated. Although specific default semantics are always predefined for each RULEML language, the intended semantics of a rule base can override it by using explicit values for corresponding semantic attributes.

(Footnote 3 continued)
made translations [66]. In turn, in [67] Wang et al. indicate that, from the perspective of modeling, RIF does not have meta-model to describe features of rules.

[4]See: http://wiki.ruleml.org.

R2ML

REWERSE Rule Markup Language R2ML [72] was developed by the *REWERSE Project Working Group 11.*[5] It is an interoperability format for rules integrating RULEML with the SWRL as well as the OCL. The main goal or R2ML is to provide a method for: rule translation between different systems and tools; enriching ontologies by rules; connecting different rule systems with R2ML-based tools for visualization, verbalization, verification and validation. It supports four rule categories: derivation rules, production rules, integrity rules and ECA/reaction rules. The concepts that are used within rules can be defined in MOF/UML. R2ML was modeled using Model-Driven Approach (MDA) and it is required to accommodate Semantic Web by: web naming concepts, such as URIs and XML namespaces; the ontological distinction between objects and data values; the datatype concepts of RDF and user-defined datatypes. On the other hand, R2ML can be used as a concrete XML syntax for Extended RDF, which extends the RDF(S) semantics. Similarly to RIF, R2ML is based upon partial logic [73]. However, R2ML is supposed to be used as an intermediate representation for rules and not as a reasoning formalism. Thus, there is no efficient tool support for R2ML.

2.6 Architectures for Integration

Historically, rule-based systems were considered as stand alone tools. This meant such a system was an independent software component (sometimes integrated into hardware system). As such, it was fully responsible for processing input data, performing processing, and then making the appropriate decision and ultimately producing output data, or carrying out control actions. Therefore, with time in classic RBS systems such as CLIPS, a number of additional libraries were created to support such an environment. However, today such an approach seems redundant, and is rather rare. RBS are considered as software components, that have to be integrated into a larger software environment using well-defined software engineering approaches [6]. Therefore, here we provide a short account of different architectures to integrate rule-based systems into a larger software environment.

The already mentioned classic approach with standalone systems can be considered a *homogeneous* one. As in such a case the RBS should be able to provide not just the decision making, but also a vital part of the interfaces on the software runtime level. An important aspect is in fact related to the rule language level. In this case, the rule language should be powerful enough to program all of these features, as it is the only language available for the system designer. This results in the design of expressive rule languages like in the case of CLIPS with additional programming libraries, or language extensions such as COOL [74].

An alternative approach is to restrict the role of the RBS only to decision making. In this case, the remaining functionality is delegated to other systems or components.

---

[5]See: http://oxygen.informatik.tu-cottbus.de/rewerse-i1/.

The RBS only need to posses interfaces allowing for lower-level integration. It also operates as intelligent middleware, not a stand-alone system. This kind of architecture can be described as *heterogeneous*.

Heterogeneous Integration Using MVC Design Pattern

The rule-based component can be then integrated with into larger software system using some of the common software design patterns [75]. An example of such an approach was previously proposed in [32]. It is related to bridging knowledge engineering with software engineering (SE). Today software engineering faces a number of challenges related to efficient and integrated design and implementation of complex systems. Historically, when systems became more complex, the engineering process became more and more declarative in order to model the systems in a more comprehensive way. It made the design stage independent of programming languages, which resulted in a number of approaches. One of the best examples is the MDA (Model-Driven Architecture) approach [76]. Since there is no direct "bridge" between declarative design and sequential implementation, a substantial amount of work is needed to turn a design into a running application. This problem is often referred to as a *semantic gap* between a design and its implementation as has been discussed by Mellor in [77]. It is worth noting that while the conceptual design can sometimes be partially formally analyzed, the full formal analysis is impossible in most cases. However, there is no way of assuring that even a fully formally correct model would translate into a correct code into a programming language. Moreover, if an application is automatically generated from a designed conceptual model then any changes in the generated code have to be synchronized with the design. Another issue is the common lack of separation between core software logic, interfaces, and presentation layers.

Some of the methodologies e.g. the MDA, and the design approaches e.g. the MVC (Model-View-Controller) [78] try to address this issue. The main goal is to avoid semantic gaps, mainly the gap between the design and implementation. In order to do so, the following elements should be developed: a rich and expressive design method, a high-level runtime environment, and an effective design process. Methodologies which embody all of these elements should eventually shorten the development time, improve software quality, and transform the "implementation" into the runtime-integration and introduce so-called "executable design".

Using these ideas the *heterogeneous integration* of a RBS may be considered on several levels:

- *Runtime level*: the application is composed of the rule-based model run by the inference engine integrated into the external interfaces.
- *Service level*: the rule-based core is exposed to external applications using a network-based protocol. This allows for an SOA (Service-Oriented Architecture)-like integration [79] where the rule-based logic is designed and deployed using an embedded inference engine.
- *Design level*: integration considers a scenario, where the application has a clearly identified logic-related part, which is designed using a visual design method for

rules (such as decision table, or decision trees), and then translated into a domain-specific representation.

• *Rule language level*: in this case rule expressions can be mixed with another programming language (most often Java). In this case both syntax and semantics is mixed. However, this allows for an easy integration of rule-based code to be easily integrated into the rich features of another programming environment (e.g. Java).

An example of integration on the first three levels will be given in Sect. 9.3.

Integration of Rules and Business Process System

A specific case of heterogeneous integration of RBS with BPMS can also be considered. The Drools 5 platform encompasses several integrated modules including *Drools Expert* and *jBPM*. The former is a business rules execution engine. The latter is a fully-fledged BP execution engine. It executes business process models encoded in Business Process Model and Notation (BPMN) [80]. This notation includes dedicated syntactic constructs, so-called rule tasks. As a result of these it is possible to delegate the execution of the details of business process logic to a rule-bases system. Practically it can be any system implemented in Drools. However, from the design transparency perspective a reasonable approach is to connect only restricted well-defined subsystems, or even single modules (tables).

Following the previously defined levels, such a scenario for integration is mainly runtime-oriented. While proper design tools are currently not available for Drools, with some extensions this integration can also be reflected on the design level. Preliminary work in this direction was presented in [81], where a web design framework for business processes with rules were presented. The Drools approach also allows for service-level integration, as the whole runtime environment is web-enabled. Drools supports the orchestration of web services using rules. The execution of such solutions is supported by the runtime environment.

We provided new results in the area of integration, including a formalized model for business process models with rules in [82]. This model will be discussed in Chap. 5. It allows for a more complete integration of rules and processes, both during design and execution, as presented in Chap. 5.

2.7 Summary

In this short chapter we provided an overview of important topics related to the development of rule-based systems, commonly simply referred to as knowledge engineering. Following classic textbooks [1, 2] we discussed the general phases of the KE process [3–5]. We emphasized aspects that are non-trivial and are explored later in this book. This includes knowledge base modeling with the use of visual methods. During this phase formalization of the representation methods can be introduced. Moreover, we discussed structuring of the rule base as an important aspect of complex real-life systems. Furthermore, we identified knowledge interoperability as playing a key role in the long-term maintenance of once acquired and modeled knowledge. Finally we explored opportunities for integration of the RBS with other systems.

In the next chapter we will discuss selected applications of RBS. Thus we will conclude the presentation of the important aspects of the state of the art in RBS needed for the discussion of original results presented later. As today's KBS are mostly software based [83] we try to identify possible bridges with KE and SE. This will provide a foundation for the subsequent parts of the book.

References

1. Guida, G., Tasso, C.: Design and Development of Knowledge-Based Systems: From Life Cycle to Methodology. Wiley, New York (1995)
2. Gonzalez, A.J., Dankel, D.D.: The Engineering of Knowledge-Based Systems: Theory and Practice. Prentice-Hall Inc, Upper Saddle River (1993)
3. Durkin, J.: Expert Systems: Design and Development. Macmillan, New York (1994)
4. Waterman, D.A.: A Guide to Expert Systems. Addison-Wesley Longman Publishing Co., Inc, Boston (1985)
5. Liebowitz, J. (ed.): The Handbook of Applied Expert Systems. CRC Press, Boca Raton (1998)
6. Sommerville, I.: Software Engineering. International Computer Science, 7th edn. Pearson Education Limited, Harlow (2004)
7. Hayes-Roth, F., Waterman, D.A., Lenat, D.B.: Building Expert Systems. Addison-Wesley Longman Publishing Co., Inc, Boston (1983)
8. Buchanan, B.G., Shortliffe, E.H. (eds.): Rule-Based Expert Systems. Addison-Wesley Publishing Company, Reading (1985)
9. von Halle, B.: Business Rules Applied: Building Better Systems Using the Business Rules Approach. Wiley, New York (2001)
10. Scott, A.C.: A Practical Guide to Knowledge Acquisition, 1st edn. Addison-Wesley Longman Publishing Co., Inc, Boston (1991)
11. van Harmelen, F., Lifschitz, V., Porter, B. (eds.) Handbook of Knowledge Representation. Elsevier Science, Amsterdam (2007)
12. Stefik, M.: Introduction to Knowledge Systems. Morgan Kaufmann Publishers, San Francisco (1995)
13. Buchanan, B.G., Wilkins, D.C.: Readings in Knowledge Acquisition and Learning: Automating the Construction and Improvement of Expert Systems. M. Kaufmann Publishers, San Francisco (1993)
14. Debenham, J.: Knowledge Engineering: Unifying Knowledge Base and Database Design, 1st edn. Springer Publishing Company, Incorporated, Berlin (2012)
15. Nalepa, G.J.: Languages and tools for rule modeling. In: Giurca, A., Gašević, D., Taveter, K. (eds.) Handbook of Research on Emerging Rule-Based Languages and Technologies: Open Solutions and Approaches, pp. 596–624. IGI Global, Hershey (2009)
16. Ligęza, A.: Logical Foundations for Rule-Based Systems. Springer, Berlin (2006)
17. Ben-Ari, M.: Mathematical Logic for Computer Science. Springer, London (2001)
18. Ignizio, J.P.: An Introduction to Model Driven Architecture. Applying MDA to Enterprise Computing. David S. Frankel, Wiley Publishing, Inc., Indianapolis 2003 Expert Systems. The Development and Implementation of Rule-Based Expert Systems. McGraw-Hill, New York (1991)
19. Graham, I.: Business Rules Management and Service Oriented Architecture. Wiley, New York (2006)

20. Jackson, P.: Introduction to Expert Systems, 3rd edn. Addison–Wesley, Boston (1999). ISBN 0-201-87686-8
21. Giarratano, J., Riley, G.: Expert Systems. Principles and Programming, 4th edn. Thomson Course Technology, Boston (2005). ISBN 0-534-38447-1
22. Sutherland, G.R.: DENDRAL, a computer program for generating and filtering chemical structures. Report Memo AI-49, Stanford University, Department of Computer Science, Stanford, California (1967)
23. Buchanan, B.G., Shortliffe, E.H.: Rule Based Expert Systems: The Mycin Experiments of the Stanford Heuristic Programming Project. The Addison-Wesley Series in Artificial Intelligence. Addison-Wesley Longman Publishing Co. Inc., Boston (1984)
24. Forgy, C.: Rete: a fast algorithm for the many patterns/many objects match problem. Artif. Intell. **19**(1), 17–37 (1982)
25. Miranker, D.P.: TREAT: a better match algorithm for AI production systems; long version. Technical report 87-58, University of Texas (1987)
26. Hanson, E.N., Hasan, M.S.: Gator: an optimized discrimination network for active database rule condition testing. Technical report 93-036, CIS Department University of Florida (1993)
27. Kaczor, K., Nalepa, G.J., Bobek, S.: Rule modularization and inference solutions – a synthetic overview. In: Schneider, A. (ed.) Crossing Borders within ABC. Automation, Biomedical Engineering and Computer Science: 55 IWK Internationales Wissenschaftliches Kolloquium: International Scientific Colloquium, Illmenau, Germany, pp. 555–560 (2010)
28. Bobek, S., Kaczor, K., Nalepa, G.J.: Overview of rule inference algorithms for structured rule bases. Gdansk Univ. Technol. Fac. ETI Ann. **18**(8), 57–62 (2010)
29. Nalepa, G., Bobek, S., Ligęza, A., Kaczor, K.: Algorithms for rule inference in modularized rule bases. In: Bassiliades, N., Governatori, G., Paschke, A. (eds.) Rule-Based Reasoning, Programming, and Applications. Lecture Notes in Computer Science, vol. 6826, pp. 305–312. Springer, Berlin (2011)
30. Vanthienen, J., Dries, E., Keppens, J.: Clustering knowledge in tabular knowledge bases. In: ICTAI, pp. 88–95 (1996)
31. Nalepa, G.J., Ligęza, A., Kaczor, K.: Formalization and modeling of rules using the XTT2 method. Int. J. Artif. Intell. Tools **20**(6), 1107–1125 (2011)
32. Nalepa, G.J.: Semantic Knowledge Engineering. A Rule-Based Approach. Wydawnictwa AGH, Kraków (2011)
33. Andert, E.P.: Integrated knowledge-based system design and validation for solving problems in uncertain environments. Int. J. Man-Mach. Stud. **36**(2), 357–373 (1992)
34. Nguyen, T.A., Perkins, W.A., Laffey, T.J., Pecora, D.: Checking an expert systems knowledge base for consistency and completeness. In: IJCAI, pp. 375–378 (1985)
35. Nazareth, D.L.: Issues in the verification of knowledge in rule-based systems. Int. J. Man-Mach. Stud. **30**(3), 255–271 (1989)
36. Preece, A.D.: Verification, validation, and test of knowledge-based systems. AI Mag. **13**(4), 77 (1992)
37. Preece, A.D.: A new approach to detecting missing knowledge in expert system rule bases. Int. J. Man-Mach. Stud. **38**(4), 661–688 (1993)
38. Preece, A.D., Shinghal, R.: Foundation and application of knowledge base verification. Int. J. Intell. Syst. **9**(8), 249–269 (1994)
39. Meseguer, P.: Incremental verification of rule-based expert systems. In: Proceedings of the 10th European Conference on Artificial Intelligence. ECAI'92. Wiley, New York, NY, USA, pp. 840–844 (1992)
40. Suwa, M., Scott, C.A., Shortliffe, E.H.: Completeness and consistency in rule-based expert system. Rule-Based Expert Systems, pp. 159–170. Addison-Wesley Publishing Company, Reading (1985)
41. Tepandi, J.: Verification, testing, and validation of rule-based expert systems. In: Proceedings of the 11-th IFAC World Congress, pp. 162–167 (1990)

42. Szpyrka, M.: Design and analysis of rule-based systems with adder designer. In: Cotta, C., Reich, S., Schaefer, R., Ligęza, A. (eds.) Knowledge-Driven Computing: Knowledge Engineering and Intelligent Computations. Studies in Computational Intelligence, vol. 102, pp. 255–271. Springer, Berlin (2008)

43. Boehm, B.W.: Verifying and validating software requirements and design specifications. IEEE Softw. **1**(1), 75–88 (1984)

44. Vermesan, A.I., Coenen, F. (eds.) Validation and Verification of Knowledge Based Systems. Theory, Tools and Practice. Kluwer Academic Publisher, Boston (1999)

45. Wentworth, J.A., Knaus, R., Aougab, H.: Verification, Validation and Evaluation of Expert Systems. World Wide Web Electronic Publication. http://www.tfhrc.gov/advanc/vve/cover.htm

46. Zacharias, V.: Development and verification of rule based systems – a survey of developers. In: Proceedings of the International Symposium on Rule Representation, Interchange and Reasoning on the Web. RuleML'08. Springer, Berlin, pp. 6–16 (2008)

47. Ligęza, A., Nalepa, G.J.: Rules verification and validation. In: Giurca, A., Gašević, D., Taveter, K. (eds.) Handbook of Research on Emerging Rule-Based Languages and Technologies: Open Solutions and Approaches, pp. 273–301. IGI Global, Hershey (2009)

48. Tsai, W.T., Vishnuvajjala, R., Zhang, D.: Verification and validation of knowledge-based systems. IEEE Trans. Knowl. Data Eng. **11**, 202–212 (1999)

49. Charles, E., Dubois, O.: Melodia: logical methods for checking knowledge bases. In: Ayel, M., Laurent, J.P. (eds.) Validation, Verification and Test of Knowledge-Based Systems, pp. 95–105. Wiley, New York (1991)

50. De Raedt, L., Sablon, G., Bruynooghe, M.: Using interactive concept-learning for knowledge base validation and verification. In: Ayel, M., Laurent, J. (eds.) Validation, Verification and Testing of Knowledge Based Systems, pp. 177–190. Wiley, New York (1991)

51. Preece, A.D., Shinghal, R., Batarekh, A.: Principles and practice in verifying rule-based systems. Knowl. Eng. Rev. **7**(02), 115–141 (1992)

52. Zhang, D., Nguyen, D.: Prepare: a tool for knowledge base verification, IEEE Trans. Knowl. Data Eng. **6**(6), 983–989 (1994)

53. Ayel, M., Laurent, J.P.: Sacco-Sycojet: two different ways of verifying knowledge-based systems. In: Ayel, M., Laurent, J.P. (eds.) Validation, Verification and Test of Knowledge-Based Systems, pp. 63–76. Wiley, New York (1991)

54. Rousset, M.C.: On the consistency of knowledge bases: the covadis system. In: ECAI, pp. 79–84 (1988)

55. Craw, S., Sleeman, D.H.: Automating the refinement of knowledge-based systems. In: ECAI, pp. 167–172 (1990)

56. Culbert, S.: Expert system verifications and validation. In: Proceedings of First AAAI Workshop on V,V and Testing (1988)

57. Preece, A.D.: A new approach to detecting missing knowledge in expert system rule bases. Int. J. Man-Mach. Stud. **38**, 161–181 (1993)

58. Vermesan, A.: Foundation and application of expert system verification and validation. The Handbook of Applied Expert Systems. CRC Press, Boca Raton (1997)

59. Genesereth, M.R., Fikes, R.E.: Knowledge interchange format version 3.0 reference manual (1992)

60. Delugach, H.: ISO/IEC 24707 information technology–common logic (CL) – A framework for a family of logic-based languages. The formally adopted ISO specification (2007)

61. Kifer, M., Boley, H.: RIF overview. W3C working draft, W3C (2009). http://www.w3.org/TR/rif-overview

62. Kifer, M.: Rule interchange format: the framework. In: Calvanese, D., Lausen, G. (eds.) Web Reasoning and Rule Systems, Second International Conference, RR 2008, Karlsruhe, Germany, October 31–November 1, 2008. Proceedings. Lecture Notes in Computer Science, vol. 5341, pp. 1–11. Springer (2008)

63. Paschke, A., Reynolds, D., Hallmark, G., Boley, H., Kifer, M., Polleres, A.: RIF core dialect. Candidate recommendation, W3C (2009). http://www.w3.org/TR/2009/CR-rif-core-20091001/

64. Kifer, M., Boley, H.: RIF basic logic dialect. Candidate recommendation, W3C (2009). http://
    www.w3.org/TR/2009/CR-rif-bld-20091001/
65. Hallmark, G., Paschke, A., de Sainte Marie, C.: RIF production rule dialect. Candidate recom-
    mendation, W3C (2009). http://www.w3.org/TR/2009/CR-rif-prd-20091001/
66. Lukichev, S.: Towards rule interchange and rule verification. Ph.D. thesis, Brandenburg Uni-
    versity of Technology (2010) urn:nbn:de:kobv:co1-opus-20772. http://d-nb.info/1013547209
67. Wang, X., Ma, Z.M., Zhang, F., Yan, L.: RIF centered rule interchange in the semantic web.
    In: Bringas, P.G., Hameurlain, A., Quirchmayr, G. (eds.) Database and Expert Systems Appli-
    cations, 21st International Conference, DEXA 2010, Bilbao, Spain, August 30–September 3,
    2010, Proceedings, Part I. Lecture Notes in Computer Science, vol. 6261, pp. 478–486. Springer
    (2010)
68. OMG: Production rule representation (OMG PRR) version 1.0 specification. Technical report
    formal/2009-12-01, Object Management Group (2009). http://www.omg.org/spec/PRR/1.0
69. Boley, H., Paschke, A., Shafiq, O.: RuleML 1.0: the overarching specification of web rules. In:
    Dean, M., Hall, J., Rotolo, A., Tabet, S. (eds.) Semantic Web Rules - International Symposium,
    RuleML 2010, Washington, DC, USA, 21–23 October 2010. Proceedings. Lecture Notes in
    Computer Science, vol. 6403, pp. 162–178. Springer (2010)
70. Boley, H., Tabet, S., Wagner, G.: Design rationale for RuleML: a markup language for semantic
    web rules. In: Cruz, I.F., Decker, S., Euzenat, J., McGuinness, D.L. (eds.) Proceedings of
    SWWS'01, The first Semantic Web Working Symposium, Stanford University, California,
    USA, July 30–August 1, 2001, pp. 381–401 (2001)
71. Wagner, G., Antoniou, G., Tabet, S., Boley, H.: The abstract syntax of RuleML - towards a
    general web rule language framework. Web Intell. IEEE Comput. Soc. 628–631 (2004)
72. Wagner, G., Giurca, A., Lukichev, S.: R2ml: a general approach for marking up rules. In:
    Bry, F., Fages, F., Marchiori, M., Ohlbach, H. (eds.) Principles and Practices of Semantic Web
    Reasoning, Dagstuhl Seminar Proceedings 05371 (2005)
73. Herre, H., Jaspars, J.O.M., Wagner, G.: Partial logics with two kinds of negation as a foundation
    for knowledge-based reasoning. Centrum voor Wiskunde en Informatica (CWI) **158**, 35 (1995)
74. Giarratano, J.C., Riley, G.D.: Expert Systems. Thomson, Toronto (2005)
75. Gamma, E., Helm, R., Johnson, R., Vlissides, J.: Design Patterns, 1st edn. Addison-Wesley
    Publishing Company, Reading (1995)
76. Miller, J., Mukerji, J.: MDA guide version 1.0.1. OMG (2003)
77. Mellor, S.J., Balcer, M.J.: Executable UML: A Foundation for Model Driven Architecture, 1st
    edn. Addison-Wesley Professional, Boston (2002)
78. Burbeck, S.: Applications programming in smalltalk-80(TM): How to use model-view-
    controller (MVC). Technical report, Department of Computer Science, University of Illinois,
    Urbana-Champaign (1992)
79. Bieberstein, N., Bose, S., Fiammante, M., Jones, K., Shah, R.: Service-Oriented Architecture
    (SOA) Compass: Business Value, Planning, and Enterprise Roadmap. IBM Press, Indianapolis
    (2006)
80. OMG: Business process model and notation (BPMN): version 2.0 specification. Technical
    report formal/2011-01-03, Object Management Group (2011)
81. Kluza, K., Kaczor, K., Nalepa, G.J.: Enriching business processes with rules using the Oryx
    BPMN editor. In: Rutkowski, L., et al. (eds.) Artificial Intelligence and Soft Computing: 11th
    International Conference, ICAISC 2012: Zakopane, Poland, April 29–May 3, 2012. Lecture
    Notes in Artificial Intelligence, vol. 7268, pp. 573–581. Springer (2012)
82. Kluza, K., Nalepa, G.J.: Business processes integrated with business rules, F.M. Inf. Syst. Front.
    (2016) submitted
83. Guida, G., Lamperti, G., Zanella, M.: Software Prototyping in Data and Knowledge Engineer-
    ing. Kluwer Academic Publishers, Norwell (1999)

Chapter 3
Selected Applications of Rules

In this chapter we discuss several applications of rules and rule-based systems. There is no doubt that such applications are numerous, and deserve a dedicated multi volume book. For a comprehensive overview see [1]. Applications discussed in this chapter are mostly related to the areas of business and software engineering. They are relevant for the applications of the SKE approach discussed in the third part of the book.

We begin with the discussion of the Business Rules Approach [2, 3] in Sect. 3.1. In our opinion this is the most important attempt to adopt and extend the existing expert systems technologies in companies and the software industry that supports them. With time rules systems had to be integrated into other business management systems using business processes. This is why, it is our second application area is this integration, see Sect. 3.2. In the first decade of the 21st century in the area of Internet-centric or Web-based applications the Semantic Web project [4, 5] played an important role. A number of past knowledge engineering experiences and achievements were placed in a completely new technological and application context. Rules were to play an important role in this enterprise, in order to provide high level inference features. However, integration of classic RBS with the Semantic Web technologies is quite challenging, due some deep conceptual assumptions, see Sect. 3.3. Furthermore, we discuss some common uses of rules in the area of software engineering. A recent emerging computing paradigm of context-aware systems is also an important area for rules as discussed in Sect. 3.4. In Sect. 3.5 we wish to demonstrate examples, where rules are commonly used in SE. Finally, in Sect. 3.6 we take a look at rules as a general programming paradigm, which is actively developed. The chapter is summarized in Sect. 3.7.

3.1 Business Rules Approach

Today, rules used in business are understood in a broad sense. In this context, rules are used for defining logical aspects of the business which involve making decisions,

© Springer International Publishing AG 2018
G.J. Nalepa, *Modeling with Rules Using Semantic Knowledge Engineering*,
Intelligent Systems Reference Library 130,
https://doi.org/10.1007/978-3-319-66655-6_3

defining behavior in the given a situation, specifying regulations or limitations. Rules that are used in this context are called *Business Rules* (BR) [2, 3, 6]. There is no single and precise definition of the BR in the literature. In [7] the BRs are described as a statement that defines or constrains some aspect of the business. They are intended to assert business structure or to control or influence the behavior of the business.

The use of rules for describing the way in which business works is currently called the *business rules approach* (BRA) [3, 8, 9]. It is a methodology – and possibly a special technology – through which one can capture, challenge, publish, automate, and change the rules from a strategic business perspective. The result is a *business rules system*, an automated system in which the rules are separated, logically and physically, from other aspects of the system and shared across data stores, user interfaces and applications [3].

Within the classic production rule systems, like those described in the Sect. 1.3, only production rules were used. The usage of rules in the business context requires a distinction between different types of rules. This is why BRA provides four types of rules:

Production rules   correspond to classic production rules that are well known from classic RBS e.g. *If it is raining then the playground is wet.*

Derivation rules   are statements that allow the generation of new knowledge based upon what is currently known e.g. *Each Female Australian is a Person who was born in Country 'Australia' and has the Gender 'Female'.*

Event-Condition-Action rules   are similar to production rules but, besides the conditional part, they provide an *event* part which defines an event that triggers rule for evaluation against satisfaction of their conditional part e.g. *If it stops raining and there is a weekend then I go play ball.*

Constraints   can be considered as rules without a conclusion part and are statement that must be always true *Person has one date of birthday.*

In general, the main goal of the BRA is to provide a clear, transparent and precise method for business description which can be easily understood and applied by non-technical people. This is why, in BRA the rule representation as well as rule expressing language play a crucial role. On the one hand, they should be easy to read and understand and on the other, they must allow for very the precise expression of knowledge. Finding a compromise between these two issues is not a trivial task because solution of the first problem usually raises the second problem.

Business rules can be expressed in a several ways: using natural language, controlled language, decision tables, decision trees or logical formulas. The way in which the BRs are expressed depends on several issues: the set of aspects that are modeled, the BRs representation that is used, the designer preferences, etc. Clearly, the most precise and versatile is the logic-based representation. Nevertheless, BRs are usually used by a non-engineer people that may not have the appropriate mathematical skills required for understanding logical formulas. In turn, a natural language-based representation can be easily understood but it is very often imprecise and inconsistent and thus very difficult for automated processing. This is why, other representations have been developed that are a fusion of the precise and consistent representations on the one hand, and on the other, are understandable for those who are not engineers. One of these representations, which is very important in the context of business rules, is the Semantics of Business Vocabulary and Business Rules (SBVR) [10].

Semantic Business Vocabulary and Business Rules

As mentioned above, providing an efficient rule representation for BRA is not a trivial task. This is why, in 2003 the Object Management Group (OMG) issued the Business Semantics of Business Rule (BSBR) Request For Proposal. As a response the Semantic Business Vocabulary and Business Rules (SBVR) [11] was developed. Currently, SBVR is an adopted OMG standard of the language allowing for the declarative description of business and what is more it is also an integral part of the OMG Model Driven Architecture (MDA).

SBVR allows business people to define the policies and rules through which they run their business in their own language, in terms of the things they deal with in the business, and to capture those rules in a way that is clear, unambiguous and readily translatable into other representations [11]. It is intended to define the meaning of concepts and rules regardless of the languages or notations used to state them. This is reached by providing rule representation meta-model which entirely abstracts from the knowledge processing, methods of inference or ways of modeling. Nevertheless, the SBVR proposal provides description of the method for expressing SBVR-based knowledge using English-based controlled natural language. This proposal also considers also other ways for expressing SBVR like RuleSpeak.

SBVR facilitates interchanging business rules among organizations by making it possible to express them in a unified and precise way. Nevertheless, the interchange process cannot be automated because of the weak tool support for SBVR which is due to the complexity of the SBVR meta-model. The weak tool support causes that SBVR can be mainly used as a method that allows the running of a business to be described in a precise, unambiguous and clear way. In turn, efficient tools support for a given rule representation brings many benefits like: automated knowledge interchange, the possibility of using of the knowledge in the computer systems, automatic verification of the rule bases, and many others. We will discuss an original design tool we developed in Sect. 15.3. However, it is worth pointing out that while SBVR is a useful language for rule acquisition and authoring, its powerful semantics does not allow for automated rule execution.

Nowadays, many dedicated tools for BR management are available. They are generally called *Business Rules Management Systems* (BRMS).

Management of Business Rules

Business Rule Management Systems (BRMS) are computer systems that are intended to provide a complete support for business logic in a given business. They provide appropriate solutions for knowledge:

- storing often in the form of centralized repository,
- modeling by providing modeling methods that are appropriate for business people,

- management by implementing user interfaces that allow a knowledge engineer to modify the knowledge repository,
- processing which makes the knowledge usable in a practical way.

What is more, they support a complete knowledge life-cycle including knowledge deployment within a company.

With all of the above mentioned systems a common knowledge life-cycle can be described [12]. It includes the main phases supported by tools:

1. Alignment and planning,
2. Capturing, Organizing, and Authoring, supported by the Rule Management Environment, and
3. Distribution, testing, and application, supported by the Implementation Environment.

Currently, there are several implementations of such systems. Most of them are parts of expensive integrated commercial solutions. One of these systems was developed by ILOG, and is now a part of IBM product line.[1] Another classic example is Pegasystems Pega 7 platform.[2] Oracle also delivers Oracle Business Rules engine.[3] Moreover, FICO BlazeAdvisor Decision Rules Management System is a also a mature solution.[4] A recent BRMS technology comes from the InRule company.[5] There are also BRMS tools that are free for use. An example is OpenRules from an independent company.[6] It is targeted at smaller applications and it is well integrated into office software suites.

One of the most commonly known free for use systems is DROOLS [7] [13] which has been already mentioned in Chap. 2. DROOLS introduces the Business Logic integration Platform which provides a unified and integrated platform for Rules, Workflow and Event Processing. It consists of several projects, among which the most important are:

Drools Expert – this constitutes a dedicated forward-chaining rule engine for DROOLS-based knowledge representation. It consists of a set of JAVA classes providing programming interface for building application that are able for reasoning. It also provides support for syntax of Drools Rule Language (DRL) which provides a native way for rules encoding.

Drools Guvnor – it is also called Business Rules Manager and provides a centralized rules repository which enables the modeling of data structure, rules and decision tables by using web-based user interface. It also supports domain specific language which allows rules to be specified natural-based language.

---

[1] See https://www-01.ibm.com/software/info/ilog.

[2] See https://www.pega.com/business-rules-engine.

[3] See http://www.oracle.com/technetwork/middleware/business-rules.

[4] See http://www.fico.com/en/products/fico-blaze-advisor-decision-rules-management-system.

[5] See http://www.inrule.com.

[6] See http://openrules.com.

[7] See http://www.jboss.org/drools/.

Drools Fusion – this is a DROOLS module which supports event processing and temporal reasoning. It is a tool that is able to support Complex Event Processing (CEP) [14] concept that deals with the task of processing multiple events with the goal of identifying the meaningful events within the event cloud.

Apart from rules, DROOLS also integrates workflow-based modeling of the processes with the help of BPM [15–17]. Workflows can be designed by using Business Process Modeling Notation (BPMN) [18] and then can be executed with the help of dedicated workflow engine.

Business Rules Approach is currently one of the most important areas where rules are applied. BRA uses rules for providing declarative specification of regulation that exist in business that can be used by business people. Besides BRA, rules also play an important role in the Semantic Web initiative, as it provides an important element which enables the performing of reasoning tasks at a new abstraction level.

Decision Model and Notation

As there is still no established standard that is widely used by the BR application vendors for decision modeling, there is an OMG attempt to provide some standardization for decision modeling and management, which is similar to Unified Modeling Language (UML) in Software Engineering or BPMN for business process modeling. The Decision Model and Notation (DMN) specification is expected to standardize some aspects of decision models [19], mainly decision modeling and management by business users (users that are at the business level i.e. business analysts rather than IT specialists). As the specification goal is to define computation independent models (those are the technology-agnostic), it should be relevant to multiple deployment technologies, e.g. Business Rule Engine, Complex Event Processing, etc.

The main DMN objectives are [19]:

- to standardize decision modeling by providing a standard framework for decision model types, decision tables, decisions in business process models, technology-agnostic notation for decision models,
- to improve the quality of decision models and decisions made in systems, by using an expert-defined standard, or by making the properties specific to decision models verifiable,
- to make decision models communicable among people and machines, and to provide intuitive user oriented decision notation,
- to encourage better control of decisions in models and applications,
- to provide interchangeable decision models between tools, assuring the independence of decision models from modeling techniques and from decision making technology.

Today business systems aim to combine different methods to capture the dynamics of business operation and the business environment. In fact, a very common approach in this area is to use workflow-based approaches. This is why business rule systems are very often integrated into business process management systems as described in following section.

3.2 Rules in Business Process Management

Business Process Management (BPM) [20, 21] is a modern holistic approach improving organization's workflow, in order to align processes with client needs, which focuses on the re-engineering of processes in order to obtain the optimization of procedures, increase efficiency and effectiveness through a constant process improvement. The key aspect of BPM is a Business Process (BP). Although there is no single definition of a Business Process, the existing definitions have many things in common [22–25]. A BP is usually described as a collection of related activities which transform different kinds of clearly specified inputs in order to produce customer value, mainly considered as products or services and organizational goals, as output.

Business Process Management requires a specification of many aspects, such as goals, inputs, outputs, used resources, activities and their order, their impact on other organizational units, customers and owners for each of the managed processes in order to enable real benefits. It unifies the previously distinct disciplines such as Process Modeling, Simulation, Workflow, Enterprise Application Integration (EAI), and Business-to-Business (B2B) integration into a single standard [26]. Therefore, BPM is often considered to be as either a legacy or the next step after workflows. Workflow Management Coalition (WfMC) [27] defines a workflow [28] in terms of automation of a business process during which documents, information or tasks are passed from one participant to another for action, according to a defined set of procedural rules.

BP Lifecycle

Although many aspects of BPM have been debated in the literature, one of the fundamental BPM issues is the repeating sequence of steps, the so-called *Business Process Management Lifecycle*. The main idea behind the BPM lifecycle is to manage and improve BPs over business changes. BPM is in fact the application of the management cycle to organization's business processes [29]. The BPM lifecycle starts with specification of organizational and process goals as well as an assessment of the environmental factors that have an effect on the organization BPs. In the following process design phase, the organization processes are to be identified or redesigned. In this phase the particular process details should be specified and the proper variables that will influence the process design should be identified as well. During the next phase the previously specified process models are implemented in the environment manually via procedure handbooks or using BPM or workflow software. Finally, the implemented process can be instantiated and executed. During execution, the performance is monitored in order to control and improve the process. Data produced during the process enactment and monitoring phases, aggregated from multiple process instances, can be used in the evaluation phase, whose purpose is to formulate the results suitable for process improvement.

From our perspective the most interesting are research directions concerning the Business Process Modeling approach in the Business Process Design phase.

Process Modeling with BPMN

The most commonly used approach for modeling BPs is one that is activity flow-oriented, which is frequently referred to as "workflow" representation. Consequently, such approaches are often referred to as workflow-oriented. BPMN [18], adopted by the OMG group, is the most widely used notation for modeling BPs. As the notation is quite complex, there are many additional documents which explain it [26, 30, 31], as well as several handbooks [24, 32, 33] and many papers devoted exclusively to this notation, e.g. [34, 35], as well as its application to various areas, e.g. [36–39].

The BPMN notation uses a set of models with predefined graphical elements to depict a process and how it is performed. The current BPMN 2.0 specification [18] defines three models to cover various aspects of Business Processes. However, in most cases, the *Business Process Model* is sufficient or even too expressive to represent complex processes. The model uses four basic categories of elements to model BPs: flow objects (activities, gateways, and events), connecting objects (sequence flows, message flows, and associations), swimlanes, and artifacts.

In the case of flow object elements, activities denote tasks that have to be performed, events denote something that happens during the lifetime of the process, and gateways determine forking and merging of the sequence flow between tasks in a process, depending on some conditions (AND, XOR, OR, event-based). The sequence flow between flow object elements is used to model the flow of control in a process. The message flow between selected elements is used to model the flow of messages between the participants of a process, which are depicted as different pools.

Integration of Business Process and Rules

While this issue is not commonly considered in classic KE handbooks, it has recently become an important aspect of KE with business rules. Although there is a difference in the abstraction levels of BP and BR, rules can be complementary to the processes. BR provide declarative specification of domain knowledge, which can be encoded into a BPMN model. On the other hand, a BPMN diagram can be used as a procedural specification of the workflow, including inference control [40–42]. The use of business rules in BP design helps to simplify complex decision modeling. Rules should describe business knowledge in a formalized way that can be further automated. Nevertheless, there is no common understanding of how process and rule models should be structured in order to be integrated [43, 44]. In BPMN models decisions are often made in process flow forks, represented by a diamond. Mostly, these are simple decisions which do not require any particular modeling or management concerns. Although some rules can be expressed this way (e.g. see Fig. 3.1), such process models quickly become overwhelmed with too much details, are hard to maintain and are considered to be poorly designed models.

There are also several problems with such ad hoc modeled decisions. Firstly, there is no explicit business logic that governs such a decision. Secondly, the logic behind the decision can be duplicated (redundant) or incomplete. Moreover, such Decision

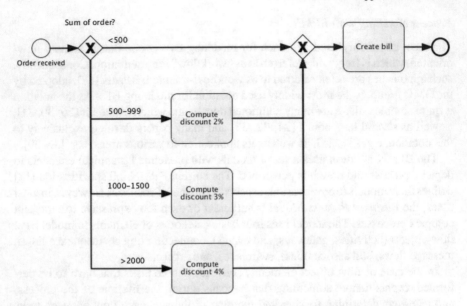

Fig. 3.1 Decision process modeled using a BPMN gateway and sequence flow guards

Fig. 3.2 Decision process of
the model modeled using a
BR task

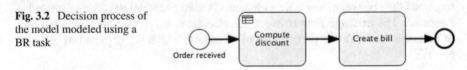

Tree or Graph represented using BPMN elements suffers from a lack of support for important modeling concepts, e.g. vocabularies/dictionaries compliance, verifying completeness, etc.

Thus, more complex business logic can be modeled using Business Rule Task and delegated to a rule engine, or a service provided by BRMS (see Fig. 3.2). The BPMN 2.0 specification defines BR tasks as elements for the association of rules, However, they are rarely used in modeling [43].

A method for graphical business rule modeling using the BPMN notations was introduced by Di Bona et al. [45]. Their approach consists of mapping a rule into a process model containing two sub-processes: the first one representing the conditional part of the rule with the activation conditions, and the second one representing the actions to be executed when the rule fires (condition and action details are specified in the DRL code and are not represented in a diagram). An example for the last rule from the decision table from Table 2.3 is presented in Fig. 3.3.

Rule-based BPMN (rBPMN) [46] constitutes one of the few examples of a coherent methodology for modeling processes integrated with rules. Milanovic et al. extended the standard of BPMN notation with the notion of rule. They introduced a new gateway symbol, called a rule gateway (a gateway with R symbol inside),

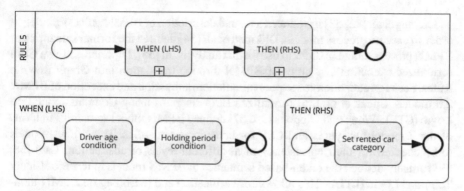

Fig. 3.3 Visualization of a rule in BPMN

which can be connected to a single rule, several rules or to a defined rule set. As they extended BPMN, their solution is not applicable for the existing BPMN models and tools.

In several approaches, rules are used to achieve flexibility in business processes. Adams et al. proposed Worklets [47], self-contained sub-processes and associated selection rules which use rules stored in the Ripple Down Rules (RDR) sets. Eijndhoven et al. [48] proposed a method which allows for production rules to be applied to several points in business processes. However, they do not use rules for specifying logic of tasks. AgentWork [49], in turn, used rules for supporting workflow adaptation. A rule-based approach is used to specify exceptions and necessary workflow adaptations. Zur Muehlen et al. [50] have considered the relationship between rules and processes. They analyzed the representation capabilities of two rule modeling languages, SRML and SBVR, in comparison to the Petri net, EPC, IDEF and BPMN 1.0 approaches. In [51] zur Muehlen et al. compared BP-oriented and BR-oriented approaches and presented a decision framework for process and rule modeling.

The above mentioned approaches do not provide any formalized specification for integration and implementation. Several of them do not concern the BPMN notation, which is *de facto* the standard for process modeling. As our interest concerns formalized and coherent approaches for modeling rules with process, we will take a look at BP formalization in the following section.

Formalization of Business Process Models

Later in this book we will discuss an approach focusing on designing the process models integrated with rules. The purpose of such models is to formally describe the integration of the BP model with rules and to provide the basis for the formal description of other integration issues. In order to provide such integrated model, we need need the formalization of Business Process models. There are several such formalizations that differ with respect to the goal with which the model semantics is specified.

Ouyang et al. [52, 53] introduced a formal description of BPMN process model, in order to execute process models. Dijkman et al. [54] defined the formal semantics of BPMN process model in order to use formal analysis. In [55], Dijkman and Van Gorp formalized execution semantics of BPMN through translation into Graph Rewrite Rules. Such formalization can support simulation, animation and execution of BPMN 2.0 models. Speck et al. [56] formalized EPC diagrams using Computational Tree Logic (CTL). Wong and Gibbons [35, 57] defined BPMN model semantics in terms of the Z model for syntax and CSP for behavioral semantics. This allows models to be checked for their consistency of at different levels of abstraction. Lam [58, 59] formally defined the token-based semantics of BPMN models in terms of Linear Temporal Logic (LTL). This allows for verification and reasoning on BPMN models, especially for checking such properties as liveness or reachability. Ligęza [60] defined a declarative model for well-defined BPMN diagrams, which allows for correct components and correct dataflow to be specified, especially by checking model consistency or the termination condition.

The above mentioned approaches were used either for formal analysis of the model or its execution. In Chap. 5 we will discuss our original approach that focuses on designing the integrated models that include processes and rules.

3.3   Rules on the Semantic Web

The Semantic Web initiative as presented by Berners-Lee, Hitzler et al. [4, 5] was based upon previous experiences and research of Knowledge Engineering in the field of Artificial Intelligence. The initiative promised to deliver interoperable web services that are able to automatically process structured data and perform simple inference tasks. It proposed a layered architecture called the Semantic Web stack. The stack provides a number of data structuring languages such as XML (eXtensible Markup Language) [61], knowledge representation languages of different expressiveness including: RDF (Resource Description Framework) [62] for metadata annotations, RDFS (RDF Schema) [63] for building a vocabulary for RDF, and OWL (Ontology Web Language) [64] for ontologies, as well as query languages including SPARQL (SPARQL Protocol and RDF Query Language) [65]. The second version of OWL [66] supports different perspectives on ontology development with OWL 2 Profiles [67]. A formal foundation for ontologies is provided by DL (Description Logics). They allow for simple inference tasks, e.g. those that correspond to concept classification. A canonical description of DL is given by Baader et al. in [68]. Considering the collective nature of the Web, these methods and technologies aim to provide solutions for distributed content and knowledge representation and basic reasoning with DL.

In order to provide more powerful reasoning capabilities, there are efforts which are focusing on developing a rule layer combined with ontologies. There exist a number of solutions that aim to integrate rules and ontologies for the Semantic Web e.g. SWRL (Semantic Web Rule Language) [69], or DLP (Description Logics

Programs) [70]. However, they were not able to overcome all of the challenges as pointed out by Horrocks et al. [71, 72] of integration of rules and DL. RBS technologies are a prime example of efficient reasoning systems based upon rules. Formal description of these systems is based upon propositional calculus, or restricted form of predicate logic – like in the case of Prolog [73].

As there still is a need to find a rule representation method appropriate for the Web (see [74]), we provide a short overview of selected research results on rules and reasoning.

Challenges for Integrating Rules with Ontologies

Rules and ontologies are complimentary approaches to knowledge representation and reasoning, see a classic paper [71, 72]. In ontologies one can capture class properties and define complex classes. Rule languages are designed mainly to define how to synthesize new facts from those stored in the knowledge base. There are things that are not easily expressed using rules, e.g. inferencing complex relationships among classes. Generally, asserting negation (complement of classes), disjunctive information or existential quantification is not possible [75]. On the other hand, there are things that cannot be expressed in ontologies or only in a complicated manner e.g. complex Horn rules [76]. Various case studies demonstrated that applications often require both approaches.

Important differences exist between ontologies based upon Description Logics and Rule-Based Systems. Description Logics and LP (Logic Programming) [76] are orthogonal in the sense that neither of them is a subset of the other. The UNA (Unique Name Assumption) in logic programming does not hold in ontologies and DL, where the same resource may be referenced to by different names and descriptions. Databases and logic programming systems use the CWA (Closed World Assumption), whereas in the Semantic Web standards there is a tendency to use OWA (Open World Assumption). Various proposals for rule representation for the Semantic Web have been formulated. The approaches to the integration of rules and ontologies may be generally divided into hybrid (heterogeneous) and homogeneous systems.

The heterogeneous approaches offer loose integration through strict semantic separation between the ontology and rule components. The resulting systems consist of an ontology component based upon a DL version, and a rule one, which usually is a dialect of Datalog. Datalog is a syntactically restricted subset of logic programming, originally defined as a query and rule language for deductive databases [77]. Homogeneous approaches result in the designing of a single logical language. Neither syntactic nor semantic distinctions are made between the ontology and the rule part, and both can be interpreted by the same reasoning engine. The language is typically either an expressive union of the component languages or an intersection of them. The union of the entire LP and DL fragments within FOL (First Order Logic) is undecidable, in general. The semantics of Semantic Web languages is based upon various logical foundations, including subsets of FOL, and F-Logic (Frame Logic) [78]. A number of languages are based upon the Datalog core. Hybrid solutions also include $\mathcal{AL}$-log introduced by Donini et al. in [79] and CARIN proposed by Levy and Rousset in [80], which integrate selected DL into Datalog rules. Integrating closed-

and open-world reasoning is an aim of the Hybrid MKNF Knowledge Bases, which has been considered by Motik et al. [81]. A MKNF knowledge base consists of a description in DL and a set of MKNF rules.

Homogeneous solutions include DLP, SWRL and ELP. DLP (Description Logics Programs) are based upon the intersection of a Description Logic with Horn Clause rules. The result is a decidable language, which is necessarily less expressive than both the DL and rules language from which it is formed. DLP, as proposed by Grosof et al. in [70], has standard First Order semantics and does not support CWA. However, it is possible to treat DLP rules as having Datalog semantics based upon CWA. In this case they are no longer semantically compatible with OWL, nor even with RDF (see the discussion of this by Horrocks et al. [71]).

Next we describe some basic yet widely used approaches for providing rules for Semantic Web.

Selected Rule Languages for the Semantic Web

SWRL (Semantic Web Rule Language) [69, 82, 83] is based upon the union of the function-free Horn logic and OWL-DL [84]. It includes a high-level abstract syntax, a model-theoretic semantics, and an XML syntax based upon RuleML (Rule Markup Language) [85]. The language enables Datalog-like rules to be combined with an OWL knowledge base. Concepts and roles are used in rules as unary and binary atoms. Subsumption and query answering with respect to knowledge bases and programs is undecidable.

In order to regain tractability, subsets of SWRL were proposed. For instance, *DL-safe* rules as proposed by Motik et al. [86] are applicable only to explicitly named objects. Another decidable fragment of SWRL is ELP proposed by Krötzsch et al. [87], a language based upon a tractable DL $\mathcal{EL}^{++}$ augmented with DL Rules.[8] The authors call it a "rule-based tractable knowledge representation" which allows reasoning in polynomial time. ELP supports inferencing in OWL 2 EL and OWL 2 RL Profiles [67].

SWRL is based upon a high-level abstract syntax and model-theoretic semantics which is built on the same Description Logics foundation as OWL. It can be serialized using an XML syntax based on RuleML. This syntax is supported by several tools like: Protégé,[9] KAON2,[10] and RacerPro.[11] SWRL was an important step in the rule-related investigation for the Semantic Web. However, it is now deprecated due to its intrinsic design assumptions which were incompatible with the further development of DL.

OWL 2 RL language[12] is a syntactic subset (also called profile) of OWL 2 that is amenable to implementation using rule-based technologies together with a partial axiomatization of the OWL 2 RDF-based semantics in the form of first-order

---

[8] See http://korrekt.org/page/Description_Logic_Rules.

[9] See: http://protege.stanford.edu.

[10] See: http://kaon2.semanticweb.org.

[11] See: http://www.racer-systems.com/products/racerpro.

[12] See: http://www.w3.org/TR/rif-owl-rl.

implications that can be used as the basis for such an implementation. OWL 2 RL is aimed at applications that require scalable reasoning in return for some restriction on expressive power. These restrictions are designed in order to avoid the need for inferencing of the knowledge existence that is not explicitly present in the knowledge base, and to avoid the need for non deterministic reasoning. This is achieved by restricting the use of constructs to certain syntactic positions. OWL 2 RL is often considered to be a default straightforward solution for providing rules similar to forward chaining rule-based systems. However, there are still are some important differences related to the assumptions of DL systems. These include open vs. closed world assumption, as well as a lack of capabilities for dynamic modification of the knowledge base used in RBS.

In last years the research on rules and OWL has been limited. For the most recent progress see [88–91].

3.4   Rules in Context-Aware Systems

The notion of context has been important in the conceptualization of computer systems for many years. However, providing its constructive and precise definition has proven to be a non-trivial task. A general observation is that *context is about evolving, structured, and shared information spaces, and that such spaces are designed to serve a particular purpose* [92]. The traditional approach for building context-aware systems (CAS) assumes that the context is obtained in the acquisition phase, modeled (automatically or by knowledge engineer) in the modeling phase, and finally executed in the processing phase.

Context representation

Contextual information delivered during the acquisition phase is hard to process without a prior interpretation. Thus, important challenge in CAS concerns context *representation*. This can be divided into two categories:

- Knowledge engineering approach – in which a full model is given a priori. Knowledge management is reduced to storing a model in a system memory and executing it during the processing phase.
- Machine learning approach – in which a model is not given, but rather learned from data. In this case, knowledge management is a non-trivial task, as it is responsible for shaping a model and thus providing correct behavior of a system.

There is a gap between these two approaches. The model based approach can provide reliability and – in some cases – allows for system verification, but the model itself is static and does not change over time. Machine learning provides adaptability, but the model is usually hidden in semantic-less numbers, which makes it difficult to adjust and understand by the user. Here we mostly focus on the KE approach as being more relevant for our discussion.

One of the simplest way to define context is with *key-value representation* to use pairs of a form: `key-value`. The `key` is usually a name that defines a context property. For instance it can denote location (e.g. room) or time (e.g. daytime). The complex contexts can be represented as a union of several `keys`. The `value` represents current state of the context property (e.g. kitchen, etc.)

The ActiveBadges [93] based system called *Watchdog* described in [94] uses the key-value context representation. A simple example might be:

```
badge location event-type action
```

The reasoning in key-values models is usually supported by a simple matching engine. When the keys match the actual context values, an action is triggered. The key-value model does not provide formalization and visualization of the model, nor provide design tools. It does not incorporate hierarchy or any sophisticated structure into model which is flat. The inference is supported usually by very simple matching algorithm that does not allow form more sophisticated reasoning. However, it is very simple to implement.

Logic-based approaches are widely used in context-aware systems. They enable automated inductive and deductive reasoning to be undertaken on contextual information and due to their strong formalization, allow for verification and validation of context models. There are several approaches that use logic to represent context. First order logic allows for an expressive description of context using Boolean operators and existential and universal quantifiers [95, 96]. Fuzzy logic and probabilistic logic is used to handle the uncertainty of the environment and to deal with the imperfections of the data [97]. Description logic is usually used in combination with ontologies. It models concepts, roles and individuals, and the relationships between them. It also provides simple reasoning capabilities that resolves classification tasks [98]. An example of a system that uses first order logic to describe context can be found in [96]. Logic-based models provide strong formalization, though their flexibility might be limited. There is also a lack of tools that provide the visualization of models defined in logic languages. A lot of programming languages and reasoners exist for expressing and processing knowledge encoded with logic based languages. However, dedicated reasoners are rarely available for mobile platforms.

Object oriented models play an important role in software engineering. In context aware systems, a Context Modeling Language (CML) [99] developed by Henricksen et al. is an example of such an approach. CML is based upon Object-Role Modeling language which was developed for conceptual modeling of databases. It provides a graphical representation of different classes and sources of context facts, relations between them and the uncertainty of information. The underlying formalism is based upon first order predicate logic and set theory. The CML approach allows for reasoning about situations that are derived from simple facts. Although the representation is human readable and expressive, it can be very complicated, especially when the number of entities and relations between them grows.

The main advantage of *ontological* context models is that they form a separate, independent layer in context-aware system. Due to standardized languages for serializing ontologies like OWL, or RDF, it is possible to reuse some well-defined models in many context-aware applications. There are many frameworks that provide such ready-to-use ontological models. One of these is CONtext ONtology (CONON) [100] which provides an upper context ontology for smart home environments. It captures general concepts about basic context like users, locations, activities. The other example is SOUPA [101] – an ontology for modeling context in pervasive environments. Ontologies have been successfully incorporated into various context-aware systems like CoBrA [102] for building smart meeting rooms, GAIA [103] for active spaces or SOCAM [104] – a middleware architecture for building context-aware mobile services. Ontologies become very popular due to the formalization and hierarchization of knowledge they provide. However, design and implementation are usually far more difficult and time consuming than other approaches. Although ontologies fall into the set of static models, there were attempts to combine them with methods that support the adaptability of the model. An interesting approach is the MUSIC [105] framework – an open platform for development of self-adaptive mobile applications.

Processes are one of the most popular methods for modeling the flow of information and/or control within a sequence of activities, actions or tasks. Jaroucheh et al. model contextual data with processes [106], which he defines as directed graphs of states. Those states denotes user current, past and possible future context. Brezillon et al. [107] presented a different approach for modeling and executing context models through the usage of processes. They proposed a structure called contextual graph. It is a directed acyclic graph that represents the actions to undertake according to the context. Processes can be automatically obtained from sensors logs with a process mining techniques [108]. However, it is a non-trivial task, and requires a lot of tuning of process mining algorithm which can be considered to be an unnecessary workload when compared to other methods. Processes provide limited reasoning support, which focus on simple tracking of a user current and future state according to the learned model.

Rules in Context Aware Systems

Rule-based systems have been in use for several decades in various branches of engineering. Hence, they have also been used in context-aware applications, both as a representation of models and as a support for reasoning [109–111]. One of the most popular tools for context-aware applications that implements a rule-based approach is Context Toolkit [112]. An example of a rule written in a Context Toolkit notation looks as follows:

```
<Reference name="Off">
   <Query name="lightOff">
      (OR
          (EQUAL presence 0)
          (GREATER brightness brightnessThreshold))
   </Query>
```

```
<Outcome outAttribute="light">0</Outcome>
    <ServiceInput service="LightService"function="lightOff" />
</Reference>
```

The example rule can be read as follows: *If there is no person in a room or brightness exceeds some threshold, then turn out the light in the room.*

Context Toolkit uses custom rule language and inference engine. However, there are several commonly used rule-based environments that provide advanced reasoning mechanisms and complex rule languages. Although there were attempts to use these tools for context-aware applications [110, 113], they are still not popular in this area.

Rules incorporates more powerful reasoning mechanisms than those available in the key-value approach. They allow for assertions of new facts into the knowledge base that can be later used as an input for other rules making the knowledge base more dynamic. They provide self-explanation mechanism that is crucial for implementing ineligibility of a system [114]. RBS provide more advanced methods for selecting rules that should be processed, improving the efficiency of the system.

The modeling phase provides a model that is used in the processing phase. This stage is responsible for context-based reasoning, which output is presented to the user. The execution mechanism is determined by the modeling approach chosen for the modeling phase and partially by the architecture of the context providers' layer. Usually every modeling language comes with a runtime environment which allows for the execution of the models created with it. Rule-based models are executed by the rule-based inference engines like ContextToolkit [114], sometimes Jess or Drools.

Context-Aware Recommender Systems

Recommender systems are an important and rapidly growing class of software tools to provide decision support in form of suggestions. These suggestions are for users to make specific decisions, usually certain choices, and selections. Very often such systems suggest certain "items" that can be of use for the user. A typical example is a web-based system that recommends items to be bought in an on-line store. The development of such systems is a mature and active field of research [115, 116].

There are a number of important classes of recommendation techniques. Some of them include: content-based, collaborative filtering, community-based, and knowledge-based [115]. The last class is especially important here, as it uses methods and models from knowledge engineering. They offer deterministic recommendations, have assured quality and do not suffer from the cold start problem.[13] Such systems can use different forms of knowledge, including case bases, constraints or rules.

Today recommender systems face a growing amount of information about user preferences, needs, but also environmental constraints. This is why, it is common to develop them in context-aware paradigm as context-aware recommender systems [117]. In this paradigm it is much easier to recommend items to the user

---

[13]This is a common problem in some recommender systems, especially based upon user-provided feedback. When the system is being bootstrapped there is no data from users, thus it is hard to assure reliable operation of the system [116].

in certain circumstances (context). Moreover, context models are often adaptable, because they use different data mining techniques. Finally, for context-aware recommender systems it is easy to acquire additional context from a range of sensors. In fact, very often such systems use mobile devices such as smart phones, smart watches, or tablets. The use of mobile systems is an opportunity but also imposes some constraints that we will discuss in the following section.

Challenges and Opportunities in Using Mobile Devices

While context-aware systems can be implemented using several approaches, recently mobile device applications become the most important. This is why, in our work in this area we focus on the aspect of designing *mobile* context-aware systems. The nature of mobile environment in which such systems are immersed, implies important assumptions regarding process of their development. Most of the existing solutions were crafted for the purpose of stationary context-aware systems, which assume that the model of the environment and the user preferences are well defined a priori and do not change over time. In mobile systems this assumption does not hold, exposing the *evolutionary* nature of the models. The models are influenced by constantly streaming data, which additionally are neither certain nor always available. Therefore, such raw contextual data needs to be appropriately collected, processed and applied to the model iteratively, which influences upon a classic three-phased approach for designing context-aware systems. In fact, this approach needs to be redefined to meet the requirements of mobile context-aware systems.

A deep analysis of the literature allowed us to formulate four main requirements that should be met by every mobile context-aware system in order to assure its high quality and to cope with such drawbacks [118, 119]. These four requirements are:

1. Intelligibility – system should allow the user to understand and modify its performance.
2. Robustness – system should be adaptable to changing user habits or environment conditions, and should be able to handle uncertain and incomplete data.
3. Privacy – system should assure the user that his or her sensitive data are secured and not accessible by the third party.
4. Efficiency – a system should be efficient both in terms of resource efficiency and high responsiveness.

We will discuss how solutions meeting these requirements were developed as part of our research on rules. Specifically in Chap. 7 we will discuss a models extending XTT2 rules with uncertainty handling. Based upon these in Chap. 17 we will present the KNOWME framework providing practical implementation of software tools supporting the development of such systems. The toolset includes HEARTDROID which is a rule-based inference engine for Android mobile devices that is based upon HEART inference engine. HEARTDROID implements the uncertainty-related extensions of XTT2.

3.5   Rules in Software Engineering

Rules are a very successful knowledge specification and system programming technique not just in KE. In fact, they find many applications in the everyday practice of *Software Engineering* [120]. In some cases they are just convenient tools to solve some engineering problems; in some other software engineers are in fact developing some KBS using SE techniques.[14] In this section we identify only several such cases. Some of them will be partially addressed in the third part of the book.

Rules in Databases

SQL is the paradigm-setting language for databases. It provides several constructs for expressing various kinds of rules: constraints, derivation and reaction. In SQL databases, integrity rules may occur in various places, most notably at the level of attribute definitions in the form of SQL check,[15] which allows a wide range of integrity rules for tables to be specified, such as a range of values and list of values, at the level of table definitions in the form of constraints, and at the database schema level in the form of assertions [122]. Derivation rules may, in turn, occur in the form of views that define a derived table by means of a query whereas reaction rules may occur in the form of triggers that define a reaction in response to update events of a certain type.

Furthermore, rules are often related to *deductive databases* [123]. They combine classic relational databases with logic programming. In this kind of database rule-based querying and deductions are possible, thus creating new facts. To maintain speed the logic programming language is usually restricted as compared to PROLOG. Commonly DATALOG and its extensions such as DATALOG* are used [124].

Rules in Software Modeling with UML

The Unified Modeling Language (UML) [125, 126] may be viewed as the paradigm-setting language for software and information systems modeling. UML allows for rules specification by using Object Constraint Language (OCL) [127]. OCL is a complementary part of the UML specification providing a dedicated language for rules definitions that are applied to UML models. It allows integrity constraints to be expressed as invariants in a formal way. It also allows for inclusion of derived attributes, classes or associations in a class diagram. The derived concepts are defined by means of derivation rules [128].

Although UML is a *de facto* standard software modeling language, there are no straightforward way to use it for modeling of rules. UML-Based Rule Modeling Language (URML) introduced by Lukichev and Wagner in [129] is one of the few

---

[14]As KBS are mostly software systems, in the 21st century the development methodologies for KBS seem to be more and more absorbed by diverse software engineering tools and techniques. However, for many KE experts it is clear that KE is distinct from SE [121] There are many reasons for this distinction. Some common arguments include the fact that KE is much more focused on declarative and reusable knowledge than dynamics or operation of the system.

[15]See: http://www.w3schools.com/sql/.

modeling languages that allows users to design rules in a visual way in UML. It is based upon UML class diagrams and extends UML meta-model by adding a rule notion and its visual notation. It supports different kinds of rules. In URML, every rule is expressed graphically as a circle with a rule identifier. Incoming arrows can express rule conditions or triggering events, and outgoing arrows express rule conclusions or produced actions. Another approach for defining the rules in UML is a UML profile for rules redefining the original UML semantics for a specific application area. Brockmans et al. in [130] has provided an example of such a profile, which is a profile for modeling SWRL [69] rules. When utilizing this profile, it is possible to model rule-extended ontologies. Although this approach does not extend the UML notation, it extends the UML semantics and enforces users to use particular stereotypes. The observation that existing solutions to rule visualization with UML fail to scale up with the growing number of rules led us to the development of an original approach in this area that will be discussed in Chap. 11.

Automation of Software Building

Build automation is one of the best examples of applications of RBS in SE. A prime example is the classic *Make* system [131]. It is a *de facto* standard in the Unix environment, especially with the use of the GNU Make implementation. The systems uses a file called simply a *Makefile* that defines a series of targets to be built, e.g. a binary program file, documentation, etc. For every target, a number of prerequisites or build dependencies are given. Furthermore, a series of actions that are needed to build that target is given. Hence, *Makefile* is composed of a series of rules in the form "to build a target, check if conditions are met, and then conduct given actions". A similar software called *Apache Ant* was provided for the Java programming environment.[16] Ant is implemented in Java, and it uses XML to encode the make file. A more flexible tool implementing similar goal-driven rule-based concepts is *Apache Maven*.[17] The most recent tool of this kind is *Gradle*[18] that uses the Groove language instead of XML to describe the make process.

Rules in Software Testing

A software testing process is an important activity in the software engineering process. There are a large number of types of tests corresponding to the phases of the software lifecycle. One of these types is black-box testing techniques that do not take the internal structure of the system into account but are based upon the system specification. Among black-box techniques, the *decision tables* based technique can be distinguished. In this technique decision tables (DTs) are used for testing system response for a given input. The content of DT consists of rules and corresponds to the possible combinations of the values of tested attributes [120]. Each rule in DT defines an input for a system as its premise and expected system response as its conclusion. The rule is satisfied when system response is the same as assumed in the DT. In fact,

---

[16]See https://ant.apache.org.

[17]See https://maven.apache.org.

[18]See https://gradle.org.

the generation of decision tables as simple rule sets can be automated [132]. Selected results in this domain will be discussed in Chap. 12.

Web Services and Rules

A Web service supports interoperable machine-to-machine interaction over the Internet [133]. It has a well-defined interface, which is described in Web Services Description Language (WSDL). Based upon the specific WSDL definition other systems interact with the service using messages in Simple Object Access Protocol (SOAP). At a high level, such a services can provide a communication interface and a workflow management protocol for business systems. The objective of such an approach is set to change the Web from a static collection of information, to a dynamic infrastructure where distributed components can be integrated to deliver a specific business logic. Clearly, rule-based systems can be used to support this operation. When typical business rules are used in a networked business, they are often too complex. Moreover, integrating RBS in a service-oriented environment [134, 135] is a challenging, as both worlds have their own paradigms [136].

In [137], authors use the MDE principles to enable for rule-based modeling of Web services. They propose the use of UML-based Rule Language (URML) and REWERSE Rule Markup Language (R2ML), together with reaction rules (Event-Condition-Action rules) for modeling Web services in terms of message exchange patterns. They also build upon W3C's Semantic Annotations for WSDL (SAWSDL) recommendation [138]. They demonstrate transformations from R2ML reaction rule-based models to production rule selected languages, including Drools, and Jess. A new generation of web services, simpler and stateless, is also proposed. They are based upon the REST (representational state transfer) protocol, and are commonly referred to as WebAPI [139]. Recent example of a rule-based service that supports intelligent automation of tasks based upon common web services is *If This Then That (IFTTT)* system.[19]

Rules in Security Systems

Computer security is yet another aspect of software engineering where RBS find applications. In fact, many security systems, mostly related to access/resource control share a similar, knowledge-based mechanism. Examples include network firewalls or intrusion detection systems. In such systems the so-called "security policy" is the core component. It is composed of different conditional statements, that are easily translatable into rules. In firewall systems the reasoning process is mostly data driven (forward), where as in intrusion detection systems backward-chaining is also often used for identification.

The XTT2 representation discussed in the second part of the book was used in such applications. In [140, 141] a rule-based approach for building GNU/Linux firewall policies was proposed. Later on that model was extended in [142] to support an application layer firewall and intrusion detection system. Furthermore, in [143] the application of this approach to the verification of security systems was discussed.

---

[19] See https://ifttt.com.

Complex Event Processing with Rules

Event driven applications rely on processing streams of events that occurs in the environment they work in. They originated from event monitoring and handling in active data bases, where ECA rules where used [144]. In last decades they acquired an increasing amount of interest in the industry as they address the challenges that complex business systems face. The term *Complex Event Processing* (CEP) was introduced to describe large scale often real-time systems that process multiple streams of events and perform some kind of reasoning based upon them [145]. Usually, a CEP system provides a certain kind of decision support. Today's distributed service oriented environments require agile and flexible software and hardware infrastructures. For this purpose high-level Event-Driven Architectures (EDA) provide proactive real-time reactions to events and relevant situations.

A common approach when it comes to building such systems is to use a dedicated kind of rule-processing [146]. A number of rule languages for CEP have been proposed [147]. A recent language oriented on interoperability and integration with Web systems is Reaction RuleML [148]. It is a general and compact serialized sublanguage RuleML that supports the family of reaction rules. It incorporates various kinds of rules including action, reaction, and temporal/event/action logic. Furthermore, it incorporates complex event/action messages into the native RuleML syntax using several extensions [149].

3.6 Rules as Programming Paradigm

Conditional statements are one of the crucial constructs in relation to encoding algorithms. Therefore, they are also commonly used in most of programming languages supporting the procedural paradigm. Moreover, some popular languages offering rich programming platforms often support dedicated means for expressing and embedding rules. In this short section we provide an overview of selected programming languages for expressing rules.

It is also worth noting, that for ten year a well-respected workshop on rule programming called "International Workshop on Rule-Based Specification and Programming (RULE)" was held (2000–2009).[20] The workshop gathered an international community interested in current applications of rule programming. The pre-proceedings of the workshop are available online.

Classic production systems

Rule-based expert systems shells described earlier in Sect. 1.3, descendants of the OPS5 system provide expressive and dedicated rule programming languages. Originally, the syntax of these languages was modeled after Lisp; this was true of OPS5. CLIPS was developed in the 1980s with speed and portability in mind, so ANSI C

---

[20]See http://twiki.di.uminho.pt/twiki/bin/view/Events/RuleWorkshop/WebHome for the workshop webpage.

was used. While it did not change the Lisp-like syntax of the language, it provided an opportunity to offer programming extensions, especially a more robust type system and object-oriented programming. A decade later Jess was developed in Java. It kept the CLIPS syntax, but opened up possibilities of easy Java integration, where Java objects could be used, and called from rules. In fact, about a decade after Jess, Drools became mature and well supported. Drools provided its own rule language DRL. Drools is implemented in Java, so processing of DRL is tightly integrated into Java runtime. In a way, DRL is built on top of Java objects. Therefore it offers the best Java integration so far. Moreover, systems implemented in Drool are Java software, so the rule-based processing can be just one several components of a larger application. To summarize, the above mentioned systems are the dedicated generic rule-based programming tools.

Prolog programming

Prolog [73] is one of the best examples of the implementation of the logic programming paradigm [76]. It can also be considered to be a rule-based programming language. The knowledge base in Prolog is encoded with the use of clauses. Simple clauses are often called facts, and compound clauses correspond to rules. As Prolog uses goal-oriented reasoning, rules are typically written with the head (goal) on the left hand side, and premises, or conditions on the right hand side. A rule is fired if the premises are met, i.e. formulae on the right hand side satisfied. Prolog searches recursively for formulae the satisfaction. Facts can be considered as condition-less rules. Thus, a Prolog knowledge base is a rule base.

However, it is worth emphasizing, that when compared to the RBS shells, Prolog is a much more powerful high-level and general purpose programming language. Besides declarative programming it supports other programming styles [150]. Moreover, as a result of to meta programming it is possible to write meta interpreters supporting other inference schemas, e.g. forward-chaining, or even interpreters for other languages. Prolog can also be used to encode and process business rules in many forms [151]. In fact in our research Prolog was used for implementing the HEART rule engine for XTT2 as discussed in Chap. 9. Moreover, it was used for implementing the knowledge processing layer of LOKI, the semantic wiki platform described in Chap. 15.

AWK

The name of this classic programming language comes from the names of its creators: Aho, Weinberger, and Kernighan. AWK was created in the 1970s and became an integral part of the Unix environment [152]. It is a very original programming language oriented at data (mostly text data) processing. In fact it is often defined as data-driven programming tool, sometimes also a stream editor. AWK programs are composed of rules. Every rule has a basic `pattern/action` scheme. Patterns in rule are matched against the input data stream. Once a pattern is matched, a corresponding rule is fired. A rule usually performs some transformation on the current portion of data (usually a single line in a text file). AWK programs scale up from handy single liners to long scripts. There are several implementations of the original AWK tool. The most popular one is GNU AWK (gawk) [153] which is present essentially in most Unix and GNU/Linux systems. In a way AWK is one of the most popular generic rule based programming tools.

Java Rules

As Java became one of the most widely used programming languages for enterprise applications, there was a push to provide rule-based programming extensions to the language. In fact one of the very first steps was the *Java Specification Request (JSR) 94*.[21] This specification is targeted both at the J2EE and J2SE platforms. It describes an API to generic operations of rule engine used from a Java application. The set of supported operations includes the execution of a rule-based engine and running an inference cycle, including parsing rules, firing rules, adding objects to an engine and getting objects from it. The engine uses a collection of rules. The rules in a rule set are expressed in any rule language, as the API itself does not define a rule language standard. The specification was supported by rule-related companies such as BEA Systems, Blaze Software, or ILOG. It is now supported by many third party rule programming solutions for Java. Besides this generic API there are a number of basic frameworks for using rules with Java. A good example is *Easyrules*[22] that allows the use of Java objects as rules using a dedicated `Rule` class, or specific Java annotations. Finally there are many approaches for bridging Java and Prolog [154].

XSLT

Extensible Stylesheet Language Transformations or XSLT for short is an XML based language for transforming XML documents into other documents [155]. It is a declarative programming language, although its syntax requires the use of XML. While inspired by functional programming languages, XSLT mostly is a pattern-driven rule-based programming language. The main statements in XML are templates and transformation rules that are fired when certain patterns in the input XML file are matched. The XML file is interpreted by an XSLT processor that runs the rules and produces an output XML file in result.

CHR

Constraint Handling Rules (CHR) is a modern concurrent declarative programming language [156]. In fact, it is a rule-based computational formalism which is very versatile. A CHR program is composed of rules that define the set of constraints (constraint store). Execution of rules may add or remove formulas from the store, thus changing the state of the program. The main types of rules include: simplification rules, propagation rules and simpagation rules (that combine the two previous types). There is a dedicated inference engine that runs the rules in a forward chaining manner. In general, the execution is non-deterministic, as fired rules can modify (rewrite) the contents of the constraint store. CHR is often used an extension of a host language, e.g.

---

[21] See https://jcp.org/en/jsr/detail?id=94.

[22] See http://www.easyrules.org.

Prolog. Thus implementations for different programming languages are available, also including Java, and Haskell.[23] Several editions of an international workshop on CHR were also held.[24]

CSP languages

As the CHR language has been mentioned a much wider class of systems and languages also needs to be recognized. Solving of Constraint Satisfaction Problems (CSP) is one of the classic areas of AI [157]. Basically, a CSP can be defined using a set of variables, along with their domains, and constraints. Practically different classes of CSP may be considered depending on the nature of the constraints and expressiveness of a language used to describe them. These languages are declarative ones, often based upon logic programming [158]. Solving a CSP can be interpreted by a special type of search with a number of diverse techniques, form basic backtracking to a number of complex constraint propagation techniques [159]. The relation of rules and constraints is such, that the definition of constraints, especially (but not only) in the case of *Constraint Logic Programming* (CLP), can be rule based. A classic CSP solving system is *ECLiPSe* [160].[25] A new and flexible language for modeling of CSP as well as optimization problems is *MiniZinc*.[26]

3.7 Summary

The topics described in this chapter clearly deserve a dedicated multi volume book. We focused only on selected applications that are somehow relevant to the scope of this book. This mostly includes the area of business applications, software engineering and Semantic Web. For the most recent in-depth study see an edited volume [1]. Some of the application areas briefly introduced in this chapter will be developed in the subsequent chapters in relation to the Semantic Knowledge Engineering approach.

As today KBS are mostly software-based, the main knowledge-based component, such as RBS is often embedded into a larger application and developed using SE techniques. Returning to the rule classification given in [161], (see Sect. 1.1), rules are considered at the lowest level of MDA, which is the PSM (Platform Specific) level. That paper provides interesting hints about programming languages considered in today's software engineering that use rules.[27] Another perspective on the use of rules, or decision tables can be found in [162]. There, rules are considered to be a special and important case of a *Domain Specific Language* (DSL).

---

[23] See http://constraint-handling-rules.org.

[24] See https://dtai.cs.kuleuven.be/CHR.

[25] See http://eclipseclp.org.

[26] See http://www.minizinc.org.

[27] For purists: these are not always programming languages in the sense of expressing algorithms, or object-oriented models. SQL is mostly defined as a query language and its core is separated from procedural extensions such as PL/SQL. Similarly OCL is not a general programming language.

This chapter concludes our presentation of the state of the art in RBS. Clearly our focus has been on issues important from the perspective needed for the discussion of original results presented in the following parts of the book. In fact, in the second part we will provide several formalized models for rules addressing the issues of knowledge base modeling, formalization of representation methods, structuring of the knowledge base, knowledge interoperability, and integration of RBS with other software systems. The third part will demonstrate the use of these models in a series of case studies.

References

1. Giurca, A., Gašević, D., Taveter, K. (eds.) Handbook of Research on Emerging Rule-Based Languages and Technologies: Open Solutions and Approaches. Information Science Reference. Hershey, New York (May (2009)
2. Ambler, S.W.: Business Rules (2003). http://www.agilemodeling.com/artifacts/businessRule.htm
3. von Halle, B.: Business Rules Applied: Building Better Systems Using the Business Rules Approach. Wiley, New York (2001)
4. Berners-Lee, T., Hendler, J., Lassila, O.: The Semantic Web. Scientific American (2001)
5. Hitzler, P., Krötzsch, M., Rudolph, S.: Foundations of Semantic Web Technologies. Chapman & Hall/CRC, Bocca Raton (2009)
6. Hay, D., Kolber, A., Healy, K.A.: Defining Business Rules - what they really are. Final Report. Technical report, Business Rules Group (2000)
7. Burns, A., Dobbing, B., Vardanega, T.: Defining business rules. what are they really? Technical Report revision 1.3, The Business Rules Group (2000)
8. Nalepa, G.J.: Business rules design and analysis approaches. In: Presentation given at the 6th European Business Rules Conference (2007)
9. Ross, R.G.: Principles of the Business Rule Approach, 1st edn. Addison-Wesley Professional, Boston (2003)
10. Object Management Group (OMG): Semantics of Business Vocabulary and Business Rules (SBVR) — Version 1.0, Framingham, Massachusetts (2008)
11. Object Management Group (OMG): Business Semantics of Business Rules – Request for Proposal (2004)
12. Nelson, M.L., Rariden, R.L., Sen, R.: A Lifecycle Approach towards Business Rules Management. In: Proceedings of the 41st Annual Hawaii International Conference on System Sciences, pp. 113–113 (2008)
13. Browne, P.: JBoss Drools Business Rules. Packt Publishing, Birmingham (2009)
14. Luckham, D.: Complex event processing (CEP). Softw. Eng. Notes 25(1), 99–100 (2000)
15. van der Aalst, W.M.P., ter Hofstede, A.H.M., Weske, M.: Business process management: A survey. In: Proceedings of Business Process Management: International Conference, BPM. Lecture Notes in Computer Science, vol. 2678, pp. 1–12. Springer, Eindhoven, The Netherlands, 26–27 June 2003 (2003)
16. Knolmayer, G., Endl, R., Pfahrer, M.: Modeling processes and workflows by business rules. In: Business Process Management, Models, Techniques, and Empirical Studies, pp. 16–29. Springer, London, UK (2000)
17. Lee, R., Dale, B.: Business process management: a review and evaluation. Bus. Process Manag. J. 4(3), 214–225 (1998)
18. OMG: Business Process Model and Notation (BPMN): Version 2.0 specification. Technical Report formal/2011-01-03, Object Management Group (2011)

19. Object Management Group (OMG): Decision model and notation request for proposal. Technical Report bmi/2011-03-04, Object Management Group, 140 Kendrick Street, Building A Suite 300, Needham, MA 02494, USA (2011)
20. Dumas, M., La Rosa, M., Mendling, J., Reijers, H.A.: Fundamentals of Business Process Management. Springer, Berlin (2013)
21. Weske, M.: Business Process Management: Concepts, Languages, Architectures, 2nd edn. Springer, Berlin (2012)
22. Davenport, T.H.: Process Innovation: Reengineering Work Through Information Technology. Harvard Business School Press, Boston (1993)
23. Hammer, M., Champy, J.: Reengineering the Corporation: A Manifesto for Business Revolution. Harper Business, New York (1993)
24. White, S.A., Miers, D.: BPMN Modeling and Reference Guide: Understanding and Using BPMN. Future Strategies Inc., Lighthouse Point, Florida, USA (2008)
25. Lindsay, A., Dawns, D., Lunn, K.: Business processes – attempts to find a definition. Inf. Softw. Technol. **45**(15), 1015–1019 (2003). Elsevie
26. Owen, M., Raj, J.: BPMN and Business Process Management. Introduction to the new business process modeling standard. Technical report, OMG (2006)
27. WfMC: Workfow Management Coalition. http://www.wfmc.org/
28. Lawrence, P. (ed.): Workflow Handbook. Wiley, New York (1997)
29. zur Muehlen, M., Ho, D.T.Y.: Risk management in the BPM lifecycle. In: Business Process Management Workshops, pp. 454–466 (2005)
30. OMG: BPMN 2.0 by Example. Technical Report dtc/2010-06-02, Object Management Group (2010)
31. White, S.: Introduction to BPMN (2004). http://www.bpmn.org/Documents/Introduction20to20BPMN.pdf
32. Allweyer, T.: BPMN 2.0. Introduction to the Standard for Business Process Modeling. BoD, Norderstedt (2010)
33. Silver, B.: BPMN Method and Style. Cody-Cassidy Press (2009)
34. Chinosi, M., Trombetta, A.: BPMN: an introduction to the standard. Comput. Stand. Interfaces **34**(1), 124–134 (2012)
35. Wong, P.Y.H., Gibbons, J.: Formalisations and applications of bpmn. Sci. Comput. Program. **76**(8), 633–650 (2011)
36. Krużel, T., Werewka, J.: Application of BPMN for the PMBOK standard modelling to scale project management efforts in IT enterprises. In: et al., Z.W., ed.: Information systems architecture and technology: information as the intangible assets and company value source, pp. 171–182. Oficyna Wydawnicza Politechniki Wrocławskiej, Wrocław (2011)
37. Ligęza, A.: A note on a logical model of an inference process : from ARD and RBS to BPMN. In: Małgorzata Nycz, M.L.O. (ed.) Knowledge acquisition and management. Research Papers of Wrocław University of Economics. 232 edn, pp. 41–49. Wrocław: Publishing House of Wrocław University of Economics (2011). ISSN 1899-3192
38. Lubke, D., Schneider, K., Weidlich, M.: Visualizing use case sets as bpmn processes. In: Requirements Engineering Visualization, 2008. REV '08, pp. 21–25 (2008)
39. Szpyrka, M., Nalepa, G.J., Ligęza, A., Kluza, K.: Proposal of formal verification of selected BPMN models with Alvis modeling language. In: Brazier, F.M., Nieuwenhuis, K., Pavlin, G., Warnier, M., Badica, C. (eds.) Intelligent Distributed Computing V. Proceedings of the 5th International Symposium on Intelligent Distributed Computing – IDC 2011. Studies in Computational Intelligence, vol. 382, pp. 249–255. Springer, Delft, The Netherlands (2011)
40. Kluza, K., Maślanka, T., Nalepa, G.J., Ligęza, A.: Proposal of representing BPMN diagrams with XTT2-based business rules. In: Brazier, F.M.T., Nieuwenhuis, K., Pavlin, G., Warnier, M., Badica, C. (eds.) Intelligent Distributed Computing V. Proceedings of the 5th International Symposium on Intelligent Distributed Computing – IDC 2011. Studies in Computational Intelligence, vol. 382, pp. 243–248. Springer, Delft, The Netherlands (2011)

41. Kluza, K., Nalepa, G.J., Łysik, Ł.: Visual inference specification methods for modularized rulebases. Overview and integration proposal. In: Nalepa, G.J., Baumeister, J. (eds.) Proceedings of the 6th Workshop on Knowledge Engineering and Software Engineering (KESE6) at the 33rd German Conference on Artificial Intelligence September 21, 2010, Karlsruhe, Germany, Karlsruhe, Germany, pp. 6–17 (2010)

42. Nalepa, G.J., Kluza, K., Ernst, S.: Modeling and analysis of business processes with business rules. In: Beckmann, J. (ed.) Business Process Modeling: Software Engineering, Analysis and Applications. Business Issues, Competition and Entrepreneurship, pp. 135–156. Nova Science Publishers (2011)

43. Hohwiller, J., Schlegel, D., Grieser, G., Hoekstra, Y.: Integration of bpm and brm. In: Dijkman, R., Hofstetter, J., Koehler, J. (eds.) Business Process Model and Notation. Lecture Notes in Business Information Processing, vol. 95, pp. 136–141. Springer, Berlin (2011)

44. Kluza, K.: Modeling of business processes consistent with business rules. PAR Pomiary Automatyka Robotyka 15(12), 194–195 (2011). ISSN 1427-9126

45. Di Bona, D., Lo Re, G., Aiello, G., Tamburo, A., Alessi, M.: A methodology for graphical modeling of business rules. In: 2011 5th UKSim European Symposium on Computer Modeling and Simulation (EMS), pp. 102–106 (2011)

46. Milanovic, M., Gašević, D.: Towards a language for rule-enhanced business process modeling. In: Proceedings of the 13th IEEE international conference on Enterprise Distributed Object Computing, EDOC'09, pp. 59–68. IEEE Press, Piscataway, NJ, USA (2009)

47. Adams, M., ter Hofstede, A.H.M., Edmond, D., van der Aalst, W.M.P.: Worklets: A service-oriented implementation of dynamic flexibility in workflows. In: OTM Conferences (1), pp. 291–308 (2006)

48. van Eijndhoven, T., Iacob, M.E., Ponisio, M.: Achieving business process flexibility with business rules. In: Proceedings of the 12th International IEEE Enterprise Distributed Object Computing Conference, 2008 EDOC '08, pp. 95–104 (2008)

49. Müller, R., Greiner, U., Rahm, E.: Agent work: a workflow system supporting rule-based workflow adaptation. Data Knowl. Eng. 51(2), 223–256 (2004)

50. zur Muehlen, M., Indulska, M., Kamp, G.: Business process and business rule modeling languages for compliance management: a representational analysis. In: Tutorials, posters, panels and industrial contributions at the 26th international conference on Conceptual modeling, vol. 83. ER '07, pp. 127–132. Darlinghurst, Australia, Australia, Australian Computer Society, Inc (2007)

51. zur Muehlen, M., Indulska, M., Kittel, K.: Towards integrated modeling of business processes and business rules. In: 19th Australasian Conference on Information Systems ACIS 2008. Christchurch, New Zealand (2008)

52. Ouyang, C., Dumas, M., ter Hofstede, A.H., van der Aalst, W.M.: From bpmn process models to bpel web services. In: IEEE International Conference on Web Services (ICWS'06) (2006)

53. Ouyang, C., Wil M.P. van der Aalst, M.D., ter Hofstede, A.H.: Translating BPMN to BPEL. Technical report, Faculty of Information Technology, Queensland University of Technology, GPO Box 2434, Brisbane QLD 4001, Australia Department of Technology Management, Eindhoven University of Technology, GPO Box 513, NL-5600 MB, The Netherlands (2006)

54. Dijkman, R.M., Dumas, M., Ouyang, C.: Formal semantics and automated analysis of BPMN process models. preprint 7115. Technical report, Queensland University of Technology, Brisbane, Australia (2007)

55. Dijkman, R.M., Gorp, P.V.: Bpmn 2.0 execution semantics formalized as graph rewrite rules. In: Mendling, J., Weidlich, M., Weske, M. (eds.) Proceedings from the Business Process Modeling Notation – Second International Workshop, BPMN 2010. Lecture Notes in Business Information Processing, vol. 67, pp. 16–30. Springer, Potsdam, Germany 13–14 Oct 2010 (2011)

56. Speck, A., Feja, S., Witt, S., Pulvermüller, E., Schulz, M.: Formalizing business process specifications. Comput. Sci. Inf. Syst./ComSIS 8(2), 427–446 (2011)

57. Wong, P.Y.H., Gibbons, J.: A process semantics for bpmn. In: Liu, S., Maibaum, T.S.E., Araki, K. (eds.) ICFEM 2008 Proceedings from the 10th International Conference on Formal

Engineering Methods. Lecture Notes in Computer Science, vol. 5256, pp. 355-374. Springer, Kitakyushu-City, Japan, 27–31 Oct 2008 (2008)

58. Lam, V.S.W.: Equivalences of BPMN processes. Serv. Oriented Comput. Appl. **3**(3), 189–204 (2009)

59. Lam, V.S.W.: Foundation for equivalences of BPMN models. Theor. Appl. Inf. **24**(1), 33–66 (2012)

60. Ligęza, A.: BPMN - a logical model and property analysis. Decis. Making Manuf. Serv. **5**(1–2), 57–67 (2011)

61. Bray, T., Paoli, J., Sperberg-McQueen, C.M., Maler, E. (eds.): Extensible Markup Language (XML) 1.0, 2nd edn, Technical report, World Wide Web Consortium, W3C Recommendation (2000). http://www.w3.org/TR/REC-xml

62. Lassila, O., Swick, R.R.: Resource description framework (RDF) model and syntax specification. Technical report, World Wide Web Consortium, W3C Recommendation (1999). http://www.w3.org/TR/REC-rdf-syntax

63. Brickley, D., Guha, R.V.: RDF vocabulary description language 1.0: RDF schema. W3C recommendation, W3C (2004). http://www.w3.org/TR/2004/REC-rdf-schema-20040210/

64. Dean, M., Schreiber, G.: OWL Web Ontology Language reference. W3C recommendation, W3C (2004). http://www.w3.org/TR/2004/REC-owl-ref-20040210/

65. Seaborne, A., Prud'hommeaux, E.: SPARQL query language for RDF. W3C recommendation, W3C (2008). http://www.w3.org/TR/2008/REC-rdf-sparql-query-20080115/

66. Hitzler, P., Krötzsch, M., Parsia, B., Patel-Schneider, P.F., Rudolph, S.: OWL 2 Web Ontology Language – primer. W3C recommendation, W3C (2009)

67. Motik, B., Grau, B.C., Horrocks, I., Wu, Z., Fokoue, A., Lutz, C.: OWL 2 Web Ontology Language: Profiles. W3C recommendation, W3C (2009)

68. Baader, F., Calvanese, D., McGuinness, D.L., Nardi, D., Patel-Schneider, P.F. (eds.): The Description Logic Handbook: Theory, Implementation, and Applications. Cambridge University Press, Cambridge (2003)

69. Horrocks, I., Patel-Schneider, P.F., Boley, H., Tabet, S., Grosof, B., Dean, M.: SWRL: A semantic web rule language combining OWL and RuleML, W3C member submission 21 May 2004. Technical report, W3C (2004)

70. Grosof, B.N., Horrocks, I., Volz, R., Decker, S.: Description Logic Programs: combining logic programs with description logic. In: Proceedings of the Twelfth International World Wide Web Conference, WWW, vol. 2003, pp. 48–57 (2003)

71. Horrocks, I., Parsia, B., Patel-Schneider, P., Hendler, J.: Semantic web architecture: stack or two towers? In: Fages, F., Soliman, S. (eds.) Principles and Practice of Semantic Web Reasoning. Lecture Notes in Computer Science, vol. 3703, pp. 37–41. Springer (2005)

72. Eiter, T., Ianni, G., Polleres, A., Schindlauer, R., Tompits, H.: Reasoning with rules and ontologies. In: Proceedings of Summer School Reasoning Web 2006 REWERSE (2006). Lecture Notes in Computer Science, vol. 4126, pp. 93–127. Lisbon, Portugal (4–8 Sept 2006)

73. Bratko, I.: Prolog Programming for Artificial Intelligence, 3rd edn. Addison Wesley, Upper Saddle River (2000)

74. Adrian, W.T., Nalepa, G.J., Kaczor, K., Noga, M.: Overview of selected approaches to rule representation on the Semantic Web. Technical Report CSLTR 2/2010, AGH University of Science and Technology (2010)

75. Antoniou, G., van Harmelen, F.: A Semantic Web Primer. The MIT Press, Cambridge (2008)

76. Nilsson, U., Małuszyński, J.: Logic, Programming and Prolog, 2nd edn. Wiley, New York (2000). http://www.ida.liu.se/~ulfni/lpp

77. Ullman, J.D.: Principles of Database and Knowledge-Base Systems, vol. I. Computer Science Press, New York (1988)

78. Kifer, M., Lausen, G., Wu, J.: Logical foundations of object-oriented and frame-based languages. J. ACM **42**(4), 741–843 (1995)

79. Donini, F.M., Lenzerini, M., Nardi, D., Schaerf, A.: $\mathcal{AL}$-log: integrating datalog and description logics. J. Intell. Coop. Inf. Syst. **10**, 227–252 (1998)

80. Levy, A.Y., Rousset, M.C.: Combining horn rules and description logics in CARIN. Artif. Intell. **104**(1–2), 165–209 (1998)
81. Motik, B., Horrocks, I., Rosati, R., Sattler, U.: Can OWL and logic programming live together happily ever after? Semant. Web - ISWC **2006**, 501–514 (2006)
82. W3C Working Group: SWRL: A Semantic Web Rule Language Combining OWL and RuleML (2004). http://www.w3.org/Submission/SWRL
83. Parsia, B., Sirin, E., Grau, B.C., Ruckhaus, E., Hewlett, D.: Cautiously Approaching SWRL. Technical report, Technical report, University of Maryland (2005)
84. McGuinness, D.L., Welty, C., Smith, M.K.: OWL Web Ontology Language guide. W3C recommendation, W3C (2004). http://www.w3.org/TR/2004/REC-owl-guide-20040210
85. Boley, H., Tabet, S., Wagner, G.: Design rationale for RuleML: A markup language for semantic web rules. In: Cruz, I.F., Decker, S., Euzenat, J., McGuinness, D.L. (eds.) SWWS, pp. 381–401 (2001)
86. Motik, B., Sattler, U., Studer, R.: Query answering for OWL-DL with rules. In: Journal of Web Semantics, pp. 549–563. Springer (2004)
87. Krötzsch, M., Rudolph, S., Hitzler, P.: ELP: Tractable rules for OWL 2. In: 7th International Semantic Web Conference (ISWC2008) (2008)
88. Knorr, M., Hitzler, P., Maier, F.: Reconciling OWL and non-monotonic rules for the semantic web. In: Raedt, L.D., Bessière, C., Dubois, D., Doherty, P., Frasconi, P., Heintz, F., Lucas, P.J.F. (eds.) ECAI 2012 - 20th European Conference on Artificial Intelligence. Including Prestigious Applications of Artificial Intelligence (PAIS-2012) System Demonstrations Track. Frontiers in Artificial Intelligence and Applications, vol. 242, pp. 474–479. IOS Press, Montpellier, France, 27–31 Aug 2012 (2012)
89. Martínez, D.C., Hitzler, P.: Extending description logic rules. In: Simperl, E., Cimiano, P., Polleres, A., Corcho, Ó., Presutti, V. (eds.) Proceedings of the Semantic Web: Research and Applications - 9th Extended Semantic Web Conference, ESWC 2012. Lecture Notes in Computer Science, vol. 7295, pp. 345–359. Springer, Heraklion, Crete, Greece, 27–31 May 2012 (2012)
90. Hitzler, P.: Recent advances concerning owl and rules. In: Invited Talk To 29th International Conference On Logic Programming (2013). http://www.iclp2013.org/files/downloads/iclp13_pascal.pdf
91. Mutharaju, R., Mateti, P., Hitzler, P.: Towards a rule based distributed OWL reasoning framework. In: Tamma, V.A.M., Dragoni, M., Gonçalves, R., Lawrynowicz, A. (eds.) Ontology Engineering - 12th International Experiences and Directions Workshop on OWL, OWLED 2015, co-located with ISWC 2015, Revised Selected Papers. Lecture Notes in Computer Science, vol. 9557, pp. 87–92. Springer, Bethlehem, PA, USA, 9–10 Oct. 2015 (2015)
92. Coutaz, J., Crowley, J.L., Dobson, S., Garlan, D.: Context is key. Commun. ACM **48**(3), 49–53 (2005)
93. Want, R., Falcao, V., Gibbons, J.: The active badge location system. ACM Trans. Inf. Syst. **10**, 91–102 (1992)
94. Schilit, B.N., Adams, N., Want, R.: Context-aware computing applications. In: Proceedings of the Workshop on Mobile Computing Systems and Applications. IEEE Computer Society, Washington, DC, USA, pp. 85–90 (1994)
95. Loke, S.W.: Representing and reasoning with situations for context-aware pervasive computing: a logic programming perspective. Knowl. Eng. Rev. **19**(3), 213–233 (2004)
96. Ranganathan, A., Campbell, R.H.: An infrastructure for context-awareness based on first order logic. Personal Ubiquitous Comput. **7**(6), 353–364 (2003)
97. Ranganathan, A., Al-Muhtadi, J., Campbell, R.H.: Reasoning about uncertain contexts in pervasive computing environments. IEEE Pervasive Comput. **3**(2), 62–70 (2004)
98. Hu, B., Wang, Z., Dong, Q.: A modeling and reasoning approach using description logic for context-aware pervasive computing. In: Lei, J., Wang, F., Deng, H., Miao, D. (eds.) Emerging Research in Artificial Intelligence and Computational Intelligence. Communications in Computer and Information Science, pp. 155–165. Springer, Berlin (2012)

99. Henricksen, K., Indulska, J.: Developing context-aware pervasive computing applications: models and approach. Pervasive Mob. Comput. **2**(1), 37–64 (2006)
100. Wang, X., Zhang, D., Gu, T., Pung, H.K.: Ontology based context modeling and reasoning using OWL. In: 2nd IEEE Conference on Pervasive Computing and Communications Workshops (PerCom 2004 Workshops), pp. 18–22. Orlando, FL, USA, 14–17 March 2004 (2004)
101. Chen, H., Perich, F., Finin, T.W., Joshi, A.: SOUPA: Standard ontology for ubiquitous and pervasive applications. In: 1st Annual International Conference on Mobile and Ubiquitous Systems (MobiQuitous 2004), Networking and Services, pp. 258–267. IEEE Computer Society, Cambridge, MA, USA, 22–25 Aug 2004 (2004)
102. Chen, H., Finin, T.W., Joshi, A.: Semantic web in the context broker architecture. In: PerCom, pp. 277–286. IEEE Computer Society (2004)
103. Ranganathan, A., McGrath, R.E., Campbell, R.H., Mickunas, M.D.: Use of ontologies in a pervasive computing environment. Knowl. Eng. Rev. **18**(3), 209–220 (2003)
104. Gu, T., Pung, H.K., Zhang, D.Q.: A middleware for building context-aware mobile services. In: 2004 IEEE 59th Vehicular Technology Conference, VTC 2004, vol. 5, pp. 2656–2660. Springer (2004)
105. Floch, J., Fra, C., Fricke, R., Geihs, K., Wagner, M., Lorenzo, J., Soladana, E., Mehlhase, S., Paspallis, N., Rahnama, H., Ruiz, P.A., Scholz, U.: Playing music – building context-aware and self-adaptive mobile applications. Softw.: Pract. Exp. **43**(3), 359–388 (2013)
106. Jaroucheh, Z., Liu, X., Smith, S.: Recognize contextual situation in pervasive environments using process mining techniques. J. Ambient Intelligence and Humanized Comput. **2**(1), 53–69 (2011)
107. Brezillon, P., Pasquier, L., Pomerol, J.C.: Reasoning with contextual graphs. Eur. J. Operation. Res. **136**(2), 290–298 (2002)
108. van der Aalst, W.M.P.: Process Mining - Discovery, Conformance and Enhancement of Business Processes. Springer, Berlin (2011)
109. Dey, A.K.: Understanding and using context. Pers. Ubiquitous Comput. **5**(1), 4–7 (2001)
110. Etter, R., Costa, P.D., Broens, T.: A rule-based approach towards context-aware user notification services. In: 2006 ACS/IEEE International Conference on Pervasive Services, pp. 281–284 (2006)
111. Wang, H., Mehta, R., Chung, L., Supakkul, S., Huang, L.: Rule-based context-aware adaptation: a goal-oriented approach. Int. J. Pervasive Comput. Commun. **8**(3), 279–299 (2012)
112. Dey, A.K.: Providing architectural support for building context-aware applications. Ph.D. thesis, Atlanta, GA, USA (2000) AAI9994400
113. Biegel, G., Cahill, V.: A framework for developing mobile, context-aware applications. In: 2004 Proceedings of the Second IEEE Annual Conference on Pervasive Computing and Communications, PerCom 2004, pp. 361–365 (2004)
114. Dey, A.K.: Modeling and intelligibility in ambient environments. J. Ambient Intell. Smart Environ. **1**(1), 57–62 (2009)
115. Ricci, F., Rokach, L., Shapira, B., Kantor, P.B.: Recommender Systems Handbook, 1st edn. Springer, New York (2010)
116. Jannach, D., Zanker, M., Felfernig, A., Friedrich, G.: Recommender Systems An Introduction. Cambridge University Press, Cambridge (2011)
117. Adomavicius, G., Tuzhilin, A.: In: Context-Aware Recommender Systems, pp. 217–253. Springer, Boston(2011)
118. Bobek, S., Nalepa, G.J., Ślażyński, M.: Challenges for migration of rule-based reasoning engine to a mobile platform. In: Dziech, A., Czyżewski, A. (eds.) Multimedia Communications, Services and Security. Communications in Computer and Information Science, vol. 429, pp. 43–57. Springer, Berlin, Heidelberg (2014)
119. Bobek, S., Nalepa, G.J.: Uncertain context data management in dynamic mobile environments. Future Gener. Comput. Syst. **66**, 110–124 (2017)
120. Sommerville, I.: Software Engineering, 7th edn. International Computer Science, Pearson Education Limited (2004)

121. Guida, G., Lamperti, G., Zanella, M.: Software Prototyping in Data and Knowledge Engineering. Kluwer Academic Publishers, Norwell (1999)
122. Wagner, G.: How to design a general rule markup language. In: XML Technology for the Semantic Web (XSW 2002). Lecture Notes in Informatics, pp. 19–37. HU, Berlin (2002)
123. Ceri, S., Gottlob, G., Tanca, L.: Logic Programming and Databases. Springer, New York (1990)
124. Seipel, D.: Practical applications of extended deductive databases in Datalog*. In: Proceedings of the 23rd Workshop on Logic Programming (WLP 2009) (2009)
125. OMG: Unified Modeling Language (OMG UML) version 2.2. superstructure. Technical Report formal/2009-02-02, Object Management Group (2009)
126. Pilone, D., Pitman, N.: UML 2.0 in a Nutshell. O'Reilly (2005)
127. Robin, J.: The object constraint language (OCL) (2007). http://www.cin.ufpe.br/~if710/2007/slides/OCL.ppt
128. Cuadra, D., Aljumaily, H., Castro, E., de Diego, M.V.: An OCL-based approach to derive constraint test cases for database applications. Int. J. Softw. Eng. Knowl. Eng. **21**(5), 621–645 (2011)
129. Lukichev, S., Wagner, G.: Visual rules modeling. In: Sixth International Andrei Ershov Memorial Conference Perspectives of System Informatics. Lecture Notes in Computer Science. Springer, Novosibirsk, Russia (2005)
130. Brockmans, S., Haase, P., Hitzler, P., Studer, R.: A metamodel and UML profile for rule-extended OWL DL ontologies. Lect. Notes Comput. Sci. **4011**, 303–316 (2006)
131. Stallman, R.M.: GNU Make Reference Manual. Samurai Media Limited (2015)
132. Nalepa, G.J., Kaczor, K.: Proposal of a rule-based testing framework for the automation of the unit testing process. In: Proceedings of the 17th IEEE International Conference on Emerging Technologies and Factory Automation ETFA 2012. Kraków, Poland, 28 Sept 2012 (2012)
133. W3C Working Group: Web services architecture w3c working group note 11 february 2004. Technical report, W3C (2004). https://www.w3.org/TR/ws-arch
134. Erl, T.; Service-Oriented Architecture (SOA): Concepts, Technology, and Design. Prentice Hall PTR, Upper Saddle River (2005)
135. Pant, K., Juric, M.: Business Process Driven SOA using BPMN and BPEL: From Business Process Modeling to Orchestration and Service Oriented Architecture. Packt Publishing, Birmingham (2008)
136. Rosenberg, F., Dustdar, S.: Business rules integration in bpel - a service-oriented approach. In: Seventh IEEE International Conference on E-Commerce Technology (CEC'05), pp. 476–479 (2005)
137. Ribarić, M. et al.: Modeling of Web Services using Reaction Rules. Information Science Reference. In: Handbook of Research on Emerging Rule-Based Languages and Technologies: Open Solutions and Approaches. IGI Global (2009)
138. Semantic Annotations for WSDL Working Group: Semantic annotations for WSDL and XML schema. W3C recommendation 28 Aug 2007. Technical report, W3C (2007). http://www.w3.org/TR/sawsdl
139. Benslimane, D., Dustdar, S., Sheth, A.: Services mashups: the new generation of web applications. IEEE Internet Comput. **12**(5), 13–15 (2008)
140. Nalepa, G.J., Ligęza, A.: Designing reliable Web security systems using rule-based systems approach. In: Menasalvas, E., Segovia, J., Szczepaniak, P.S. (eds.) Advances in Web Intelligence. First International Atlantic Web Intelligence Conference AWIC 2003. Lecture Notes in Artificial Intelligence, vol. 2663, pp. 124–133. Springer, Berlin, Heidelberg, Madrid, Spain, 5-6 May 2003 (2003)
141. Nalepa, G.J., Ligęza, A.: Security systems design and analysis using an integrated rule-based systems approach. In: Szczepaniak, P.S., Kacprzyk, J., Niewiadomski, A. (eds.) Advances in Web Intelligence: 3rd international Atlantic Web Intelligence Conference AWIC 2005. Lecture Notes in Artificial Intelligence, vol. 3528, pp. 334–340. Springer, Berlin, Heidelberg, New York, Lodz, Poland, 6-9 June 2005 (2005)

142. Nalepa, G.J.: A unified firewall model for web security. In: Węgrzyn-Wolska, K.M., Szczepaniak, P.S. (eds.) Advances in Intelligent Web Mastering, Proceedings of the 5th Atlantic Web Intelligence Conference – AWIC'2007. Advances in Soft Computing, vol. 43, pp. 248–253. Springer, Berlin, Heidelberg, New York, Fontainebleau, France (2007)

143. Nalepa, G.J.: Application of the XTT rule-based model for formal design and verifcation of internet security systems. In Saglietti, F., Oster, N. (eds.) Computer safety, reliability, and security: 26th international conference, SAFECOMP 2007. Lecture Notes in Computer Science, vol. 4680, pp. 81–86. Springer, Berlin, Heidelberg, Nuremberg, Germany, 18–21 Sept 2007 (2007)

144. Paton, N.W., Díaz, O.: Active database systems. ACM Comput. Surv. **31**(1), 63–103 (1999)

145. Luckham, D.: The power of events: an introduction to complex event processing in distributed enterprise systems. Addison Wesley Professional, Boston (2002)

146. Paschke, A., Boley, H.: Rules Capturing Events and Reactivity. Information Science Reference. In: Handbook of Research on Emerging Rule-Based Languages and Technologies: Open Solutions and Approaches. IGI Global (2009)

147. Paschke, A., Vincent, P., Alves, A., Moxey, C.: Tutorial on advanced design patterns in event processing. In: Bry, F., Paschke, A., Eugster, P.T., Fetzer, C., Behrend, A. (eds.) Proceedings of the Sixth ACM International Conference on Distributed Event-Based Systems, DEBS 2012, pp. 324–334. ACM, Berlin, Germany, 16–20 July 2012 (2012)

148. Paschke, A.: Reaction ruleml 1.0 for rules, events and actions in semantic complex event processing. In: Bikakis, A., Fodor, P., Roman, D. (eds.) Proceedings of Rules on the Web. From Theory to Applications - 8th International Symposium, RuleML 2014, Co-located with the 21st European Conference on Artificial Intelligence, ECAI 2014. Lecture Notes in Computer Science, vol. 8620, pp. 1–21. Springer, Prague, Czech Republic, 18–20 Aug 2014 (2014)

149. Paschke, A., Boley, H., Zhao, Z., Teymourian, K., Athan, T.: Reaction ruleml 1.0: Standardized semantic reaction rules. In: Bikakis, A., Giurca, A. (eds.) Proceedings of Rules on the Web: Research and Applications - 6th International Symposium, RuleML 2012. Lecture Notes in Computer Science, vol. 7438, pp. 100-119. Springer, Montpellier, France, 27–29 Aug 2012 (2012)

150. Covington, M.A., Nute, D., Vellino, A.: Prolog Programming in Depth. Prentice-Hall, Upper Saddle River (1996)

151. Ostermayer, L., Seipel, D.: A prolog framework for integrating business rules into java applications. In: Nalepa, G.J., Baumeister, J. (eds.) Procccdings of 9th Workshop on Knowledge Engineering and Software Engincering (KESE9) co-located with the 36th German Conference on Artificial Intelligence (KI2013). CEUR Workshop Proceedings, vol. 1070, Koblenz, Germany, 17 Sept 2013 (2013). http://CEUR-WS.org

152. Aho, A.V., Kernighan, B.W., Weinberger, P.J.: The AWK Programming Language. Addison-Wesley, Boston (1988)

153. Robbins, A.D.: GAWK: Effective AWK Programming. Free Software Foundation (2016)

154. Ostermayer, L.: Seamless Cooperation of Java and Prolog with CAPJA – A Connector Architecture for Prolog and Java. Ph.D. thesis, Univeristy of Würzburg (2017)

155. Clark, J.: XSL Transformations (XSLT) version 1.0 W3C recommendation 16 november 1999. Technical report, World Wide Web Consortium (W3C) (1999)

156. Frhwirth, T.: Constraint Handling Rules, 1st edn. Cambridge University Press, New York (2009)

157. Russell, S., Norvig, P.: Artificial Intelligence: A Modern Approach, 3rd edn. Prentice-Hall, Upper Saddle River (2009)

158. Hentenryck, P.V.: Constraint Satisfaction in Logic Programming. The MIT Press, Cambridge (1989)

159. Dechter, R.: Constraint Processing. The Morgan Kaufmann Series in Artificial Intelligence. Morgan Kaufmann (2003)

160. Apt, K.R.: Principles of Constraint Programming. Cambridge University Press, Cambridge (2003)

161. Wagner, G., Damásio, C.V., Antoniou, G.: Towards a general web rule language. Int. J. Web Eng. Technol. **2**(2/3), 181–206 (2005)
162. Fowler, M.: Domain-Specific Languages. Addison Wesley, Boston (2011)

Part II

Formal Models for Rules

Starting from the motivations and state of the art presented in first part of the book, this one introduces number of formalized models related to rules. We begin with a formalized model for decision rules called XTT2 introduced in Chap. 4. This model was first introduced in [1] and described in [2], then presented in [3]. It allows us to capture in a formalized way the interpretation of single rules. It also makes possible to describe the inference of the rule level using the ALSV(FD) logic [1, 4]. Moreover, the model assumes that rules are grouped into decision tables that are then used as decision units during the inference tasks. These tasks can be modeled as specific inference processes.

Inference processes in a decision network such as XTT2 can be complex. Therefore, in some cases it is desirable to delegate high-level inference control to a dedicated process engine. Moreover, this can allow integration with business process models. To meet these objectives, in Chap. 5 a formalized model for integrating the XTT2 decision units with business process models is presented. The XTT2 model is the foundation for a formalized rule-based design method called ARD+ and is discussed in Chap. 6. Moreover, it allows to automatically prototype a structure of a business logic model where high-level inference is provided by a business process engine, and XTT2 rule tasks are executed by a rule engine. Using this method, the structure of the linked decision tables is built in an iterative method based on the concept of the functional dependencies between attributes used in rules.

Two important extensions of the XTT2 representation are considered in the following chapters. The first, presented in Chap. 7 allows for the handling of uncertainty in rules. Consequently, XTT2 models can be easily used in highly dynamic environments such as context-aware systems on mobile platforms. Finally, the second one supports interoperability between different rule representations. In Chap. 8 an extended meta-model is presented. It enables the sharing of the same rule bases between the systems based on the XTT2 model discussed here as well as commonly used rule engines such as CLIPS or Drools.

Based on these models, in the third part of the book the SKE approach will be presented. Then, several case studies using the approach and the models from part two will be put froward. As such, the second part provides a formal foundation for the practical applications discussed in the third part of the book.

References

1. Nalepa, G.J., Ligęza, A.: HeKatE methodology, hybrid engineering of intelligent systems. Int. J. Appl. Math. Comput. Sci. **20**(1), 35–53 (2010)
2. Nalepa, G.J.: Semantic Knowledge Engineering. A Rule-Based Approach. Wydawnictwa AGH, Kraków (2011)
3. Vanthienen, J., Dries, E., Keppens, J.: Clustering knowledge in tabular knowledge bases. In: Proceedings Eighth IEEE International Conference on Tools with Artificial Intelligence, pp. 88–95 (1996)
4. David, J.M., Krivine, J.P., Simmons, R. (eds.): Second Generation Expert Systems. Springer, Secaucus (1993)

Chapter 4
Formalism for Description of Decision Rules

In this chapter we discuss the *eXtended Tabular Trees* (XTT2). It is a knowledge representation method for rules, but also a visual modeling method for RBS. It uses strict formalization of rule syntax and improves design and verification of RBS. It is the core of the *Semantic Knowledge Engineering* (SKE) approach. The formalization of XTT2 was introduced in [1], and the assumptions of SKE itself were first presented in [2].

This chapter discusses the core features of XTT2. In fact, the work on the representation lasted about a decade. In [3] some concepts of formal representation with the so-called Ψ-trees were introduced. This framework was described at a conceptual level in [4, 5]. Then the knowledge base design using the XTT2 method based upon the ALSV(FD); was partially presented in [6, 7] and finally in [8]. This work follows the comprehensive overview of the formal foundations of Rule-Based Systems discussed in [9]. Finally, the formalization of XTT2 was introduced in [1]. Some of the benefits of such formalized methods were mentioned in Sect. 1.6.

The discussion is partitioned as follows. We begin with the formalization of single rules with the ALSV(FD) logic in Sect. 4.1. This approach extends the logical description first discussed in Sect. 1.5. Inference with ALSV(FD) is discussed in Sect. 4.2. Based on this, the formalization of rule bases with the XTT2 method is discussed in Sect. 4.3. Modularization described in Sect. 4.4 follows the ideas discussed in Sect. 2.3. Such rule bases need custom inference algorithms, as described below in Sect. 4.5. Finally, the formalization allows for the XTT2 rule bases, as presented in Sect. 4.6, to be verified. The chapter is concluded in Sect. 4.7.

4.1 Attributive Logic with Set Values over Finite Domains

Here we consider an improved and extended version of Set Attributive Logics, namely ALSV(FD) (Attributive Logic with Set Values over Finite Domains), previously discussed in [6–8]. For the sake of simplicity, no objects are specified in an explicit

© Springer International Publishing AG 2018   85
G.J. Nalepa, *Modeling with Rules Using Semantic Knowledge Engineering*,
Intelligent Systems Reference Library 130,
https://doi.org/10.1007/978-3-319-66655-6_4

way. The formalism is oriented towards Finite Domains (FD) and its expressive power is increased through the introduction of new relational symbols enabling definitions of atomic formulae. Moreover, ALSV(FD) introduces a formal specification for the partitioning of the attribute set needed for its practical implementation, and a more coherent notation.

Simple and Generalized Attributes

Let $\mathbb{A}$ denote the set of all attributes used to describe the system. Each attribute has a set of admissible values that it takes (a domain). $\mathbb{D}$ is the *set of all possible attributes' values*:

$$\mathbb{D} = \mathbb{D}_1 \cup \mathbb{D}_2 \cup \cdots \cup \mathbb{D}_n \tag{4.1}$$

where $\mathbb{D}_i$ is the domain of attribute $A_i \in \mathbb{A}, i = 1 \ldots n$. Any domain $\mathbb{D}_i$ is assumed to be a *finite*,[1] discrete set. In a general case, a domain can be ordered, partially ordered or unordered (this depends on the specification of an attribute, see Sect. 9.6).

In ALSV(FD) (as in SAL) two types of attributes are identified: *simple* – those which take only one value at a time, and *generalized* – those that take multiple values at a time. Therefore, we introduce the following partitioning of the set of all attributes:

$$\mathbb{A} = \mathbb{A}^s \cup \mathbb{A}^g, \ \mathbb{A}^s \cap \mathbb{A}^g = \varnothing \tag{4.2}$$

where:

- $\mathbb{A}^s$ is the set of simple attributes, and
- $\mathbb{A}^g$ is the set of generalized attributes.

A *simple attribute* A_i is a function (or a partial function) of the form:

$$A_i : \mathbb{O} \to \mathbb{D}_i \tag{4.3}$$

where:

- $\mathbb{O}$ is a set of objects,
- $\mathbb{D}_i$ is the domain of attribute A_i.

The definition of *generalized attribute* is as follows:

$$A_i : \mathbb{O} \to 2^{\mathbb{D}_i} \tag{4.4}$$

where:

- $\mathbb{O}$ is a set of objects,
- $2^{\mathbb{D}_i}$ is the set of all possible subsets of the domain $\mathbb{D}_i$.

---

[1] This assumption has an engineering motivation. In most of the practical applications, the domains of attributes can be defined in a satisfactory way as finite. This is due to a number of physical limitations of the modeled systems.

Attribute A_i denotes a property of an object. The expression $A_i(o)$, where $o \in \mathbb{O}$, denotes the value of property A_i of object o. However, here we assume that only one object (in this case it is the system being described) with a specific property name exists. This is why, in the remaining part of this discussion, the following *notational convention* is used: the formula A_i simply denotes a *value* of the attribute A_i.

Let us consider a basic introductory Example 4.1.1,[2] called *Bookstore*. It is the first of several examples found in this book.

> **Example 4.1.1** (**BOOKSTORE: attributes**) *Consider a system for recommending books to different groups of people depending on their age and reading preferences. The age of a reader and his/her preference could be represented by the following attributes:* $\mathbb{A} = \{fav\_genres, age, age\_filter, rec\_book\}$. *In this case we assume that the second attribute is a simple one whereas the others are generalized. The fourth attribute contains book titles that can be recommended to a reader. The attributes have the following domains:* $\mathbb{D} = \mathbb{D}_{fav\_genres} \cup \mathbb{D}_{age} \cup \mathbb{D}_{age\_filter} \cup \mathbb{D}_{rec\_book}$, *where:*
>
> - $\mathbb{D}_{fav\_genres} = \{horror, handbook, fantasy, science, historical, poetry\}$,
> - $\mathbb{D}_{age} = \{1 \ldots 99\}$,
> - $\mathbb{D}_{age\_filter} = \{young\_horrors, young\_poetry, adult\_horrors, adult\_poetry\}$,
> - $\mathbb{D}_{rec\_book} = \{'It', 'Logical\ Foundations\ for\ RBS', 'The\ Call\ of\ Cthulhu'\}$.

The rule-based system is described using attributes. The system is in a certain state specified by the values of these attributes.

State Representation

The current values of all attributes are specified within the contents of the knowledge base. From logical point of view the *state* of the system is represented as a logical formula of the form:

$$s: (A_1 = S_1) \wedge (A_2 = S_2) \wedge \ldots \wedge (A_n = S_n) \tag{4.5}$$

where A_i are the attributes and S_i are their current values. Note that $S_i \in \mathbb{D}_i$ for simple attributes and $S_i \subseteq \mathbb{D}_i$ for generalized ones.

An explicit notation for covering unspecified, unknown values is proposed: $A_i = null$ means that the value of A_i is unspecified.

---

[2]The examples of books given here are as follows: *It* by Stephen King (considered to be a horror book), *Logical Foundations for RBS* by Antoni Ligęza (a science book), and *The Call of Cthulhu* by Howard P. Lovecraft (a fantasy book), where *Cthulhu* is the name of an extraterrestrial entity.

Example 4.1.2 (**BOOKSTORE: state**) *Following the example, an exemplary state can be defined as:*

$$(age = 16) \wedge (fav\_genres = \{horror, fantasy\})$$

This means that a given person is 16 years old and she or he likes reading horror and fantasy books. In fact, it is a partial state where only the values of the input attributes are defined. In this example it will be sufficient to start the inference process. To specify the full state, the values of the remaining attributes should be defined as null.

Classes of Attributes

Considering the practical implementation of the communication architecture of a rule-based system,[3] where several attribute classes are identified, the following partitioning of the set of attributes is introduced:

$$\mathbb{A}^s = \mathbb{A}^s_{in} \cup \mathbb{A}^s_{int} \cup \mathbb{A}^s_{out} \cup \mathbb{A}^s_{io} \tag{4.6}$$
$$\mathbb{A}^g = \mathbb{A}^g_{in} \cup \mathbb{A}^g_{int} \cup \mathbb{A}^g_{out} \cup \mathbb{A}^g_{io} \tag{4.7}$$

where all these sets are pairwise disjoint:

- $\mathbb{A}^s_{in}$, $\mathbb{A}^g_{in}$ are the sets of *input* attributes,
- $\mathbb{A}^s_{int}$, $\mathbb{A}^g_{int}$ are the sets of *internal* attributes,
- $\mathbb{A}^s_{out}$, $\mathbb{A}^g_{out}$ are the sets of *output* attributes, and
- $\mathbb{A}^s_{io}$, $\mathbb{A}^g_{io}$ are the sets of attributes that can be simultaneously input and output (*communication* attributes).

These *attribute classes* (i.e. input, internal, output, and communication) are used in the rule specification to support the interaction of the system with environment. They are handled by dedicated *callbacks*. These callbacks are procedures providing means for reading and writing attribute values (see Sect. 9.6).

Example 4.1.3 (**BOOKSTORE: Classes of attributes**) *In the example, both fav_genres and age attributes are input, age_filter is internal, and rec_book is an output one.*

The ALSV(FD) has been developed to describe rules. In order to do so, it provides certain expressions to represent the conditions and actions of rules. These expressions are the atomic formulae of ALSV(FD). Their syntax is presented in the following section.

---

[3] An example of this kind of architecture for the SKE approach will be given later on in Sect. 9.3.

Table 4.1 Simple attribute formulae syntax

Syntax	Meaning	Relation
$A_i = d_i$	The value of A_i is precisely defined as d_i	eq
$A_i \in V_i$	The current value of A_i belongs to V_i	in
$A_i \neq d_i$	Shorthand for $A_i \in (\mathbb{D}_i \setminus \{d_i\})$	neq
$A_i \notin V_i$	Shorthand for $A_i \in (\mathbb{D}_i \setminus V_i)$	notin

Table 4.2 Generalized attribute formulae syntax

Syntax	Meaning	Relation
$A_i = V_i$	A_i equals to V_i (and nothing more)	eq
$A_i \neq V_i$	A_i is different from V_i (at least one element)	neq
$A_i \subseteq V_i$	A_i is a subset of V_i	subset
$A_i \supseteq V_i$	A_i is a superset of V_i	supset
$A_i \sim V_i$	A_i has a non-empty intersection with V_i	sim
$A_i \nsim V_i$	A_i has an empty intersection with V_i	notsim

Atomic Formulae Syntax

Let A_i be an attribute from $\mathbb{A}$, and $\mathbb{D}_i$ the domain related to it. Let V_i denote an arbitrary subset of $\mathbb{D}_i$ and let $d_i \in \mathbb{D}_i$ be a single element of the domain. The legal atomic formulae of ALSV(FD) along with their semantics are presented in Tables 4.1 and 4.2, for simple and general attributes respectively.

If V_i is an empty set (the attribute takes no value), we shall write $A_i = \varnothing$. In the case when the value of A_i is unspecified, we shall write $A_i = null$. If the current attribute value is of no importance, we shall write $A_i = any$.

More complex formulae can be constructed with the *conjunction* ($\wedge$) and *disjunction* ($\vee$); both of these have classical interpretation. For enabling efficient verification, there is no explicit use of negation in the formulae. The proposed set of relations has been selected for convenience and they are not completely independent.[4] The meaning of these formulae is presented below.

Formulae Semantics

The semantics of the atomic formulae is as follows:

- If $V_i = \{d_1, d_2, \ldots, d_k\}$, then $A_i = V_i$ means that the attribute takes as its value the set of all the values specified with V_i (and nothing more).
- $(A_i \subseteq V_i) \equiv (A_i = U_i)$ for some U_i such that $U_i \subseteq V_i$, i.e. A_i takes *some* of the values from V_i (and nothing out of V_i),
- $(A_i \supseteq V_i) \equiv (A_i = W)$, for some W such that $V_i \subseteq W$, i.e. A_i takes *all* of the values from V_i,

---

[4]For example, $A_i = V_i$ can be defined as $A_i \subseteq V_i \wedge A_i \supseteq V_i$, but it is much more concise and natural to use just "=" directly.

- $(A_i \sim V_i) \equiv (A_i = X_i)$, for some X_i such that $V_i \cap X_i \neq \varnothing$, i.e. A_i takes *some* of the values from V_i.

> **Example 4.1.4** (**BOOKSTORE: atomic formulae**) *In the example the follow-ing atomic formulae could be present:* $age \in \{1, 2, 3, 4, 5, 6, 7, 8, 9, 10, 11, 12, 13, 14, 15, 16, 17\}$ *which could also be denoted as* $age < 18$ *and* $fav\_genres \subseteq$ *{science, fantasy, horror}. The interpretation of the second one is: the person likes a subset of science, fantasy, horror books.*

ALSV(FD) is an expressive language that allows for the building of conditional statements of rules. It can handle complex conditions, e.g. where formulae may not be logically independent. For example imagine two formulae: the first: $age \in \{1, 2, 3, 4\}$ and the second one: $age \in \{1, 2, 3, 4, 5, 6, 7, 8\}$. If these formulae are used in the conditions of two different rules, the first rule is not needed, and as such can be removed from the knowledge base. This is because the second condition is more general.[5] This is a case of rule subsumption. In order to be able to analyze such a case of a logical relationship between rules, as well as other cases, the inference rules for the atomic formulae are considered in the next section.

4.2 Inference in ALSV(FD) Formulae

The summary of the inference rules for atomic formulae with simple attributes (where an atomic formula is the logical consequence ($\models$) of another atomic formula) is presented in Table 4.3. The summary of the inference rules for atomic formulae with generalized attributes is presented in Table 4.4. The tables are to be read as follows: if an atomic formula in the leftmost column holds and a condition stated in the same row is true then the corresponding atomic formula in the topmost row is also true. In other words, the formula in the topmost row is a logical consequence ($\models$) of the one from the leftmost column provided the condition is fulfilled.

For example the first row of the second table should be read: if $A_i = V_i$ (see the leftmost column) and provided that $V_i \subseteq W_i$ (the same row, the fourth column) we can conclude that $A_i \subseteq W_i$ (the topmost row), where $V_i \subseteq \mathbb{D}_i$, $W_i \subseteq \mathbb{D}_i$.

The conditions in Tables 4.3 and 4.4 are *satisfactory*. The mark "—" has the interpretation "does-not-apply/no-definition". Rules in Tables 4.3 and 4.4 can be used to verify rule *subsumption* (see section "Subsumption of a Pair of Rules").

Furthermore, to introduce more complex procedures for the logical analysis of formulae e.g. analysis of intersection (overlapping of rule preconditions), one may

---

[5]This is a simplified case, where these formulae are the only conditions. Moreover, the decisions of both rules must be the same. For a detailed discussion refer to section "Subsumption of a Pair of Rules".

Table 4.3 Inference rules for atomic formulae, simple attributes

$\models$	$A_i = d_j$	$A_i \neq d_j$	$A_i \in V_j$	$A_i \notin V_j$
$A_i = d_i$	$d_i = d_j$	$d_i \neq d_j$	$d_i \in V_j$	$d_i \notin V_j$
$A_i \neq d_i$	—	$d_i = d_j$	$V_j = \mathbb{D}_i \setminus \{d_i\}$	$V_j = \{d_i\}$
$A_i \in V_i$	$V_i = \{d_j\}$	$d_j \notin V_i$	$V_i \subseteq V_j$	$V_i \nsim V_j$
$A_i \notin V_i$	$\mathbb{D}_i \setminus V_i = \{d_j\}$	$V_i = \{d_j\}$	$V_j = \mathbb{D}_i \setminus V_i$	$V_j \subseteq V_i$

Table 4.4 Inference rules for atomic formulae, generalized attributes

$\models$	$A_i = W_i$	$A_i \neq W_i$	$A_i \subseteq W_i$	$A_i \supseteq W_i$	$A_i \sim W_i$	$A_i \nsim W_i$
$A_i = V_i$	$V_i = W_i$	$V_i \neq W_i$	$V_i \subseteq W_i$	$V_i \supseteq W_i$	$V_i \sim W_i$	$V_i \nsim W_i$
$A_i \neq V_i$	—	$V_i = W_i$	$W_i = \mathbb{D}_i$	—	$W_i = \mathbb{D}_i$	—
$A_i \subseteq V_i$	—	$V_i \subset W_i$	$V_i \subseteq W_i$	—	$W_i = \mathbb{D}_i$	$V_i \nsim W_i$
$A_i \supseteq V_i$	—	$W_i \subset V_i$	$W_i = \mathbb{D}_i$	$V_i \supseteq W_i$	$V_i \sim W_i$	—
$A_i \sim V_i$	—	$V_i \nsim W_i$	$W_i = \mathbb{D}_i$	—	$V_i = W_i$	—
$A_i \nsim V_i$	—	$V_i \sim W_i$	$W_i = \mathbb{D}_i$	—	$W_i = \mathbb{D}_i$	$V_i = W_i$

Table 4.5 Inconsistency conditions for atomic formulae pairs

INCONS	$A_i = W_i$	$A_i \subseteq W_i$	$A_i \supseteq W_i$	$A_i \sim W_i$
$A_i = V_i$	$W_i \neq V_i$	$V_i \nsubseteq W_i$	$W_i \nsubseteq V_i$	$V_i \nsim W_i$
$A_i \subseteq V_i$	$W_i \nsubseteq V_i$	$V_i \nsim W_i$	$W_i \nsubseteq V_i$	$V_i \nsim W_i$
$A_i \supseteq V_i$	$V_i \nsubseteq W_i$	$V_i \nsubseteq W_i$	—	—
$A_i \sim V_i$	$V_i \nsim W_i$	$V_i \nsim W_i$	—	—

be interested if two atomic formulae cannot simultaneously be true, and if so – under what conditions. Table 4.5 specifies the conditions for mutual exclusion. The interpretation of the table is straightforward: if the condition specified at the intersection of a row and column holds then the atomic formulae labeling this row and column cannot hold simultaneously. However, this can only be viewed as a satisfactory condition.

For example, formula $A_i \subseteq V_i \wedge A_i \subseteq W_i$ is *inconsistent* if $V_i \cap W_i = \varnothing$.

The ALSV(FD) has been introduced with practical applications for rule languages in mind. In fact, the primary aim of the presented language is to formalize and extend the notational possibilities and expressive power of rule languages for modularized RBS. The analysis and formalization presented in this section started from the basic concept of an attribute. The ALSV(FD) logic formulae correspond to simple statements (facts) about attribute values. These formulae are then used to express certain conditions. Using this formalism, a complete solution that allows for decision rules to be build is discussed in the following section.

4.3 Formulation of XTT2 Rules

The rule formalism considered here is called XTT2. It provides a formalized rule specification, based upon some preliminary results presented in [4, 5]. This section starts from single rule formulation using the ALSV(FD) concepts. Then provides definitions for grouping similar rules into decision components (tables) linked into an inference network.

Let us consider a rule: *IF a person is younger than 18 and she or he likes horror books THEN recommend the book "It"*. It would be formalized in ALSV(FD) as:

$$r : [age < 18 \wedge fav\_genres = \{\text{horror}\}] \longrightarrow [rec\_book := \{\text{'It'}\}]$$

The $\longrightarrow$ symbol separates the *condition* part of a rule (on the left) from the *decision* part (on the right). Both parts are built using *triples* composed of an attribute name, operator, and a value. In the decision part of the rule, only simple assignment operator ($:=$) of a new attribute value is permitted. In the condition a number of expressions using the atomic formulae are permitted. In the above rule, there are two triples in the condition part of the rule. A rule can be *fired* if its conditions are satisfied. Hence, considering the exemplary state defined on p. 88, a person is in fact younger than 18, and she or he likes horror books. Moreover, we consider a concept of rule *schema*. A schema contains attributes used in a rule. In this case the schema is $h = (\{age, fav\_genres\}, \{rec\_book\})$.

Schemas are used to identify rules working in the same context, and group them into *tables*. Tables can be *linked* to hint the inference control mechanism. Firing a rule from a given table may result in switching the rule interpretation to another table. A set of rules, grouped in tables linked into a network forms an XTT2 *knowledge base*.

In the following subsections a number of *structural definitions* are given. Their goal is to formalize the structure of the knowledge base. Moreover, they are used to organize the process of the design and possible translation of the knowledge base.

Conclusion and Decision

Let us consider the following convention, where two identifiers will be used to denote attributes as well as operators in rule parts:

- *cond* corresponds to the *conditional* part of a rule, and
- *dec* corresponds to the *decision* part of a rule.

Using it, two subsets of the attribute set can be identified. $\mathbb{A}^{cond}$ is a subset of attributes taken from set $\mathbb{A}$ that contains attributes present in the conditional part of a rule. $\mathbb{A}^{dec}$ is a subset of attributes taken from set $\mathbb{A}$ that contains attributes present in the decision part.

Relational Operators in Rules

Considering the syntax of the legal ALSV(FD) formulae previously presented in Tables 4.1 and 4.2, the legal use of the relational operators in rules is specified. With respect to the previously identified attribute classes, not every operator can be used with any attribute or in any rule part. Hence, the set of all operators has been divided into smaller subsets that contain all the operators, which can be used at the same time.

The set of all relational *operators* that can be used in rules is defined as follows:

$$\mathbb{F} = \mathbb{F}^{cond} \cup \mathbb{F}^{dec} \qquad (4.8)$$

where:

- $\mathbb{F}^{cond}$ is a set of all operators that can be used in the conditional part of a rule.

$$\mathbb{F}^{cond} = \mathbb{F}_a^{cond} \cup \mathbb{F}_s^{cond} \cup \mathbb{F}_g^{cond} \qquad (4.9)$$

where:

- $\mathbb{F}_a^{cond}$ contains operators, that can be used in the rule conditional part with all attributes. The set is defined as:

$$\mathbb{F}_a^{cond} = \{=, \neq\} \qquad (4.10)$$

- $\mathbb{F}_s^{cond}$ is the set that contains the operators, which can be used in the rule conditional part with simple attributes. The set is defined as:

$$\mathbb{F}_s^{cond} = \{\in, \notin\} \qquad (4.11)$$

ALSV(FD) also allows for the use of the following operators $<, >, \leq, \geq$ which provide only a variation for $\in$ operator. These operators can be used only with attributes whose domains are ordered sets.

- $\mathbb{F}_g^{cond}$ contains operators that can also be used in the rule conditional part with generalized attributes. The set is defined as:

$$\mathbb{F}_g^{cond} = \{\subseteq, \supseteq, \sim, \nsim\} \qquad (4.12)$$

- $\mathbb{F}^{dec}$ is a set of all operators that can be used in the rule decision part.

$$\mathbb{F}^{dec} = \{:=\} \qquad (4.13)$$

The operator $:=$ allows a new value to be assigned to an attribute.[6]

---

[6]Here only the assignment operator is used. In fact some RBS provide more robust operators, such as *restrict* in Kheops [10]. It allows the set of values of a given attribute to be narrowed. While such

To specify in the rule condition that the value of an attribute is to be *null* (unknown) or *any* (unimportant) the operator $=$ is used. To specify in the same rule part the value of an attribute is not *null* the operator $\neq$ is used.

ALSV(FD) *Triples*

Let us consider the set $\mathbb{E}$ that contains all the triples that are legal atomic formulae in ALSV(FD). The triples are built using the operators:

$$\mathbb{E} =\{(A_i, \propto, d_i), A_i \in \mathbb{A}^s, \propto \in (\mathbb{F}\backslash\mathbb{F}_g^{cond}), d_i \in \mathbb{D}_i\} \cup \qquad (4.14)$$
$$\{(A_i, \propto, V_i), A_i \in \mathbb{A}^g, \propto \in (\mathbb{F}\backslash\mathbb{F}_s^{cond}), V_i \in 2^{\mathbb{D}_i}\}$$

The ALSV(FD) triples are the basic components of rules.

XTT2 *Rule*

Let us consider the set of all rules defined in the knowledge base denoted as $\mathbb{R}$. A single XTT2 rule is a triple:

$$r = (\mathbb{COND}, \mathbb{DEC}, \mathbb{ACT}) \qquad (4.15)$$

where:

- $\mathbb{COND} \subseteq \mathbb{E}$,
- $\mathbb{DEC} \subseteq \mathbb{E}$, and
- $\mathbb{ACT}$ is a set of actions to be executed when a rule is fired.

A rule can be written using *LHS* (*Left Hand Side*) and *RHS* (*Right Hand Side*)[7]:

$$LHS(r) \rightarrow RHS(r), DO(\mathbb{ACT}) \qquad (4.16)$$

where *LHS(r)* and *RHS(r)* correspond respectively to the condition and decision parts of the rule r, and $DO(\mathbb{ACT})$ involves executing actions from a predefined set. Actions are not included in the *RHS* of the rule, because it is assumed that they are independent from the considered system, and the execution of actions does not change the state of the system.

The rule defined by Formula 4.15 can also be presented in the following form:

$$r: [\phi_1 \wedge \phi_2 \wedge \cdots \wedge \phi_k] \rightarrow [\theta_1 \wedge \theta_2 \wedge \cdots \wedge \theta_l], DO(\mathbb{ACT}) \qquad (4.17)$$

where:

(Footnote 6 continued)
an operator is useful in certain applications (e.g. in diagnostic systems, where the set of potential diagnoses can be incrementally narrowed), it poses practical verification problems. Therefore, it is not used in ALSV(FD).

[7]The LHS/RHS convention is common among a number of classic RBS, e.g. see [11].

- $\{1, \ldots, k\}$ and $\{1, \ldots, l\}$ are the sets of identifiers, $k \in \mathbb{N}, l \in \mathbb{N}$
- $\phi_1, \ldots, \phi_k \in \mathbb{COND}$, and
- $\theta_1, \ldots, \theta_l \in \mathbb{DEC}$.

From a logical point of view, the order of the atomic formulae in both the precondition and conclusion parts is unimportant. Moreover, rules with empty decisions are considered. They are useful in the inference process control. Rules with no conditions can be used to set the attribute value. Such rules may be used to populate the knowledge base with facts or import sets of values.

> **Example 4.3.1** (**BOOKSTORE: Rule firing**) *Consider the example of the following rules:*
>
> $r_1 : [age < 18 \wedge fav\_genres \supseteq \{horror\}] \rightarrow [age\_filter := young\_horrors]$
> $r_2 : [age = \_ \wedge fav\_genres \subseteq \{science\}] \rightarrow [age\_filter := all\_science]$
> $r_3 : [age\_filter \in \{young\_horrors, adult\_horrors\}] \rightarrow [rec\_book := 'It']$
> *Now, having the previously defined state:* $(age = 16) \wedge (fav\_genres = \{horror, fantasy\})$, *it can be observed, that rules* r_1 *and* r_3 *could be fired. The notion of rule firing is explained in the next section.*

Rule Firing

Considering the previous definitions, firing a single XTT2 rule r involves the following basic steps:

1. Checking if *all* the ALSV(FD) triples in the $\mathbb{COND}$ part are satisfied.
2. If so, changing the system state by evaluating triples (assigning new values to attributes) in the $\mathbb{DEC}$ part.
3. Executing actions defined by $\mathbb{ACT}$; actions do not change attribute values.

Having the structure of a single rule defined, the structure of the complete knowledge base is introduced. The knowledge base is composed of tables grouping rules having the same attributes lists (rule schemas).

4.4 Structure of XTT2 Rule Base

Rule Schema

Let us introduce the *trunc* function which transforms the set of atomic formulae into a set of attributes that are used in these triples. It is defined as follows:

$$\text{trunc} \colon 2^{\mathbb{E}} \rightarrow 2^{\mathbb{A}} \tag{4.18}$$

$$\text{trunc}((A, \propto, d)) = A,$$
$$\text{trunc}((A, \propto, V)) = A,$$
$$\text{trunc}(\{e_1, e_2, \ldots, e_n\}) = \text{trunc}(e_1) \cup \text{trunc}(e_2) \cup \ldots \cup \text{trunc}(e_n).$$

where $e_1, e_2 \ldots e_n \in \mathbb{E}$

Now let us introduce a concept of a *rule schema*. It can be defined as follows:

$$\forall r = (\text{COND}, \text{DEC}, \text{ACT})$$
$$h(r) = (\text{trunc}(\text{COND}), \text{trunc}(\text{DEC})) \tag{4.19}$$

Therefore, each rule has a schema that is a pair of attribute sets:

$$h = (H^{cond}, H^{dec}) \tag{4.20}$$

where H^{cond} and H^{dec} sets define the attributes occurring in the conditional and decision part of the rule. Therefore, $H^{cond} = \text{trunc}(\text{COND})$ and $H^{dec} = \text{trunc}(\text{DEC})$.

A schema is used to identify rules working in the same situation (operational context). Such a set of rules can form a decision component in the form of a decision table. A schema can also be considered a table header.

Decision Component (Table)

Let us consider a *decision component* (or table). It is an ordered set (sequence) of rules having the same rule schema, defined as follows:

$$t = (r_1, r_2, \ldots, r_n) \text{ such that } \forall i, j : r_i, r_j \in t \rightarrow h(r_i) = h(r_j) \tag{4.21}$$

where $h(r_i)$ is the schema of the rule r_i.

In XTT2 the rule schema h can also be called the schema of the component (or table). Considering the rule schema notation, a table schema has the following property:

$$\forall r \in t \; h(r) = h(t) \tag{4.22}$$

Example 4.4.1 (Basic XTT2 knowledge base) *Consider the following illustration given in Fig. 4.1. On the left table t_1 is represented. It is an example of a table having three rules: r_1, r_2, r_3. These rules have the same schema $h_1 = (\{A_1, A_2, A_3\}, \{A_4, A_5\})$. This means that respective ALSV(FD) triples contain given attribute e.g. a triple $e_{2,3}$ is a part of rule r_2 and it contains the attribute A_3. To simplify the visual representation, a convention is introduced, where the schema of a table is depicted on the top of the table.*

Inference Link

An *inference link* l is an ordered pair:

$$l = (r, t), \; l \in \mathbb{R} \times \mathbb{T} \tag{4.23}$$

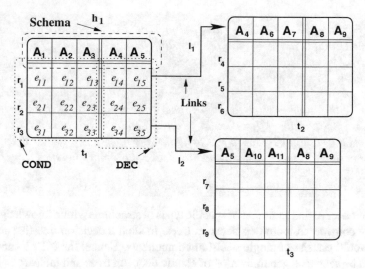

Fig. 4.1 An example of an XTT2 knowledge base

where:

- $\mathbb{R}$ is the set of rules in the knowledge base, and
- $\mathbb{T}$ is the set of tables in the knowledge base.

Components (tables) are connected (linked) in order to provide inference control. A single link connects a single rule (a row in a table) with another table. A structure composed of linked decision components is called a XTT2 knowledge base.

XTT2 *Knowledge Base*

The XTT2 *knowledge base* is the set of components connected with links. It can be defined as an ordered pair:

$$X = (\mathbb{T}, \mathbb{L}) \qquad (4.24)$$

where:

- $\mathbb{T}$ is a set of components (tables),
- $\mathbb{L}$ is a set of links, and
- all the links from $\mathbb{L}$ connect rules from $\mathbb{R}$ with tables from $\mathbb{T}$.

Links are introduced during the design process according to the specification provided by the designer. The knowledge base is a form of an inference network.

Example 4.4.2 (XTT2 inference control) *An example of a simple XTT2 knowledge base is presented in Fig. 4.1. It is composed of three tables $X_1 = (\{t_1, t_2, t_3\}$ and two links $\{l_1, l_2\}$). In the first table t_1 a rule schema can be observed. There are nine rules, three in each of the tables. In the first table,* *the conditional and decision parts are presented (the action-related part is optional). Rules in a table are fired starting from the first one. The order in which tables are fired depends on a specific inference mode (see Sect. 4.5). In the simple forward chaining (data-driven) mode, the inference process would be as follows: If rule r_1 from the table t_1 is fired the inference control would be passed to table t_2 through link l_1, otherwise fire rule r_2. If rule r_2 is fired then proceed to rule r_3. If rule r_3 is fired the inference control is be passed to table t_3 through link l_2. If it is not fired then the inference process stops (this is the case of a simple forward chaining mode; for other inference modes see Sect. 4.5).*

Let us observe that a number of specific types of structures within knowledge bases could be considered including decision trees. In such a decision tree-like structure nodes would consist of single decision components. Hence, the XTT2 knowledge base can be seen as a generalization of classic decision trees and tables.[8]

Examples of XTT2 *Knowledge Bases*

So far we have introduced the BOOKSTORE example to illustrate the basic definitions of ALSV(FD) formulae and XTT2 rules. In the rest of the book we will use several extended examples (called "system cases") to show how different features of XTT2 and related models can be used. The complete list of examples was discussed in the introduction to the book. The contents of the examples are also listed in the appendices. In this section we introduce the second example called PLI.

PLI *System Case*

The so called "Polish Liability Insurance system case" (PLI for short)[9] considers a system for determining the price of the car liability insurance in Poland, which protects against third party insurance claims. The insurance price can be determined based upon data such as the driver's age, the period of holding the license, the number of accidents in the last year, and the previous class of insurance junction. Other relevant factors when it comes to calculating the insurance price are data about the vehicle: the engine capacity, age, car seats, and its technical examination. Moreover, in the calculation, the insurance premium can be increased or decreased because of number of the payment installments, other insurances, continuity of insurance or the number of cars insured.

This illustrative example consists of 54 business rules. For a compact representation, all rules are grouped within three decision tables that correspond to three steps of insurance price calculation. The first step consists of 5 rules and it determines the basic rate based upon the vehicle engine capacity (see Table 4.6). The second

---

[8]In fact, it can also be a generalization of structures describing other inference processes such as business processes described using BPMN (Business Process Model and Notation) [12], or selected UML diagrams.

[9]http://ai.ia.agh.edu.pl/wiki/student:msc2008_bizrules_cases:hekate_case_ploc.

Table 4.6 Calculating the base rate (charge)

Car capacity [cm^3]	Base rate (charge) [PLN]
<900	537
901–1300	753
1301–1600	1050
1601–2000	1338
>2000	1536

Table 4.7 Bonus-Malus table

Customer class	Base rate percent	New customer class depending on number of accidents in the last year		
		1 Accident	2 Accidents	More than 2
M	260	M	M	M
0	160	M	M	M
1	100	0	M	M
2	90	0	M	M
3	80	1	0	M
4	70	2	1	M
5	60	3	1	M
6	60	3	2	M
7	50	4	3	M
8	50	6	3	M
9	40	7	4	0

step concerns discounts and these increase as a result of accident-free driving (see Table 4.7). Within this step one can distinguished 33 rules. The last step, consisting of 16 rules, takes discounts and other increases such as the driver age, additional insurance, etc. into account (see Table 4.8).

Example 4.4.3 (**PLI: basic decision table**) *Consider the general decision table specification given in Table 4.6. This table calculates the* baseCharge *price based upon the* carCapacity *(the capacity of the car engine). In order to define it two attributes are needed:*
$\mathbb{A} = \{carCapacity\}, \{baseCharge\}$ *with domains:*
$\mathbb{D}_{carCapacity} = \{0 \ldots 3000\}, \mathbb{D}_{baseChargeage} = \{0 \ldots 3000\}.$
the table schema would be:
$s = (\{carCapacity\}, \{baseCharge\})$
and an example of rule could be as follows:
$r_2 : [carCapacity \in [901, 1300]] \longrightarrow [baseCharge := 753.00]$

Table 4.8 Other discounts and rises

Discount/Rise (%)	Rule
−50	When the car is antique.
−20	When the customer has other insurance. When the car age do not exceed 1 year.
−15	When the car age do not exceed 2 years.
−10	When the car age do not exceed 3 years. When single payment. When the customer prolongs the insurance. When the customer buys more than one insurance. When the driver age is between 44 and 55 years.
+10	In case of installments.
+15	When the car age exceeds 10 years.
+20	When the car has more than 5 seats. When the car do not have a valid technical examination.
+30	When the driver have the driver license shorter than 3 years.
+50	When the driver age is younger than 25 years.
+60	When the car do not have an insurance history.

The PLI example will be used and extended in the following chapters to demonstrate the integration of XTT2 into business processes as well as RBS shells.

4.5 Inference Control in Structured Rule Bases

Any XTT2 table can have input links as well as output links. Links are related to the possible inference order. Tables to which no connections point are referred to as input (or start) tables. Tables with no connections pointing to other tables are referred to as output tables. All the other tables (those which have both input and output links) are referred to as middle tables.

We consider a network of tables connected according to the following general principles: there is at least one input table, there is at least one output table, there is a zero or more middle tables, and all the tables are interconnected. The aim is how to determine the inference order. The basic principle is that before firing a table, all the immediately preceding tables must have already been fired. The structure of the network imposes a partial order with respect to the order of table firing. Firing a table involves processing in a sequence all the rules in the table. Below, four possible algorithms for inference control are described. The concepts of the algorithms were presented in [8]. The preliminary formalization was introduced in [13].

The first basic algorithm consists of a hard-coded order of inference. Every table is assigned a unique integer number. The tables are fired in order from the lowest number to the highest. After starting the inference process, the predefined order of inference is followed. The inference stops after firing the last table. In case a table contains a complete set of rules (w.r.t. possible outputs generated by preceding tables) the inference process should end with all the attribute values defined by all the output tables being produced. This approach is only suitable for relatively small knowledge bases, where manual analysis is possible. Therefore, more complex modes are considered, including DDI (Data-Driven Inference), TDI (Token-Driven Inference), and GDI (Goal-Driven Inference). These modes are used with a general inference algorithm.

All complex inference modes considered here depend upon building a stack of tables to be processed. The general approach to the reasoning is as follows:

1. Based upon the dependencies between tables determine the order in which tables are supposed to be processed.
2. If the TDI mode is used determine how many tokens each table on the stack will require to be processed.
3. Process rules in tables that were put on the stack.

The term "dependency" in the first step means either a link, or the attribute dependency following from table schemas for the tables.

The *Data-Driven Inference* algorithm identifies start tables, and puts all the tables that are linked to the initial ones in the XTT2 network into a FIFO queue. When there are no more tables to be added to the queue, the algorithm fires selected tables in the order they are popped from the queue. The forward-chaining strategy is suitable for simple tree-like inference structures. However, it has limitations in a general case, because it cannot determine tables having multiple dependents.

The operation of the main algorithm using the DDI mode can be observed in Fig. 4.2. It is a simplified illustration, where table schemas are as follows: $table_1 = (\{A\}, \{B\})$, $table_2 = (\{C\}, \{D\}), \ldots, table_7 = (\{H, J\}, \{Z\})$. Within the figures triples in tables are denoted using the attribute name, so triples Ax contain attribute A. Moreover, the expression "fact $A1$ is in the knowledge base" means that triple $A1$ is satisfied. In the first step, the DDI algorithm builds $N_1 = \{table_1, table_2\}$, then $N_2 = \{table_3, table_4\}$, next then $N_3 = \{table_6, table_7\}$. So $\mathbb{U} = \{table_1, table_3, table_4, table_6, table_7\}$. Considering the example where it is assumed that only the formula labeled in the figure as $A1$ is satisfied, rules in tables 1, 3, 4, and 6 are fired. No rules in table 7 are fired due to unsatisfied conditions. In fact, in this algorithm the explicit inference links are not used, only dependencies at the table schema level are considered.

The *Token-Driven Inference* approach is based upon monitoring the partial inference order defined by the network structure with tokens assigned to tables. A table can be fired only when there is a token at each input. Intuitively, a token is a flag signaling that the necessary data generated by the preceding table is ready for use. The operation of the main algorithm using the TDI mode can be observed in Fig. 4.3. The TDI algorithm builds $\mathbb{U} = \{table_1, table_2, table_4, table_5, table_7\}$. Considering the example where it is assumed that only formulas labeled in the figure as $A1$ and $C1$ are satisfied, appropriate rules in tables 1, 2, 4, 5, and 7 are fired. In this algorithm

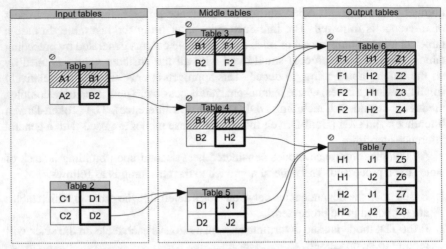

⊘ DDI with the assumption that fact **A1** is in the knowledge base and **Table 1** is a start table

Fig. 4.2 DDI inference mode for XTT2

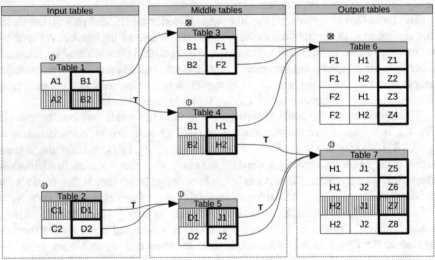

⦾ TDI with the assumption that facts **A2** and **C1** are in the knowledge base and **Table 6** and **Table 7** are goal tables
T Token sent from one table to another

Fig. 4.3 TDI inference mode for XTT2

the explicit inference links are used to represent transferring of tokens, represented in the figure as *T*s. Note that this model of inference execution covers the case of possible loops in the network. For example, if there is a loop and a table should be fired several times, the token is passed from its output to its input, and it is analyzed if it can be fired; if so, it is placed in the queue.

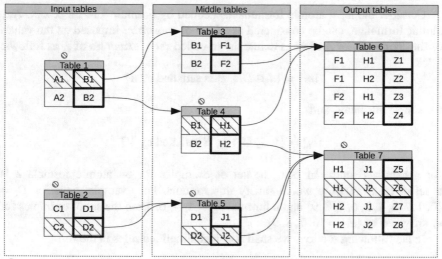

⊘ GDI with the assumption that facts **A1** and **C2** are in the knowledge base, and **Table 7** is a the goal table

Fig. 4.4 GDI inference mode for XTT2

The *Goal-Driven inference* approach works backwards with respect to selecting the tables necessary for a specific task, and then fires the tables forward in order to achieve the goal. One or more output tables are identified as those which can generate the desired goal values and are put into a LIFO queue. As a consequence, only the tables that lead to the desired solution are fired, and no rules are fired without purpose.

The operation of the main algorithm using the GDI mode can be observed in Fig. 4.4. The GDI algorithm builds $\mathbb{U} = \{table_1, table_2, table_4, table_5, table_7\}$. Considering the example where it is assumed that only formulas labeled in the figure as $A1$ and $C1$ are satisfied, and $table_7$ is selected as the goal table, appropriate rules in tables 1, 2, 4, 5, and 7 are fired. In this algorithm the explicit inference links are not used, only dependencies at the table schema level are considered. The Goal-Driven Inference may be particularly suitable for situations where the context of the operation can be defined and it is possible to clearly identify the knowledge component that needs to be evaluated.

4.6   Formalized Detection of Anomalies in Rules

One of the main objectives of the formalization of the rule language is the introduction of formalized rule verification procedures. The next section will present the most important verification tasks selected from those that had previously been identified in Sect. 2.4.

Consider the legal atomic formulae as defined by Formula 4.14 on p. xxx. Any atomic formula e_i can be considered as a kind of *constraint* imposed on the values of the domain $\mathbb{D}_i$ of A_i. Let us define the so-called *set of examples* of e_i as follows:

$$[e_i] = \{d \in \mathbb{D}_i : e_i \text{ is satisfied by } d\}$$

for simple attributes, and

$$[e_i] = \{V \subseteq \mathbb{D}_i : e_i \text{ is satisfied by } V\}$$

for generalized ones. In fact, the set of examples for an atomic formula e is a set of legal attribute which satisfy this formula. For example, consider $D_i = \{0, 1, 2, 3, 4, 5, 6, 7, 8, 9\}$ and a formula $e_i = A_i \geq 6$. Then the set of examples for e_i are given by $[e_i] = \{6, 7, 8, 9\}$.

In the following sections we shall consider simplified rules in the form:

$$e_1 \wedge e_2 \wedge \ldots \wedge e_n \longrightarrow \theta$$

where θ is an atomic formula assigning specific value to a decision attribute. Let *LHS* denote the set of preconditions of the rule, i.e. $LHS = \{e_1, e_2, \ldots, e_n\}$.

Below definitions for some of the most common anomalies to be detected and eliminated in Rule-Based Systems are considered.

Inconsistency of a Single Rule

An inconsistency of a single rule can be the result of the two following situations, where in the preconditions of the rule:

- there are two logically inconsistent atomic formulae,
- there exists an atomic formula logically inconsistent with the conclusion.

The first case occurs if there exist atomic formulae $e_i, e_j \in LHS$, such that:

$$[e_i] \cap [e_j] = \varnothing \tag{4.25}$$

The second situation takes place when:

$$[e_i] \cap [\theta] = \varnothing \tag{4.26}$$

In both cases, the formulae must define constraints referring to the same attribute.

Inconsistency of a Pair of Rules

An inconsistency within a pair of rules occurs if both rules can be simultaneously fired, and their conclusions are inconsistent. This means that a state which satisfies the preconditions of each rule, but their conclusions define different values of the same attribute. Consider two rules with preconditions given by $LHS(r_k)$ and $LHS(r_l)$.

These preconditions can be simultaneously satisfied if and only if for any pair of atomic formulae e_i, e_j, such that $e_i \in LHS(r_k)$ and $e_j \in LHS(r_l)$, e_i and e_j define the constraints over the same attribute and the following conditions are satisfied:

$$[e_i] \cap [e_j] \neq \varnothing \tag{4.27}$$

If the above holds, and simultaneously the rules define different values of the same conclusion attribute, then the rules are inconsistent.

Incompleteness of a Group of Rules

Completeness is defined as the ability to react to every admissible input value. This means that the Cartesian product of domains of attributes of a group of rules is covered by rule preconditions. If a system has an incomplete rule base, a valid system input exists for which no rule would be fired. Incompleteness is a serious anomaly that can make the system unsafe and undependable.

Consider a group of rules with preconditions defined with the use of attributes $A_1, A_2, \ldots, A_n$. Let the Cartesian product of the domains of these attributes be denoted as $\mathbb{U}$. Now, consider a single rule $e_1 \wedge e_2 \wedge \ldots \wedge e_n \longrightarrow \theta$. The Cartesian product of states covered by the preconditions of the rule is given by:

$$\mathbb{COV} = [e_1] \times [e_2] \times \ldots \times [e_n] \tag{4.28}$$

The completeness holds if and only if, for any $u \in \mathbb{U}$, a rule exists with $\approx$ such that $u \in \mathbb{COV}$.

Subsumption of a Pair of Rules

Subsumption between a pair of rules r_k and r_l occurs if and only if firing of one rule always means firing the other one and the rules have the same conclusion. The subsumption holds in fact among the joint precondition formulae. Let $\mathbb{COV}_k$, $\mathbb{COV}_l$ denote the Cartesian products of states covered by the respective formulae. Subsumption holds if and only if:

$$\mathbb{COV}_k \subseteq \mathbb{COV}_l \tag{4.29}$$

In more operational terms, the subsumption can also be defined as follows: Consider two rules with preconditions constructed over the same attributes. Rule r_l subsumes rule r_k if and only if for any $e_i \in LHS(r_k)$ there exists $e_j \in LHS(r_l)$, such that:

$$[e_i] \subseteq [e_j] \tag{4.30}$$

i.e. e_j subsumes e_i. Obviously, if e_i is satisfied, then e_j must be satisfied as well, but not vice versa. The more general rule r_l cannot have any extra atoms in preconditions, i.e. each of its preconditions must cover some atom in the preconditions of rule r_k. The subsumed rule r_k can be eliminated.

4.7   Summary

This chapter provided on overview of a logic-based formalism for decision rules. The formalism allows for modeling, verifying, and as a result of the extra tools executing modularized knowledge bases. The primary motivation for developing these was to overcome the limitations of traditional approaches, discussed in Chaps. 1 and 2. In existing systems rules are informally described and often have no formal semantics. A lack of a formal relationship between the knowledge representation language and logic leads to difficulties when it comes to understanding its expressive power. At the same time, the lack of standards for knowledge representation results in a lack of rule knowledge portability between different systems. Moreover, single rules constitute items of low knowledge processing capabilities, while for practical applications a higher level of abstraction is needed. A flat rule base includes hundreds of different rules that are considered to be equally important, and equally unrelated. The rule-centric design, with no structure explicitly identified by the user, makes the hierarchization of the logic model very difficult. Furthermore, common inference algorithms assume a flat rule base structure, where a brute force (blind) search for a solution is used, whereas number of practical applications are goal-oriented. Classic inference engines, especially those that are forward-chaining, are highly inefficient with respect to the focus on the goal to be achieved. Finally, as a result of the formalized description of the rule base it is possible to detect logical anomalies during the design.

This chapter introduced the underlying formal notation for XTT2 rules. This notation will be used in the remaining chapters. However, some simplifications of the notation will be introduced later in the book. They are mostly motivated by transparency, but also by implementation of software tools.

In the following chapter, a formalized model for business processes will be discussed. As a result of this model, an alternative inference control for rule-based knowledge bases can be achieved. These two models can be combined and supported at the design level, as discussed in Chap. 6. Furthermore, the XTT2 method was extended to support uncertain knowledge specification, as presented in Chap. 7. Finally, a supplementary formal model of rule semantics was introduced to support rule interoperability based upon XTT2. It will be presented in Chap. 8.

The third part of the book provides an overview of the XTT2-based approach: *Semantic Knowledge Engineering* (SKE) [2]. SKE delivers software tools for design, analysis and execution of rule bases. The term SKE will be used in the following chapter to denote the broader scope and use of XTT2 along with supporting software tools. Moreover, practical applications of XTT2 and other models contained in the second part will be discussed.

References

1. Nalepa, G.J., Ligęza, A., Kaczor, K.: Formalization and modeling of rules using the XTT2 method. Int. J. Artif. Intell. Tools **20**(6), 1107–1125 (2011)
2. Nalepa, G.J.: Semantic Knowledge Engineering. A Rule-Based Approach. Wydawnictwa AGH, Kraków (2011)
3. Ligęza, A., Wojnicki, I., Nalepa, G.J.: Tab-trees: a case tool for design of extended tabular systems. In: Mayr, H.C., Lazansky, J., Quirchmayr, G., Vogel, P. (eds.) Database and Expert Systems Applications. Lecture Notes in Computer Sciences, vol. 2113, pp. 422–431. Springer, Berlin (2001)
4. Nalepa, G.J.: A new approach to the rule-based systems design and implementation process. Comput. Sci. **6**, 65–79 (2004)
5. Nalepa, G.J., Ligęza, A.: A graphical tabular model for rule-based logic programming and verification. Syst. Sci. **31**(2), 89–95 (2005)
6. Nalepa, G.J., Ligęza, A.: XTT+ rule design using the ALSV(FD). In: Giurca, A., Analyti, A., Wagner, G. (eds.) ECAI 2008: 18th European Conference on Artificial Intelligence: 2nd East European Workshop on Rule-based applications, RuleApps2008: Patras, 22 July 2008, pp. 11–15. University of Patras, Patras (2008)
7. Nalepa, G.J., Ligęza, A.: On ALSV rules formulation and inference. In: Lane, H.C., Guesgen, H.W. (eds.) FLAIRS-22: Proceedings of the Twenty-Second International Florida Artificial Intelligence Research Society Conference: 19–21 May 2009, Sanibel Island, Florida, USA, Menlo Park, California, FLAIRS, pp. 396–401. AAAI Press (2009)
8. Nalepa, G.J., Ligęza, A.: HeKatE methodology, hybrid engineering of intelligent systems. Int. J. Appl. Math. Comput. Sci. **20**(1), 35–53 (2010)
9. Ligęza, A.: Logical Foundations for Rule-Based Systems. Springer, Berlin (2006)
10. Gouyon, J.P.: Kheops users's guide. Technical Report 92503, Report of Laboratoire d'Automatique et d'Analyse des Systemes, Toulouse, France (1994)
11. Giarratano, J., Riley, G.: Expert Systems. Principles and Programming, 4th edn. Thomson Course Technology, Boston (2005). ISBN 0-534-38447-1
12. OMG: Business Process Model and Notation (BPMN): Version 2.0 specification. Technical report formal/2011-01-03, Object Management Group (January 2011)
13. Nalepa, G., Bobek, S., Ligęza, A., Kaczor, K.: Algorithms for rule inference in modularized rule bases. In: Bassiliades, N., Governatori, G., Paschke, A. (eds.) Rule-Based Reasoning, Programming, and Applications. Lecture Notes in Computer Science, vol. 6826, pp. 305–312. Springer, Berlin (2011)

Chapter 5
Formalized Integration of Rules and Processes

Business Process Management (BPM) was introduced in Sect. 3.2. In BPM, a Business Process (BP) can be simply defined as a collection of related tasks which produces a specific service or product for a customer [1]. Business Rules (BR), introduced in Sect. 3.1 can be successfully used to specify process low-level logic [2, 3]. There is an important difference in abstraction levels of BP and BR. However, rules can to a certain degree be complementary to processes. BR provide a declarative specification of domain knowledge, which can be encoded into a process model. On the other hand, a process can be used as a procedural specification of the workflow, including the inference control [4]. The use of BR in BP design helps to simplify complex decision modeling. Although rules should describe business knowledge in a formalized way that can be further automated, there is no common understanding of how process and rule models should be structured in order to be integrated [5].

In this chapter a formalized model for describing the integration of BP with BR is discussed. The solution uses the existing representation methods for processes and rules, specifically the Business Process Model and Notation (BPMN) [6] for BP models, and XTT2 [7]. The proposed model deals with the integration of processes with rules in order to provide a coherent formal description, and to support the practical design. Furthermore, in such an approach, BP can be used as a high level inference control for the XTT2 knowledge base.

The chapter is organized as follows. Section 5.1 provides a formal description of a BPMN process model. It introduces the notation and its formal representation. This formalized process model is then integrated with rules, and this integration is specified as the General Business Logic Model in Sect. 5.2. In order to apply this model to a specific rule solution, the SKE-specific Business Logic Model is presented in Sect. 5.3. As an evaluation, a case study described using the proposed model is presented in Sect. 5.4. The chapter is summarized in Sect. 5.5.

© Springer International Publishing AG 2018   109
G.J. Nalepa, *Modeling with Rules Using Semantic Knowledge Engineering*,
Intelligent Systems Reference Library 130,
https://doi.org/10.1007/978-3-319-66655-6_5

5.1   Formal Description of BPMN Process Model

Let us define a BPMN 2.0 process model that describes the most important artifacts of the BPMN notation.[1] As it also focuses on several details that are key elements from the rule perspective, the process model takes into account flow objects (activities, events and gateways), sequence flows between these flow objects, as well as the set of model attributes.

BPMN Process Models

Definition 5.1 A **BPMN 2.0 process model** is a tuple $\mathcal{P} = (\mathcal{O}, \mathcal{F}, \Lambda)$, where:

- $\mathcal{O}$ is the set of flow objects, $o_1, o_2, o_3, \ldots \in \mathcal{O}$,
- Λ is the set of model attributes, $\lambda_1, \lambda_2, \lambda_3, \ldots \in \Lambda$,
- $\mathcal{F} \subset \mathcal{O} \times \mathcal{O} \times 2^{\Lambda_{\mathcal{F}}}$ is the set of sequence flows,
  where $\Lambda_{\mathcal{F}} \subset \Lambda$ is a subset of attributes that are used in sequence flows.

Moreover, the set of flow objects is divided into sets:

- $\mathcal{A}$ is a set of activities such that $\mathcal{A} = \mathcal{T} \cup \mathcal{S}, \mathcal{T} \cap \mathcal{S} = \varnothing$,
  where $\mathcal{T}$ is the set of tasks and $\mathcal{S}$ is the set of sub-processes, $\tau_1, \tau_2, \tau_3, \ldots \in \mathcal{T}$
  and $s_1, s_2, s_3, \ldots \in \mathcal{S}$,
- $\mathcal{E}$ is the set of events, $e_1, e_2, e_3, \ldots \in \mathcal{E}$
- $\mathcal{G}$ is the set of gateways, $g_1, g_2, g_3, \ldots \in \mathcal{G}$

such that $\mathcal{O} = \mathcal{A} \cup \mathcal{E} \cup \mathcal{G}$ and $\mathcal{A} \cap \mathcal{E} = \mathcal{A} \cap \mathcal{G} = \mathcal{E} \cap \mathcal{G} = \varnothing$.

A set of all the possible BPMN 2.0 process models will be denoted $\mathcal{P}$, e.g. $\mathcal{P}_1, \mathcal{P}_1, \mathcal{P}_1, \ldots \in \mathcal{P}$.

Example 5.1.1 (*Basic BPMN model used in formalization*) The process model shown below presents an example of a BPMN model consisting of two events, two gateways and two tasks.

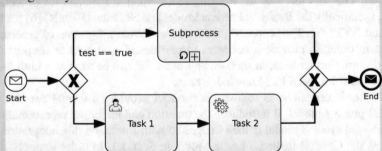

Using the presented formalization, it can be described as follows: $\mathcal{P}_1 = (\mathcal{O}_1, \mathcal{F}_1, \Lambda_1)$, where

---

[1]The model is limited to the most popular elements used in the private process model diagrams of the BPMN 2.0 notation.

- $\mathcal{O}_1 = \{\tau_1, \tau_2, s_1, e_1, e_2, g_1, g_2\}$, and $\mathcal{O}_1$ can be divided into following subsets:
- $\mathcal{O}_1 = \mathcal{A}_1 \cup \mathcal{E}_1 \cup \mathcal{G}_1$, $\mathcal{A}_1 = \mathcal{T}_1 \cup \mathcal{S}_1$, $\mathcal{T}_1 = \{\tau_1, \tau_2\}$
  and $\mathcal{S}_1 = \{s_1\}$, $\mathcal{E}_1 = \{e_1, e_2\}$, $\mathcal{G}_1 = \{g_1, g_2\}$,
- $\mathcal{F}_1 = \{(e_1, g_1, \Lambda_{e_1,g_1}), (g_1, s_1, \Lambda_{g_1,s_1}), (g_1, \tau_1, \Lambda_{g_1,\tau_1}),$
  $(\tau_1, \tau_1, \Lambda_{\tau_1,\tau_1}), (s_1, g_2, \Lambda_{s_1,g_2}), (\tau_2, g_2, \Lambda_{\tau_2,g_2}), (g_2, e_2), \Lambda_{g_2,e_2}\}$,   where
  $\Lambda_{e_1,g_1}, \Lambda_{g_1,s_1}, \Lambda_{g_1,\tau_1}, \Lambda_{\tau_1,\tau_1}, \Lambda_{s_1,g_2}, \Lambda_{\tau_2,g_2}, \Lambda_{g_2,e_2} \subset \Lambda_1$ are sets of sequence
  flows' attributes.

Definition 5.2 A **task** interpretation is a pair:

$$\tau = (type(\tau), \Lambda_\tau)$$

where:

- $type(\tau)$ determines the type of the task τ,
  $type(\tau) \in \{None, User, Manual, Send, Receive, Script, Service, BusinessRule\}$
- $\Lambda_\tau \subset \Lambda$ is the set of the task attributes,
  $\Lambda_\tau = \{id, name, documentation, markers, resources, implementation,$
  $ioSpecification, startQuantity, completionQuantity, loopCharacteristics,$
  $calledElementRef, multiInstanceLoopCharacteristics\}$,[2] some attributes can take
  set values, such as:

  – $markers \subset \{loop, parallelMI, sequentialMI, adhoc, compensation\}$,

  some of the attributes may contain other attributes, such as:

  – $ioSpecification = \{dataInputs, dataOutputs\}$.

Moreover, $\mathcal{T}_x$ will denote the set of tasks which are of the same type x:

$$\mathcal{T}_x = \{\tau \in \mathcal{T} : type(\tau) = x\}$$

For simplicity, the value of an attribute can be obtained using a function, the name of which matches the attribute name $attribute(\tau)$, e.g. $id(\tau_1)$ denotes the value of the id attribute for the task τ_1.

The tasks of a different types use the *implementation* attribute to specify the implementation technology, e.g. "$\#\#WebService$" for the Web service technology or a URI identifying any other technology or coordination protocol. The purpose of the implementation technology is different for different types of tasks, e.g.: in the case of Service tasks ($\mathcal{T}_{Service}$) it determines the service technology, in the case of Send tasks ($\mathcal{T}_{Send}$) or Receive tasks ($\mathcal{T}_{Receive}$), it specifies the technology that will be

---

[2]The set of attributes is limited to the most popular ones, for other see Sect. 10.2 Activities of the BPMN 2.0 Specification [6].

used to send or receive messages respectively, and in the case of Business Rule tasks ($\mathcal{T}_{BusinessRule}$), it specifies the Business Rules Engine technology. The *ioSpecification* attribute is used for specifying the data inputs and outputs suitable for the implementation technology. Some types of tasks can also have several additional attributes (Λ_τ) specified.

> **Example 5.1.2** (*Formalized task description*) For the process model presented in Example 5.1.1, tasks can be described as follows:
>
> - $\mathcal{T}_1 = \mathcal{T}_{User} \cup \mathcal{T}_{Service}$, $\mathcal{T}_{User} = \{\tau_1\}$, $\mathcal{T}_{Service} = \{\tau_2\}$,
> - $name(\tau_1) = $ "*Task 1*", $name(\tau_2) = $ "*Task 2*", $markers(\tau_1) = markers(\tau_1) = \varnothing$.

Definition 5.3 A **sub-process** interpretation is a triple:

$$s = (\mathcal{P}_s, type, \Lambda_s)$$

where:

- $type(s)$ determines the type of the sub-process s,
  $type(s) \in \{Sub\text{-}process, Embedded, CallActivity, Transaction, Event\}$,
- $\mathcal{P}_s \in \mathcal{P}$ is a BPMN 2.0 process model nested in the sub-process s,
- $\Lambda_s \subset \Lambda$ is the set of the sub-process attributes,
  $\Lambda_\tau = \{id, name, documentation, markers, triggeredByEvent, loopCharacteristics, multiInstanceLoopCharacteristics, calledElementRef, ioSpecification\}$,[3]
  such attributes as *markers*, *ioSpecification*, *loopCharacteristics*, and *multiInstanceLoopCharacteristics* are defined same as for the tasks (see Definition 5.2).

> **Example 5.1.3** (*Formalized subprocess description*) For the process model presented in Example 5.1.1, a sub-process $s_1 \in \mathcal{S}_1$ can be described as follows:
>
> - $name(s_1) = $ "*Subprocess*",
> - $type(s_1) = Sub\text{-}process$,
> - $markers(s_1) = \{loop\}$,
> - $triggeredByEvent(s_1) = $ `false`,
> - $loopCharacteristics(s_1) = \{loopCondition(s_1), testBefore(s_1), loopMaximum(s_1)\}$, where:

---

[3]The set of attributes is limited to those that are most popular, for others see Sect. 10.2 Activities of the BPMN 2.0 Specification [6].

- $loopCondition(s_1) \neq$ null,
- $testBefore(s_1) =$ false,
- $loopMaximum(s_1) =$ null,

where null indicates that the value does not exist.

While activities represent parts of work that are performed within a Business Process, events denote things that occur during the lifetime of a Business Process.

Definition 5.4 An **event** interpretation is a pair:

$$e = (type(e), trigger(e), \Lambda_e)$$

where:

- $type(e) \in \{Start, Intermediate, End\}$,
- $trigger(e)$ determines the trigger of the event e,
  $trigger(e) \in \{Cancel, Compensation, Conditional, Error, Escalation, Link,$
  $Message, Multiple, None, ParallelMultiple, Signal, Terminate, Timer\}$,
- $\Lambda_e \subset \Lambda$ is the set of the event attributes,
  $\Lambda_e = \{id, name, documentation, method, boundary, attachedToRef,$
  $cancelActivity\}$,[4] $method(e) \in \{catch, throw\}$.

Moreover, $\mathcal{E}^x$ will denote the set of events with the same trigger x:

$$\mathcal{E}^x = \{e \in \mathcal{E}: trigger(e) = x\}$$

$\mathcal{E}_x$ will denote the set of events of the same type x:

$$\mathcal{E}_x = \{e \in \mathcal{E}: type(e) = x\}$$

Different types of events have different event definition attributes specified, e.g.:

- $messageEventDefinition$ for $e \in \mathcal{E}^{Message}$,
  $messageEventDefinition = \{messageRef, operationRef\}$,
- $timerEventDefinition$ for $e \in \mathcal{E}^{Timer}$,
  $timerEventDefinition = \{timeCycle, timeDate, timeDuration\}$,
- $conditionalEventDefinition$ for $e \in \mathcal{E}^{Conditional}$,
  $conditionalEventDefinition = \{condition\}$,

---

[4]The set of attributes is limited to the most popular ones, for other see Sect. 10.4 Events of the BPMN 2.0 Specification [6].

Example 5.1.4 (Formalized events description) For the process model presented in Example 5.1.1, the events $e_1, e_2 \in \mathcal{E}_1$ can be described as follows:

- $name(e_1) =$ "*Start*", $name(e_2) =$ "*End*",
- $method(e_1) = catch$, $method(e_2) = throw$,
- $e_1 \in \mathcal{E}_{Start}^{Message}$ and $e_2 \in \mathcal{E}_{End}^{Message}$.
- $type(e_1) = Start$, $type(e_2) = End$,
- $trigger(e_1) = Message$, $trigger(e_2) = Message$,
- $boundary(e_1) = boundary(e_2) = \texttt{false}$
- $cancelActivity(e_1) = cancelActivity(e_2) = \texttt{false}$

For the fragment of a process model with a task τ_3 and an event e_3 presented below:

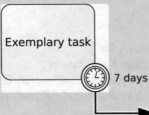

the following aspects can be described using such a model:

- $e_3 \in \mathcal{E}_{Intermediate}^{Timer}$,
- $name(e_3) =$ "*7 days*",
- $method(e_1) = catch$,
- $timeDuration(e_3) =$ "*P7D*",
- $timeCycle(e_3) = timeDate(e_3) = \texttt{null}$,
- $cancelActivity(e_3) = \texttt{true}$,
- $boundary(e_3) = \texttt{true}$,
- $attachedToRef(e_3) = id(\tau_3)$.

It is also important to note that not every *trigger(e)* is allowed for any *type(e)* of event – Table 5.1 presents the possible combinations. In the case of this formalization, the *condition* attribute for $e \in \mathcal{E}^{Conditional}$ is especially important. It defines an expression stored in *body* attribute and expressed in the *language* language: $condition(e) = \{body(e), language(e)\}$.

Gateway elements are used to control the flow of tokens through sequence flows as they converge and diverge within a process. Although, according to the BPMN 2.0 specification, a single gateway can have multiple input and multiple output flows, the formalized model that is proposed seeks to enforce the best practice of a gateway by only performing one of these functions. Thus, a gateway should have either one input or one output flow and a gateway with multiple input and output flows should be modeled with two sequential gateways, the first of which converge and the second diverge from the sequence flows.

Table 5.1   Possible combinations of $trigger(e)$ and $type(e)$

Type(e)	Start	Intermediate		End
Trigger(e)	Method(e) = catch	Method(e) = catch	Method(e) = throw	Method(e) = throw
Cancel		✓		✓
Compensation	✓	✓		✓
Conditional	✓	✓		
Error	✓	✓		✓
Escalation	✓	✓		✓
Link		✓	✓	
Message	✓	✓	✓	✓
Multiple	✓	✓	✓	✓
None	✓		✓	✓
ParallelMultiple	✓	✓		
Signal	✓	✓	✓	✓
Terminate				✓
Timer	✓	✓		

Definition 5.5   A **gateway** interpretation is a tuple:

$$g = (\mathcal{F}_g^{in}, \mathcal{F}_g^{out}, type(g), \Lambda_g)$$

where:

- $\mathcal{F}_g^{in}$ and $\mathcal{F}_g^{out}$ are sets of sequence flows (input and output flows respectively), $\mathcal{F}_g^{in}, \mathcal{F}_g^{out} \subset \mathcal{F}$,
  $\mathcal{F}_g^{in} = \{(o_i, o_j, \Lambda_{i,j}) \in \mathcal{F} : o_j = g\}$ and $\mathcal{F}_g^{out} = \{(o_i, o_j, \Lambda_{i,j}) \in \mathcal{F} : o_i = g\}$,
- $type(g)$ determines the type of the gateway g,
  $type(g) \in \{Parallel, Exclusive, Inclusive, Complex, Event\text{-}based, ParallelEvent\text{-}based\}$,
- $\Lambda_g \subset \Lambda$ is the set of the gateway attributes,
  $\Lambda_g = \{id, name, documentation, gatewayDirection\}$,[5]
  $gatewayDirection(g) \in \{converging, diverging\}$.

Furthermore, the following notational elements will be used:

- $\mathcal{G}_+ = \{g \in \mathcal{G} : type(g) = Parallel\}$,
- $\mathcal{G}_\times = \{g \in \mathcal{G} : type(g) = Exclusive\}$,
- $\mathcal{G}_\circ = \{g \in \mathcal{G} : type(g) = Inclusive\}$,
- $\mathcal{G}_* = \{g \in \mathcal{G} : type(g) = Complex\}$,
- $\mathcal{G}_\otimes = \{g \in \mathcal{G} : type(g) = Event\text{-}based\}$,
- $\mathcal{G}_\oplus = \{g \in \mathcal{G} : type(g) = ParallelEvent\text{-}based\}$.

---

[5]The set of attributes is limited to those that are most popular, for others, see Sect. 10.5 Gateways of the BPMN 2.0 Specification [6].

Some types of gateways can have several additional attributes specified, such as:

- *{instantiate, eventGatewayType}* for $g \in \mathcal{G}_\otimes \cup \mathcal{G}_\oplus$,
  $instantiate(g) \in \{\texttt{true}, \texttt{false}\}$,
  $eventGatewayType(g) \in \{Parallel, Exclusive\}$,
- *{default}* for $g \in \mathcal{G}_\times \cup \mathcal{G}_\circ \cup \mathcal{G}_*$,
  $default(g) \in \mathcal{F}_{out} \cup \{\texttt{null}\}$.

Example 5.1.5 (*Formalized Gateway description*) For the process model presented in Example 5.1.1, the gateways $g_1, g_2 \in \mathcal{G}_1$ can be described as follows:

- $type(g_1) = type(g_2) = Exclusive$, thus $g_1, g_2 \in \mathcal{G}_\times$,
- $gatewayDirection(g_1) = diverging$,
  $gatewayDirection(g_2) = converging$.
- $\mathcal{F}^{in}_{g_1} = \{(e_1, g_1, \Lambda_{e_1,g_1})\}$ and
  $\mathcal{F}^{out}_{g_1} = \{(g_1, s_1, \Lambda_{g_1,s_1}), (g_1, \tau_1, \Lambda_{g_1,\tau_1})\}$,
- $\mathcal{F}^{in}_{g_2} = \{(s_1, g_2, \Lambda_{s_1,g_2}), (\tau_2, g_2, \Lambda_{\tau_2,g_2})\}$ and
  $\mathcal{F}^{out}_{g_2} = \{(g_2, e_2, \Lambda_{g_2,e_2})\}$,
- $default(g_1) = (g_1, \tau_1, \Lambda_{g_1,\tau_1})$, $default(g_1) = \texttt{null}$,

---

Sequence Flows are used for connecting flow objects $o \in \mathcal{O}$ in the process.

Definition 5.6 A **sequence flow** interpretation is a tuple:

$$f_{o_1,o_2} = ((o_1, o_2), \Lambda_{o_1,o_2})$$

where:

- $(o_1, o_2) \in \mathcal{O} \times \mathcal{O}$ and o_1, o_2 are respectively source and target elements,
- $\Lambda_{o_1,o_2} \subset \Lambda_\mathcal{F}$ is the set of sequence flow attributes,
  $\Lambda_{o_1,o_2} = \{id, name, documentation, default, conditional, condition\}$,[6]
  $condition = \{body, language\}$.

Two Boolean attributes: *conditional* and *default* determine the conditional or default type of the flow. A conditional flow has to specify the *condition* and a default flow has no condition, i.e.

- $conditional(f) = \texttt{true} \Rightarrow condition(f) \neq \texttt{null}$,
- $default(f) = \texttt{true} \Rightarrow condition(f) = \texttt{null}$.

A subset of conditional sequence flows will be denoted $\mathcal{F}_{Conditional}$, i.e.
$\mathcal{F}_{Conditional} = \{f \in \mathcal{F}: conditional(f) = \texttt{true}\}$,

---

[6]The set of attributes is limited to those that are most popular, for others, see Sect. 10.5 Gateways of the BPMN 2.0 Specification [6].

A *condition* attribute defines an expression indicating that the token will be passed down the sequence flow only if the expression evaluates to `true`. An expression *body* is basically specified using natural-language text. However, it can be interpreted as a formal expression by a process execution engine; in such case, BPMN provides an additional *language* attribute that specifies a language in which the logic of the expression is captured.

In the presented BPMN model, the evaluation of the value can be obtained using a function *eval(value)*, e.g. for the *condition* attribute of the f sequence flow: $eval(condition(f)) \in \{$`true`, `false`$\}$. If *condition* is not explicitly defined for a particular sequence flow f, then it is implicitly always evaluated to `true`, i.e.: $condition(f) = $ `null` $\Rightarrow eval(condition(f)) \equiv $ `true`.

Example 5.1.6 (*Description of output sequence flows*) For the process model presented in Example 5.1.1, the output sequence flows from gateway g_1, i.e. $f \in \mathcal{F}_{g_1}^{out}$, can be described as follows:

- $f_{g_1,s_1} = (g_1, s_1, \Lambda_{g_1,s_1})$ and $f_{g_1,\tau_1} = (g_1, \tau_1, \Lambda_{g_1,\tau_1})$.
- $default(f_{g_1,s_1}) = $ `false` and $default(f_{g_1,\tau_1}) = $ `true`,
- $conditional(f_{g_1,s_1}) = $ `true` and $conditional(f_{g_1,\tau_1}) = $ `false`,
- $condition(f_{g_1,s_1}) = \{body(f_{g_1,s_1}), language(f_{g_1,s_1})\}$, $body(f_{g_1,s_1}) = $ "*test == true*",
- $condition(f_{g_1,\tau_1}) = $ `null`.

In this section, we presented a formalized model of a BPMN business process. This model will be used in the following section for defining a model that combines business processes with business rules.

5.2 General Business Logic Model

In this section we define a General Business Logic Model, which specifies business logic as the knowledge stored in the form of processes integrated with rules. As the model uses the abstract rule representation, it is general and can be refined into one that is more specific by adjusting it to the specific rule representation. The model uses the process model presented in the previous section and integrates it with rules.[7] As rules constitute a part of a rule base, it is defined as follows.

Definition 5.7 A **Rule Base** is a tuple $\mathbb{K} = (\mathbb{A}, \mathbb{R}, \mathbb{T})$, where:

- $\mathbb{A}$ is the set of all attributes used in the rule base,

---

[7]The proposed model focuses on selected details that are important from the rule perspective.

Fig. 5.1 Conditional
sequence flow

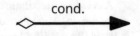

- $\mathbb{R}$ is the set of all rules, $r_1, r_2, r_3, \ldots \in \mathbb{R}$,
  and a single rule r_i contains its conditional part denoted as $cond(r_i)$.[8]
- $\mathbb{T}$ is the set of all decision components, $t_1, t_2, t_3, \ldots \in \mathbb{T}$,
  which can be rule sets or more sophisticated structures (rule sets represented as
  decision tables, trees, etc.) that organize rules in the rule base ($\mathbb{T} \subset 2^{\mathbb{R}}$),

Moreover, it is assumed that each rule base specifies a $language(r)$[9] in which the
rules are specified and provides additional methods that can be used to obtain pieces
of information from the rule base, such as $eval(r)$[10] for evaluating a conditional part
of the rule, and $infer(t)$[11] for obtaining a result of inference on a specified rule set.

Definition 5.8 A General Business Logic Model is a tuple $\mathcal{M} = (\mathcal{P}, \mathbb{K}, map)$,
where:

- $\mathbb{K}$ is a rule base containing rules (as defined in **Definition** 5.7),
- $\mathcal{P}$ is a BPMN 2.0 process model (as defined in **Definition** 5.1),
- map is a map function defined as follows:

$$
map(x) = \begin{cases}
\mathcal{F}_{Conditional} \to \mathbb{R} & \text{for } x \in \mathcal{F}_{Conditional} \\
\mathcal{E}_{Conditional} \to \mathbb{R} & \text{for } x \in \mathcal{E}_{Conditional} \\
\mathcal{T}_{BusinessRule} \to \mathbb{T} & \text{for } x \in \mathcal{T}_{BusinessRule}
\end{cases}
$$

In the following paragraphs, the mapping details for the specific BPMN elements
and more complex BPMN constructs are presented.

Conditional Sequence Flow

For a Conditional Sequence Flow $f \in \mathcal{F}_{Conditional}$ (see Fig. 5.1) the following
requirements have to be fulfilled in $\mathcal{M}$:

- All BPMN conditional sequence flows in $\mathcal{P}$ have the condition in the form of a
  conditional part of a rule from the $\mathbb{K}$ rule base assigned, formalized, the following
  holds:
  $\forall_{f \in \mathcal{F}_{Conditional}} \exists_{r \in \mathbb{R}} (map(f) = r) \wedge (body(f) = cond(r)) \wedge (language(f) = language(r))$.

---

[8]It is assumed that various rule bases can contain different kinds of rules (see categories of rules
presented in [8]). Regardless of the kind, in every rule it is possible to isolate their conditional part.
In some cases, a rule may consist only of a conditional part.

[9]$language(r)$ denotes $language(\mathbb{K})$.

[10]$eval(r)$ denotes $eval(cond(r))$ and $eval(r) \in \{\texttt{true}, \texttt{false}\}$.

[11]$infer(r)$ denotes $infer(\{r\})$.

- All condition attributes $\mathbb{A}_r \subset \mathbb{A}$ required by the rule r should be available in the $\mathcal{P}$ model, i.e.:
  $\forall_{r \in \mathbb{R}} \left((\exists_{f \in \mathcal{F}} \, map(f) = r) \Rightarrow (\forall_{\lambda \in cond(r)} \, \lambda \in \Lambda_{\mathcal{F}}) \right).$

Conditional Event

A Conditional Event $e \in \mathcal{E}_{Conditional}$ denotes that a particular *condition* specified by a rule condition is fulfilled. For Conditional Event, the following requirements have to be fulfilled in $\mathcal{M}$:

- All BPMN conditional events in $\mathcal{P}$ have the condition in the form of a conditional part of a rule from the $\mathbb{K}$ rule base assigned, i.e.:
  $\forall_{e \in \mathcal{E}_{Conditional}} \exists_{r \in \mathbb{R}} \, (map(e) = r) \wedge (body(e) = cond(r)) \wedge (language(e) = language(r)).$
- All condition attributes $\mathbb{A}_r \subset \mathbb{A}$ required by the rule r should be available in the $\mathcal{P}$ model, i.e.:
  $\forall_{r \in \mathbb{R}} \left((\exists_{e \in \mathcal{E}} \, map(e) = r) \Rightarrow (\forall_{\lambda \in cond(r)} \, \lambda \in \Lambda_e) \right).$

A Conditional Event can be used in BPMN in several structures in order to support different situations based upon the evaluation of the condition expression in the process instance, such as:

- A Simple Start and Intermediate Conditional Event can be used as conditional triggers providing the ability to trigger the flow of a token. The notation for a conditional start and intermediate events are presented in Fig. 5.2.
- Non-interruptive and Interruptive Boundary Conditional Events attached to a Task or a Subprocess can be used for interrupting a task or subprocess.
  The notation for conditional non-interruptive and interruptive boundary events are presented in Fig. 5.3.
- Event Subprocess with a Conditional Start Event can be used for interrupting the process and initiating a subprocess that is not a part of the regular control

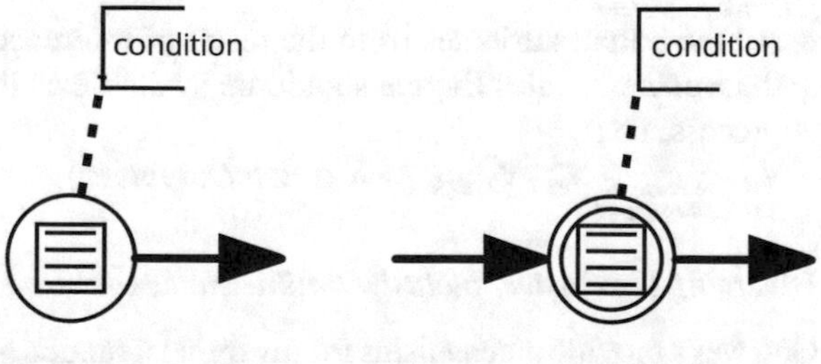

Fig. 5.2 Conditional (start and intermediate) events

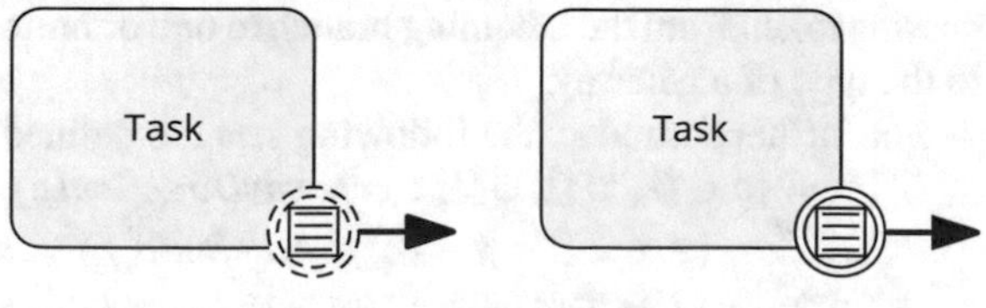

Fig. 5.3 Conditional (non-interruptive and interruptive) boundary events

Fig. 5.4 Event subprocesses
with a conditional start event

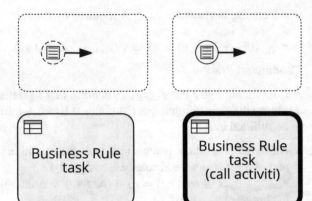

Fig. 5.5 Business rule task
(a standard and a call activity
task)

flow starting from the conditional start event. The notation for conditional non-interruptive and interruptive boundary events are presented in Fig. 5.4.

Business Rule Task

Business Rule (BR) Tasks allow for the task logic to be specified using rules and delegating work to a Business Rules Engine in order to receive calculated or inferred data. The notation for a BR task is presented in Fig. 5.5.

For the BPMN Business Rule tasks, the following formulas have to be fulfilled in $\mathcal{M}$:

- All BPMN BR tasks in $\mathcal{P}$ have the decision component from the $\mathbb{K}$ rule base assigned, i.e.:

  $\forall_{\tau \in \mathcal{T}_{BusinessRule}} \exists_{t \in \mathbb{T}} \; map(\tau) = t.$

- All the input attributes required by the Business Rules Engine for a rule set specified by the decision component should be available in the process model, i.e.:

  $\forall_{\substack{\tau \in \mathcal{T}_{BusinessRule} \\ t \in \mathbb{T}, map(\tau)=t}} \forall_{r \in t} \forall_{\lambda \in cond(r)} \; \lambda \in dataInputs(\tau).$

- All the output attributes from the result of inference on a specified rule set from the Business Rules Engine should be available as the output of a BR task in the process, i.e.:

  $\forall_{\substack{\tau \in \mathcal{T}_{BusinessRule} \\ t \in \mathbb{T}, map(\tau)=t}} \forall_{r \in t} \forall_{\lambda \in infer(r)} \; \lambda \in dataOutputs(\tau).$

Diverging (Exclusive, Inclusive/Multi-choice and Complex) Gateways

Gateways provide mechanisms for diverging a branch into two or more branches, and passing token from the incoming branch to one or more outgoing branches according to the type of a gateway.

For further formulae, the following sets are defined:

$\mathcal{G}_{div}^{cond} = \{g \in \mathcal{G}_{\times} \cup \mathcal{G}_{\circ} \cup \mathcal{G}_{*} : gatewayDirection(g) = diverging\},$

$\mathcal{F}_{g,div}^{out,cond} = \{f \in \mathcal{F}_{g}^{out} : g \in \mathcal{G}_{div}^{cond} \wedge default(f) \neq \texttt{true}\},$

$\mathcal{F}_{g,default}^{out,cond} = \{f \in \mathcal{F}_{g}^{out} : g \in \mathcal{G}_{div}^{cond} \wedge default(f) = \texttt{true}\}.$

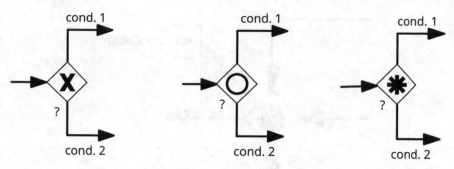

Fig. 5.6 Exclusive, inclusive (multi-choice) and complex diverging gateways

In the case of exclusive ($\mathcal{G}_\times$), inclusive ($\mathcal{G}_\circ$) and complex ($\mathcal{G}_*$) diverging gateways (see Fig. 5.6), there is a need for the model $\mathcal{M}$ to satisfy the following requirements:

- All BPMN sequence flows (apart from the default ones) outgoing from a diverging gateway have the condition in the form of a conditional part of a rule from the $\mathbb{K}$ rule base assigned, i.e.:
  $\forall_{f \in \mathcal{F}_{g,div}^{out,cond}} \exists_{r \in \mathbb{R}} (map(f) = r) \wedge (body(f) = cond(r)) \wedge (language(f) = language(r))$.
- In the case of exclusive, inclusive and complex diverging gateways, they can have a maximum of one outgoing default sequence flow, i.e.:
  $\forall_{g \in \mathcal{G}_{div}^{cond}} |\mathcal{F}_{g,default}^{out,cond}| \leq 1$.
- In the case of exclusive gateways, the evaluated conditions have to be exclusive, i.e.:
  $\forall_{f_1,f_2 \in \mathcal{F}_{g,div}^{out}} \forall_{g \in \mathcal{G}_\times} (\exists_{r_1,r_2 \in \mathbb{R}} map(f_1) = r_1 \wedge map(f_2) = r_2) \Rightarrow (eval(r_1) \neq eval(r_2))$.

Converging Complex Gateway

In the case of converging exclusive, inclusive and parallel gateways, their semantics are defined by the BPMN 2.0 specification and they do not require any rule-based description. However, a Converging Complex Gateway (see Fig. 5.7) requires an additional *activationCondition* expression which describes the precise behavior (defines the rule of passing tokens).

Thus, for BPMN Converging Complex Gateways the following requirements have to be fulfilled in $\mathcal{M}$:

- All BPMN Converging Complex Gateways in $\mathcal{P}$ specify the rule of passing tokens, i.e.:
  $\forall_{g \in \mathcal{G}_*} \exists_{r \in \mathbb{R}} (map(g) = r) \wedge (body(g) = cond(r)) \wedge (language(g) = language(r))$,
  where $activationCondition(g) = \{body(g), language(g)\}$.
- All condition attributes $\mathbb{A}_r \subset \mathbb{A}$ required by the rule r should be available in the process model, i.e.:
  $\forall_{r \in \mathbb{R}} ((\exists_{g \in \mathcal{G}_*} map(g) = r) \Rightarrow (\forall_{\lambda \in cond(r)} \lambda \in \Lambda_g))$.

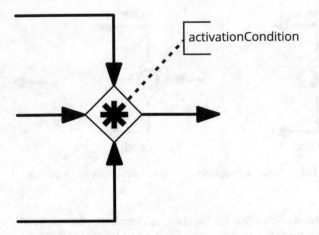

Fig. 5.7 Converging complex gateway

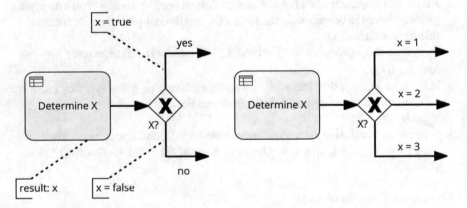

Fig. 5.8 Gateway after the BR task

Gateway Preceded by a BR Task

A special case of the two above mentioned examples occurs when a gateway is preceded by the BR task. In such a case, there is a need for the model $\mathcal{M}$ to satisfy the requirements specified for Business Rule Tasks and for Gateways, as well as the following additional requirement (Fig. 5.8):

- All BPMN sequence flows (apart from the default sequence flows) outgoing from a diverging gateway preceded by the BR task have the conditions based upon the output attributes of the BR task, i.e.:

$$\forall_{\substack{\tau \in T_{BusinessRule} \\ t \in \mathbb{T}, map(\tau)=t}} \forall_{g \in \mathcal{G}_{div}^{cond}} \left((\tau, g, \lambda_{\tau g}) \in \mathcal{F}\right) \Rightarrow \left(\forall_{f \in \mathcal{F}_g^{out}} \forall_{\lambda \in body(f)} \lambda \in infer(t)\right).$$

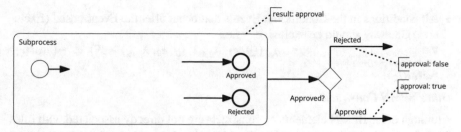

Fig. 5.9 Gateway preceded by a subprocess

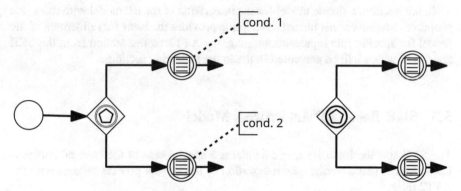

Fig. 5.10 Event-based exclusive gateways (non-instantiating and instantiating)

Gateway Preceded by a Subprocess

Another special case of using a gateway is a gateway preceded by a subprocess in which a decision is made (see Fig. 5.9). In such the case, there is a need for the model $\mathcal{M}$ to satisfy the requirements specified for Diverging Gateways, as well as the following additional requirements:

- All BPMN sequence flows (apart from the default sequence flows) outgoing from a diverging gateway preceded by a subprocess have the conditions based upon the attributes set by the preceded subprocess:
  $$\forall_{s \in \mathcal{S}} \, \forall_{g \in \mathcal{G}_{div}^{cond}} \left((s, g, \lambda_{s,g}) \in \mathcal{F} \right) \Rightarrow \left(\forall_{f \in \mathcal{F}_g^{out}} \, \forall_{\lambda \in body(f)} \, \lambda \in dataOutputs(s) \right).$$
- The number of sequence flows outgoing from a diverging gateway should be greater than or equal to the number of Message or None end events in the subprocess, i.e.:
  Let: $\mathcal{E}_{End}^s = \{ e \in \mathcal{O}_s : e \in \mathcal{E}_{End}^{None} \vee e \in \mathcal{E}_{End}^{Message} \}$, where $\mathcal{P}_s = (\mathcal{O}_s, \mathcal{F}_s, \Lambda_s)$.
  $$\forall_{s \in \mathcal{S}} \, \forall_{g \in \mathcal{G}_{div}^{cond}} \left((s, g, \lambda_{s,g}) \in \mathcal{F} \right) \Rightarrow \left(|\mathcal{F}_g^{out}| >= |\mathcal{E}_{End}^s| \right).$$

Event-Based Gateway

The use of Event-based (Exclusive) Gateway extends the use of Conditional Events (see Fig. 5.10). Thus, in this case, there is a need for the model $\mathcal{M}$ to satisfy the requirements specified for Conditional Events, as well as the following additional requirements:

- All conditions in the Conditional Events that occur after the Event-based (Exclusive) Gateway should be exclusive,[12] i.e.:

$$\forall_{\substack{e_1,e_1 \in \mathcal{E}_{Conditional} \\ r_1,r_1 \in \mathbb{R}, map(e_1)=r_1, map(e_2)=r_2}} \forall_{g \in \mathcal{G}_\otimes} \left((g, e_1, \lambda_{g,e_1}), (g, e_2, \lambda_{g,e_2}) \in \mathcal{F} \right) \Rightarrow \neg \big(eval(r_1)$$
$$\wedge eval(r_2)\big).$$

Other BPMN Constructs

Although other BPMN elements or constructs are not directly associated with rules from the rule base, they can be described by rules. However, such a representation of rules is not formally defined in the model presented here.

In this section, a simple model for the integration of the BP model with rules was proposed. Moreover, this formal description provides the basis for refinement of the model for specific rule representation, e.g. the XTT2 representation from the SKE approach, which will be presented in the in the following sections.

5.3 SKE-Based BPMN Process Model

The SKE-specific Business Logic Model is a special case of the General Business Logic Model that describes the integration of the BPMN process models with the XTT2 rules.

Definition 5.9 SKE-specific Business Logic Model is a tuple: $\mathcal{M}_{SKE} = (\mathcal{P}, \mathbb{K}_{SKE}, map)$, where:

- $\mathbb{K}_{SKE} = (\mathbb{A}_{SKE}, \mathbb{R}_{SKE}, \mathbb{T}_{SKE}) = (A, R, T_\mathcal{X})$ is an SKE-specific rule base, where:

  - $T_\mathcal{X}$ is a set of the XTT2 decision components,
  - R is a set of the XTT2 rules, such as:
    $R = \{r_i \in t : t \in T_\mathcal{X}\}$,
    $\forall_{r_i \in R} schema(r_i) = schema(t)$,
    and the conditional $cond(r_i)$ part of a rule is defined as follows:
    $cond(r_i) = E_i^{cond}$,

    where $r_i = (E_i^{cond}, E_i^{dec}, ACT_i)$,
  - A is a set of the attributes used in the XTT2 rule base, i.e.[13]:
    $A = \{a_i : \exists_{r_i \in R} a_i \in A_i^{cond} \vee a_i \in A_i^{dec}\}$.

- $\mathcal{P}$ is a BPMN 2.0 process model,
- map is a mapping function between the elements of the $\mathcal{P}$ process model and the elements of the $\mathbb{K}_{SKE}$ rule base.

---

[12]In fact, the exclusive relation here applies only to evaluation to `true` values. Thus, both conditions can not be fulfilled at the same time.

[13]Note that every rule in the XTT2 representation belongs to a particular decision table. Thus, there is no rule which would not be an element of a decision table. However, it is possible that a decision table can consist of a single rule.

The $\mathbb{K}_{SKE}$ rule base specifies the value of *language*, such as:
$\forall_{r_i \in R} \ language(r) = $ "XTT2". Moreover, the *infer(t)* method is defined as follows:
$infer(t) = A_t^{dec}$. This stems from the fact that in the SKE-specific Business Logic
Model, every decision component $t \in T_\chi$ is an XTT2 decision table. Thus, the result
of the inference is the set of decision attributes within this decision table.[14]

In the following paragraphs, the integration details are specified.[15]

Conditional Sequence Flow

For the Conditional Sequence Flows $f \in \mathcal{F}_{Conditional}$ the following hold:

- All BPMN conditional sequence flows in $\mathcal{P}$ have the condition in the form of a
  conditional part of a rule from the $\mathbb{K}_{SKE}$ rule base assigned, formalized, the following holds:
  $\forall_{f \in \mathcal{F}_{Conditional}} \exists_{r_i \in R} \ (map(f) = r_i) \wedge (body(f) = E_i^{cond}) \wedge (language(f) = $
  "XTT2").
- Values of the condition attributes required by the rule are mapped to the values of
  corresponding attributes in the rule base:
  $\forall_{\substack{f \in \mathcal{F}_{Conditional} \\ r_i \in R \\ map(f)=r_i}} \forall_{\lambda \in body(f)} \exists_{a_i \in E_i^{cond}} \ \lambda(f) \in \mathbb{D}_i \wedge \lambda(f) = a_i.$

Conditional Event

For the Conditional Events the following hold:

- All BPMN conditional events in $\mathcal{P}$ have the condition in the form of a conditional
  part of a rule from the $\mathbb{K}_{SKE}$ rule base assigned, i.e.:
  $\forall_{e \in \mathcal{E}_{Conditional}} \exists_{r_i \in R} \ (map(e) = r_i) \wedge (body(e) = E_i^{cond}) \wedge (language(e) = $
  "XTT2").
- Values of the condition attributes required by the rule are mapped to the values of
  corresponding attributes in the rule base:
  $\forall_{\substack{e \in \mathcal{E}_{Conditional} \\ r_i \in R \\ map(e)=r_i}} \forall_{\lambda \in body(e)} \exists_{a_i \in E_i^{cond}} \ \lambda(e) \in \mathbb{D}_i \wedge \lambda(e) = a_i.$

Business Rule Task

For the BPMN Business Rule tasks, the following formulae have to be fulfilled:

- All BPMN BR tasks in $\mathcal{P}$ have the decision component from the $\mathbb{K}_{SKE}$ rule base
  assigned:
  $\forall_{\tau \in T_{BusinessRule}} \exists_{t \in T_\chi} \ map(\tau) = t.$
- All the input attributes required by the HEART rule engine[16] for a rule set specified
  by the decision component should be available in the process model, i.e.:
  $\forall_{\substack{\tau \in T_{BusinessRule} \\ t \in T_\chi \\ map(\tau)=t}} \forall_{a_i \in A_t^{cond}} \exists_{\lambda \in dataInputs(\tau)} \ \lambda(\tau) \in \mathbb{D}_i \wedge \lambda(\tau) = a_i.$

---

[14]More precisely: attributes and their values that are set by a particular rule. An XTT2 decision
table is a first hit table [9], therefore it returns the output of a single rule (the first hit).

[15]If for a particular element, there are no additional requirements or conditions to specify, the
formulae from General Business Logic can be used.

[16]HEART is an inference engine that is used in the SKE approach, see Sect. 9.7.

- All the output attributes from the result of inference on a specified rule set from the HEART rule engine should be available as the output of a BR task in the process, i.e.:

$$\forall_{\substack{\tau \in T_{BusinessRule} \\ t \in T_X \\ map(\tau)=t}} \forall_{\lambda \in dataOutputs(\tau)} \exists_{a_i \in A_t^{dec}} \lambda(\tau) \in \mathbb{D}_i \wedge \lambda(\tau) = a_i.$$

Diverging Gateways

For the Diverging (Exclusive, Inclusive/Multi-choice and Complex) Gateways the following hold:

- All BPMN sequence flows (apart from the default ones) outgoing from a diverging gateway have the condition in the form of a conditional part of a rule from the $\mathbb{K}_{SKE}$ rule base assigned, i.e.:

$$\forall_{f \in \mathcal{F}_{g,div}^{out,cond}} \exists_{r \in R} (map(f) = r_i) \wedge (body(f) = E_i^{cond}) \wedge (language(f) =$$
"XTT2).

- In the case of exclusive gateways, the evaluated conditions have to be exclusive, i.e.:

$$\forall_{f_1,f_2 \in \mathcal{F}_{g,div}^{out}} \forall_{g \in \mathcal{G}_\times} (\exists_{r_1,r_2 \in R} map(f_1) = r_1 \wedge map(f_2) = r_2) \Rightarrow (eval(r_1) \neq eval(r_2)).$$

Gateway Preceded by a BR Task

For the Gateways preceded by a BR task the following hold:

- All BPMN sequence flows (apart from the default sequence flows) outgoing from a diverging gateway preceded by the BR task have the conditions based upon the output attributes of the BR task, i.e.:

$$\forall_{\substack{\tau \in T_{BusinessRule} \\ t \in T \\ map(\tau)=t}} \forall_{g \in \mathcal{G}_{div}^{cond}} ((\tau, g, \lambda_{\tau g}) \in \mathcal{F}) \Rightarrow$$
$$\left(\forall_{f \in \mathcal{F}_g^{out}} \forall_{\lambda \in body(f)} \exists_{a_i \in A_t^{dec}} \lambda(\tau) \in \mathbb{D}_i \wedge \lambda(\tau) = a_i\right).$$

The whole specification of the BP Model Integrated with the XTT2 Rules with constraints defining the connections between process elements and rules was presented in [10]. This simple notation will be used in the following section for description of the case study example.

5.4 Description of Example Using the Model

In order to evaluate the proposed model, we used selected case studies which show its feasibility and efficiency. The described models are executable[17] in the provided runtime environment [11].

To clarify the model, let us present an illustrative example of the Polish Liability Insurance (PLI) case study, previously introduced in Sect. 4.4. An excerpt of the most relevant formulae of the $\mathcal{M}_{SKE}^{PLI}$ model is as follows:

---

[17]The models consist of the BPMN 2.0 elements from the Common Executable Conformance Sub-Class [6].

$$\mathcal{M}_{SKE}^{PLI} = (\mathcal{P}^{PLI}, \mathbb{K}_{SKE}^{PLI}, map^{PLI})$$

$$\mathcal{P}^{PLI} = (\mathcal{O}, \mathcal{F}, \Lambda)$$

$$\mathcal{O} = \mathcal{A} \cup \mathcal{E} \cup \mathcal{G}$$

$$\mathcal{A} = \mathcal{T}_{Business\ Rule} \cup \mathcal{T}_{User}$$

$$\mathcal{T}_{Business\ Rule} = \{\tau_{Determine\ clientclass}, \tau_{Calculate\ base\ charge},$$
$$\tau_{Calculate\ driver\ discount\ base}, \tau_{Calculate\ car\ discount\ base},$$
$$\tau_{Calculate\ other\ discount\ base}, \tau_{Calculate\ driver\ discount}, \tau_{Calculate\ car\ discount},$$
$$\tau_{Calculate\ other\ discount}, \tau_{Calculate\ payment}\},$$

$$\mathcal{T}_{User} = \{\tau_{Enter\ car\ capacity\ information}, \tau_{Enter\ Bonus\ Malus\ information},$$
$$\tau_{Enter\ Premium\ information}, \tau_{Display\ payment\ result}\},$$

$$\mathcal{E} = \{e_{Start}, e_{End}\},$$

$$|\mathcal{G}| = 4.$$

The process model, presented in Fig. 5.11, consists of: 4 User tasks, 9 Business Rule tasks, start and end events, as well as 4 parallel gateways.

Suitable forms for acquiring data from the user are specified within the user tasks. The forms consists of the G_A^{PLI} attributes. As the attributes have suitable types defined, it is possible for the execution environment to adapt the user interface in order to support predefined form fields for particular types as well as to impose constraints on input according to the attribute domain.[18] Thus, the following forms are defined:

$$form(\tau_{Enter\ car\ capacity\ information}) = \{carCapacity\},$$
$$form(\tau_{Enter\ Bonus\ Malus\ information}) = \{clientClass, accidentNo\},$$
$$form(\tau_{Enter\ Premium\ information}) = \{driverAge, driverLicenceAge, carAge, antiqueCar,$$
$$seatsNo, technical, installmentsNo, insuranceCont,$$
$$insuranceCarsNo, otherInsurance, insuranceHistory\},$$
$$form(\tau_{Display\ payment\ result}) = \{payment\}.$$

This model can be integrated with rules from the $\mathbb{K}_{SKE}^{PLI}$ rule base. In such a case, the Business Rule tasks have to be associated with the decision tables from the T_χ set containing the proper XTT2 rules. Below, the specification of the decision tables is presented (it provides decision table schemas which have to be complemented with XTT2 rules).

$$\mathbb{K}_{SKE}^{PLI} = (A, R, T_\chi), \text{ where:}$$

$$T_\chi = \{t_{Determine\ client\ class}, t_{Calculate\ base\ charge}, t_{Calculate\ driver\ discount},$$
$$t_{Calculate\ car\ discount}, t_{Calculate\ car\ discount\ base}, t_{Calculate\ driver\ discount\ base},$$
$$t_{Calculate\ other\ discount\ base}, t_{Calculate\ other\ discount}, t_{Calculate\ payment}\},$$

$$schema(t_{Determine\ client\ class}) = (\{accidentNo, clientClass\}, \{clientClass\}),$$
$$schema(t_{Calculate\ base\ charge}) = (\{carCapacity\}, \{baseCharge\}),$$
$$schema(t_{Calculate\ driver\ discount}) = (\{driverAge, driverLicenceAge,$$
$$driverDiscountBase\}, \{driverDiscount\}),$$
$$schema(t_{Calculate\ car\ discount}) = (\{seatsNo, technical, antiqueCar,$$

---

[18] As ARD does not the specification of various types of attributes, by default all the attributes are of the default (symbolic) type, without a predefined domain. In such a case all form fields will be used default text input field.

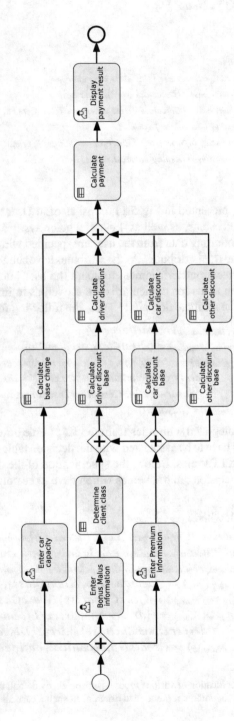

Fig. 5.11 The BPMN model for the PLI case study

Table 5.2   "Calculate other discount base" XTT2 decision table

(?) installmentNo	(?) insuranceCont	(?) insuranceCarsNo	(–>) otherDiscountBase
1	1	1	−10
1	0	1	0
1	1	>= 1	−20
1	0	>= 1	−10
2	1	1	0
2	0	1	10
2	1	>= 1	−10
2	0	>= 1	0

$$car\,Discount\,Base\}, \{car\,Discount\}),$$
$$schema(t_{Calculate\ car\ discount\ base}) = (\{car\,Age\}, \{car\,Discount\,Base\}),$$
$$schema(t_{Calculate\ driver\ discount\ base}) = (\{client\,Class\},$$
$$\{driver\,Discount\,Base\}),$$
$$schema(t_{Calculate\ other\ discount\ base}) = (\{installment\,No, insurance\,Cont,$$
$$insurance\,Cars\,No\}, \{other\,Discount\,Base\}),$$
$$schema(t_{Calculate\ other\ discount}) = (\{insurance\,History, other\,Insurance,$$
$$other\,Discount\,Base\}, \{other\,Discount\}),$$
$$schema(t_{Calculate\ payment}) = (\{base\,Charge, driver\,Discount, car\,Discount,$$
$$other\,Discount\}, \{payment\}).$$
$$map^{PLI} = \{ \ (\tau_{Determine\ class}, t_{Determine\ class}),$$
$$(\tau_{Calculate\ base\ charge}, t_{Calculate\ base\ charge}),$$
$$(\tau_{Calculate\ driver\ discount}, t_{Calculate\ driver\ discount}),$$
$$(\tau_{Calculate\ car\ discount}, t_{Calculate\ car\ discount}),$$
$$(\tau_{Calculate\ driver\ discount\ base}, t_{Calculate\ driver\ discount\ base}),$$
$$(\tau_{Calculate\ car\ discount\ base}, t_{Calculate\ car\ discount\ base}),$$
$$(\tau_{Calculate\ other\ discount\ base}, t_{Calculate\ other\ discount\ base}),$$
$$(\tau_{Calculate\ other\ discount}, t_{Calculate\ other\ discount}),$$
$$(\tau_{Calculate\ payment}, t_{Calculate\ payment}) \ \}.$$

The decision table "Calculate other discount base" related to the Business Rule task in the process model is filled in with suitable rules and is presented in Table 5.2. In the model, this decision table is represented as follows:

$$t_{Calculate\ other\ discount\ base} = (r_1, r_2, \ldots, r_8)$$
$$r_1 = (\{(installment\,No, =, 1), (insurance\,Cont, =, 1), (insurance\,Cars\,No, =, 1)\},$$
$$\{(other\,Discount\,Base, :=, -10)\}),$$
$$r_2 = (\{(installment\,No, =, 1), (insurance\,Cont, =, 0), (insurance\,Cars\,No, =, 1)\},$$
$$\{(other\,Discount\,Base, :=, 0)\}),$$
$$\ldots$$
$$r_8 = (\{(installment\,No, =, 2), (insurance\,Cont, =, 0), (insurance\,Cars\,No, >=, 1)\},$$
$$\{(other\,Discount\,Base, :=, 0)\}).$$

The practical implementation of this approach will be discussed in Sect. 3.5.

XTT2 rules can be verified in the dedicated environment that will be described in Chap. 9. The formal verification of each part of the integrated model (process or rules) is in fact possible. Because of the use of the XTT2 rule representation, it is possible to use the existing methods for the formal verification of rules [12], especially for the formal verification of decision components, such as inconsistency, completeness, subsumption or the equivalence [13]. In the case of the process model itself, one can use the existing verification methods [14], specifically those methods that can take into account the task logic [15].

5.5   Summary

The chapter presented a formal model for the integration of Business Processes with Business Rules. This model uses the existing representation methods for processes and rules, such as the BPMN notation for process models, and the XTT2 representation for rules. Such a model can be treated as a structured rule base that provides explicit inference flow determined by the process control flow. The evaluation we provided [10, 16] demonstrated that the presented model provides a sufficient formal means for describing a process model integrated with rules.

The presented model can be used for a clear description of a process model, especially for the practical design, specification of integration issues and ensuring data types consistency. The model can also be used as a specification of the constraints required for the execution purposes. As the BPMN models are executable in process engines and rules in the XTT2 representation can be executed in the HEART rule engine, such integrated models can be executed in the hybrid runtime environment [11].

This model can be treated as an alternative for the dedicated inference strategies described in Sect. 4.5. Moreover, it can enrich the design of a complex business oriented rule-based system. It was used in specification of the algorithm for generation of the integrated models from the ARD+ diagrams [17] that support the prototyping of the structure of the XTT2 knowledge bases. These prototyping and design issues will be discussed in the following chapter.

References

1. Lindsay, A., Dawns, D., Lunn, K.: Business processes – attempts to find a definition. Inf. Software Technol. **45**(15), 1015–1019 (2003). Elsevier
2. Charfi, A., Mezini, M.: Hybrid web service composition: business processes meet business rules. In: Proceedings of the 2nd International Conference on Service-Oriented Computing, ICSOC 2004, New York, NY, USA. ACM, pp. 30–38 (2004)

3. Knolmayer, G., Endl, R., Pfahrer, M.: Modeling processes and workflows by business rules. In: Business Process Management, Models, Techniques, and Empirical Studies, pp. 16–29. Springer, London (2000)
4. Kluza, K., Nalepa, G.J., Łysik, Ł.: Visual inference specification methods for modularized rule-bases. Overview and integration proposal. In: Nalepa, G.J., Baumeister, J. (eds.) Proceedings of the 6th Workshop on Knowledge Engineering and Software Engineering (KESE6) at the 33rd German Conference on Artificial Intelligence, 21 September 2010, Karlsruhe, Germany, pp. 6–17 (2010)
5. Hohwiller, J., Schlegel, D., Grieser, G., Hoekstra, Y.: Integration of bpm and brm. In: Dijkman, R., Hofstetter, J., Koehler, J. (eds.) Business Process Model and Notation. Lecture Notes in Business Information Processing, vol. 95, pp. 136–141. Springer, Berlin (2011)
6. OMG: Business Process Model and Notation (BPMN): Version 2.0 specification. Technical report formal/2011-01-03, Object Management Group, January 2011
7. Nalepa, G.J., Ligęza, A., Kaczor, K.: Formalization and modeling of rules using the XTT2 method. Int. J. Artif. Intell. Tools **20**(6), 1107–1125 (2011)
8. Wagner, G., Giurca, A., Lukichev, S.: R2ml: a general approach for marking up rules. In: Bry, F., Fages, F., Marchiori, M., Ohlbach, H. (eds.) Principles and Practices of Semantic Web Reasoning, Dagstuhl Seminar Proceedings 05371 (2005)
9. Object Management Group (OMG): Decision model and notation request for proposal. Technical report bmi/2011-03-04, Object Management Group, 140 Kendrick Street, Building A Suite 300, Needham, MA 02494, USA, March 2011
10. Kluza, K.: Methods for Modeling and Integration of Business Processes with Rules. Ph.D. thesis, AGH University of Science and Technology, Supervisor: Grzegorz J. Nalepa, March 2015
11. Nalepa, G.J., Kluza, K., Kaczor, K.: Proposal of an inference engine architecture for business rules and processes. In: Rutkowski, L., et al. (eds.) Artificial Intelligence and Soft Computing: 12th International Conference, ICAISC 2013, Zakopane, Poland, 9–13 June 2013, vol. 7895, Lecture Notes in Artificial Intelligence, pp. 453–464. Springer, Berlin (2013)
12. Ligęza, A.: Intelligent data and knowledge analysis and verification; towards a taxonomy of specific problems. In: Vermesan, A., Coenen, F. (eds.) Validation and Verification of Knowledge Based Systems, pp. 313–325. Springer, US (1999)
13. Nalepa, G., Bobek, S., Ligęza, A., Kaczor, K.: HalVA – rule analysis framework for XTT2 rules. In: Bassiliades, N., Governatori, G., Paschke, A. (eds.) Rule-Based Reasoning, Programming, and Applications, vol. 6826. Lecture Notes in Computer Science, pp. 337–344. Springer, Berlin (2011)
14. Wynn, M., Verbeek, H., van der Aalst, W., Hofstede, A., Edmond, D.: Business process verification - finally a reality!. Bus. Process Manag. J. **1**(15), 74–92 (2009)
15. Szpyrka, M., Nalepa, G.J., Ligęza, A., Kluza, K.: Proposal of formal verification of selected BPMN models with Alvis modeling language. In: Brazier, F.M., Nieuwenhuis, K., Pavlin, G., Warnier, M., Badica, C. (eds.) Intelligent Distributed Computing V. Proceedings of the 5th International Symposium on Intelligent Distributed Computing – IDC 2011, Delft, The Netherlands – October 2011, vol. 382. Studies in Computational Intelligence, pp. 249–255. Springer, Berlin (2011)
16. Kluza, K., Nalepa, G. J.: Formal Model of Business Processes Integrated with Business Rules. Information Systems Frontiers (2016, submitted)
17. Kluza, K., Nalepa, G.J.: Automatic generation of business process models based on attribute relationship diagrams. In: Lohmann, N., Song, M., Wohed, P. (eds.) Business Process Management Workshops, vol. 171. Lecture Notes in Business Information Processing, pp. 185–197. Springer International Publishing, Berlin (2014)

Chapter 6
Prototyping Structure of Rule Bases

In Chap. 2 we introduced the main issues and challenges related to the process of building RBS. As we pointed out, designing a knowledge base for a RBS is a tedious task. The main issue concerns the identification of system properties on which the rules are based. This is an iterative process that needs proper support from the design method as well as computer tools. Furthermore, this is especially hard in the case of structured knowledge bases, as discussed in Sect. 2.3. Unfortunately, there are few well-established tools for providing a transition from vague concepts provided by the user or expert to actual rules. This stage is often generally referred to as the conceptual design. It is addressed by some recent representations such as the SBVR [1] by the OMG and Business Rules communities, see Sect. 3.1.

The XTT2 knowledge representation introduced in Chap. 4 relies on structured knowledge bases. While this solution has many advantages, it also introduces challenges for the design and inference. High level inference control can be handled by dedicated inference engine, or delegated to a process engine, as discussed in Chap. 5. However, the proper design of a structured knowledge base remains complicated.

The focus of the ARD+ method presented in this chapter is the initial transition from user-provided specification (often in natural language) that includes general concepts, to the rule specification that ties rules with these concepts. Moreover, the semi-automated prototyping of the structure of the knowledge bases is possible. The schemas of XTT2 decision tables can be obtained, along with default inference links between tables. In Sect. 6.1 we introduce the main intuitions behind the ARD+ method. The formalization of the method is provided in Sect. 6.2. Then an algorithm for prototyping the structure of the XTT2 rule base is given in Sect. 6.3. In fact, the design with ARD+ can be also used to generate a business process with rules, as discussed in Sect. 6.4 conforming to the model introduced in Chap. 5. We conclude the chapter in Sect. 6.5.

© Springer International Publishing AG 2018   133
G.J. Nalepa, *Modeling with Rules Using Semantic Knowledge Engineering*,
Intelligent Systems Reference Library 130,
https://doi.org/10.1007/978-3-319-66655-6_6

6.1   Main Concepts of ARD+

Attribute Relationship Diagrams (ARD+) was inspired by ERD (Entity Relationship Diagrams) [2]. It as a method for prototyping a knowledge base structure in a similar way as the relational data base structure (tables) is generated from the ERD diagrams, and as a simple alternative to the classic approaches [3]. The preliminary concepts of the ARD+ method were introduced in [4], then in [5]. Here, an extended version called ARD+ is discussed. The ARD+ method supports the rule designer at a very general design level as described in [6, 7]. The extended version of the method called ARD+ was proposed in [8]. It enables hierarchical rule prototyping that supports the gradual design process. The main goal of ARD+ is to identify rule schemas, thus supporting the practical prototyping of the structure of the XTT2 knowledge base. Its input is a general systems description in natural language, and output is a model capturing knowledge about relationships among the attributes describing system properties. The model is subsequently used in the next design stage where the logical design with rules is carried out.

The main concepts behind ARD+ are:

- *functional dependency* – a general relationship between two or more attributes (or attribute sets called *properties*) which are called *dependent* and *independent*. The dependency means that in order to determine the values of the dependent attributes, the values of those that are independent are needed, and
- *derivation* – a relationship between two properties denoting a *refinement* operation, from a more general property, to one that is more specific. In ARD+ there are two refinement operations: *finalization* and *split*.

The ARD+ design is being specified in a number of steps, using structural transformations where each step is more detailed than the previous one. During the design, a hierarchical model is built. It captures all of the subsequent design steps, by holding knowledge about the system on the different abstraction levels.

6.2   ARD+ Method Formalization

This formalization of ARD+ is based upon the version of the formalization presented in [6], and was partially presented in [9]. However, several new definitions were introduced. In the discussion we will use the previously introduced PLI example.

Let us begin with the general idea of attributes for characterizing the system. Let us consider the set C of conceptual attributes, the set A of physical attributes, the set P of properties, the set D of dependencies and the set Q consisting of derivations, defined as follows.

Definition 6.1 A **conceptual attribute** $c \in C$ is an attribute describing some general, abstract aspect of the specified system.

> *Example 6.2.1* Conceptual attribute name starts with a capital letter, e.g.
> `BaseRate`.

During the design process, conceptual attributes are being *finalized* into, possibly multiple, physical attributes (see Definition 6.10 for the ***finalization*** description).

Definition 6.2 A **physical attribute** $a \in A$ is an attribute describing a specific well-defined, atomic aspect of the specified system.

> *Example 6.2.2* Names of physical attributes are not capitalized, e.g.
> `payment`.

A physical attribute originates from one or more (indirectly) conceptual attributes and can not be further *finalized*.

Definition 6.3 A **property** $p \in P$ is a non-empty set of attributes ($p \subseteq A \cup C$) describing the property and representing a piece of knowledge about a certain part of the system being designed.

A set of properties P can be partitioned into disjoint sets of simple and complex properties:

$$P = P^s \cup P^c, \ P^s \cap P^c = \varnothing$$

Definition 6.4 A **simple property** $p \in P^s$ is a property consisting of a single attribute ($|p| = 1$).

Definition 6.5 A **complex property** $p \in P^c$ is a property consisting of multiple attributes ($|p| > 1$).

Definition 6.6 A **dependency** $d \in D$ is an ordered pair of properties (f, t), where $f \in P$ is the **independent property** and $t \in P$ is the **dependent property** that depends on f. If $f = t$ the property is called **self-dependent**. For notational convention $d = (f, t), d \in D, D \subseteq P \times P$ will be presented as: $d(f, t)$.[1]

A (functional) dependency is a relationship between two properties that shows that in order to determine the dependent property attribute values, values of the attributes of the independent property are needed.

Definition 6.7 A **derivation** $q \in Q$ is an ordered pair of properties (f, t), where $t \in P$ is derived from $f \in P$ upon a transformation. Similarly to dependency $Q \subseteq P \times P$, however $D \cap Q = \varnothing$.

---

[1]$d(f, t)$ denotes a dependency d from a property f to a property t.

Definition 6.8 A **Design Process Diagram** G_D is a triple (P, D, Q), where:

- P is a set of properties,
- D is a set of dependencies,
- Q is a set of derivations.

The DPD diagram is a directed graph with properties as nodes and both dependencies and derivations as edges. Between two properties only a single dependency or derivation is allowed, i.e. the following hold:

$$\forall_{d_1, d_2 \in D} \ (d_1 = (f, t) \land d_2 = (f, t)) \Rightarrow (d_1 = d_2)$$
$$\forall_{d_1, d_2 \in D} \ (d_1 = (f, t) \land d_2 = (t, f)) \Rightarrow ((t = f) \land (d_1 = d_2))$$
$$\forall_{q_1, q_2 \in Q} \ (q_1 = (f, t) \land q_2 = (f, t)) \Rightarrow (q_1 = q_2)$$
$$\forall_{q \in Q} \ (q = (f, t)) \Rightarrow (t, f) \notin Q$$

Definition 6.9 An **Attribute Relationship Diagram** G_A is a pair (P_{ARD}, D), where:

- there is a $G_D = (P, D, Q)$,
- P_{ARD} is a subset of G_D properties ($P_{ARD} \subseteq P$) such that
  $P_{ARD} = \{p_i \in P : \forall_{p_j \in P} \ (p_i, p_j) \notin Q\}$,
- and D is a set of dependencies.

An ARD diagram can be depicted as a graph with the properties represented as nodes and dependencies represented as edges.

Example 6.2.3 To illustrate the ARD concepts, a fragment of an exemplary ARD diagram with two properties and the dependency between them is presented in the figure below. The diagram should be interpreted in the following way: payment depends on carCapacity and baseCharge (either on value or existence). Note that in other modeling languages the dependency is often modeled inversely, i.e. as an arrow pointing from a dependent object to one that is independent, e.g. UML [10].

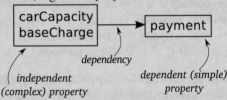

Example 6.2.4 To clarify the ARD method, let us present an illustrative example of the PLI case study. This is a case, in which the price for the liability
insurance for protecting against third party insurance claims is to be calculated.
The price is calculated based upon various reasons, which can be obtained from
the insurance domain expert. The main factors in calculating the liability insurance premium are data about the vehicle: the car engine capacity, the car age,
seats, and a technical examination. Additionally, the impact upon the insurance price have the driver's age, the period of holding the license, the number
of accidents in the last year, and the previous class of insurance. In the calculation, the insurance premium can be increased or decreased because of the
number of payment installments, other insurances, continuity of insurance or
the number of cars insured. All these pieces of data, obtained from an expert,
can be specified using the ARD method and presented using the ARD diagram
(see the figure below).

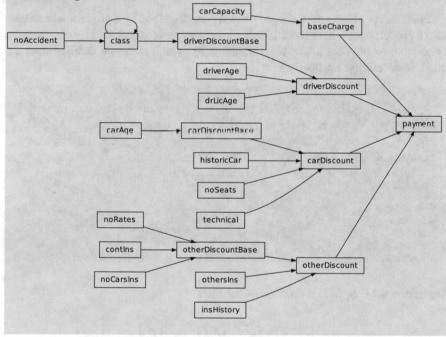

Specification of ARD is an iterative process, with regard to properties and serve as
a tool for diagram specification and development. The two diagram transformations
(*finalization* and *split*) constitute the core aspects of the ARD method. They transform
a property into one or more properties, specifying new derivations and dependencies
into a G_D model. These transformations are also required in order to introduce new
attributes for the system.

For the transformation of properties from the diagram G_A^1 into the properties in diagram G_A^2, the properties in the G_A^2 diagram are more specific than in the G_A^1.

Definition 6.10 Finalization is a function of the form:

$$finalization : p_1 \xrightarrow{Q_f} p_2$$

where:

- p_1 and p_2 are properties, such that:

  $p_1 \in P^s \wedge p_1 = \{c_i : c_i \in C\},$
  $p_2 \in P,$
  $p_1 \cap p_2 = \emptyset,$

- $Q_f \subseteq Q$ is a subset of new derivations, such that: $Q_f = \{(p_1, p_2)\}$

Finalization transforms a simple property p_1 described by a conceptual attribute into a property p_2, where the attribute describing p_1 is substituted by one or more conceptual or physical attributes describing p_2, which are more detailed than the attribute describing a property p_1.

Example 6.2.5 The figure below presents an exemplary ***finalization*** transformation. It shows that the simple property `BaseRate` (described by a single conceptual attribute) is finalized into a new complex property described by two physical attributes `carCapacity` and `baseCharge`.

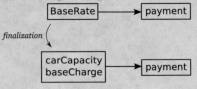

Definition 6.11 Split is a function of the form:

$$split : p_c \xrightarrow{Q_s, D_s} \{p^1, p^2, \ldots, p^n\}$$

where:

- p_c and $p^1, p^2, \ldots, p^n$ are properties, such that:

  $p_c \in P^c,$
  $p^1 \cup p^2 \cup \ldots \cup p^n = p_c$
  $\forall_{i,j \in \{1,\ldots,n\}, i \neq j}\ p^i \cap p^j = \emptyset,$

- $Q_s \subseteq Q$ is the subset of new derivations, such that: $Q_s = \{q_i \in Q : q_i = (p_c, p^i)\}$
- $D_s \subseteq D$ is the subset of new dependencies defined by a designer.

In the split transformation a complex property p_c is replaced by n properties, each of them described by one or more attributes originally describing p_c. Since p_c may depend on some other properties $p_o^1 \ldots p_o^m$, dependencies between these properties and $p^1 \ldots p^n$ have to be defined.

Example 6.2.6 An example of ***split*** transformation is illustrated in the figure below. The complex property described by two physical attributes (carCapacity and baseCharge) is split into two simple properties described by these attributes.

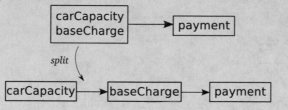

During the design process, upon splitting and finalizing, the ARD model is more and more specific. Thus, after n split and m finalization transformations, the DPD diagram consists of the introduced properties (and attributes), as well as derivations and dependencies, such that:

- $Q = Q_s^1 \cup Q_s^2 \cup \ldots \cup Q_s^n \cup Q_f^1 \cup Q_f^2 \cup \ldots \cup Q_f^m$,
- $D = D_s^1 \cup D_s^2 \cup \ldots \cup D_s^n$.

The transformations can be depicted in a hierarchical Transformation Process History (TPH) model as defined below and presented in Example 6.2.7. However, the consecutive ARD levels of more and more detailed diagrams describing the designed system can be also presented separately as depicted in Example 6.2.8.

Definition 6.12 A **Transformation Process History** G_T is a pair (P, Q), where P and Q are properties and dependencies respectively from the existing $DPD = (P, D, Q)$.

A TPH diagram forms a tree with properties as nodes and derivations as edges. It denotes what particular property is split into or what attributes a particular property attribute is finalized into. Such a diagram stores the history of the design process.

Example 6.2.7 An example of a TPH model is presented below. The diagram corresponds to the ARD model presented in Example 6.2.4.

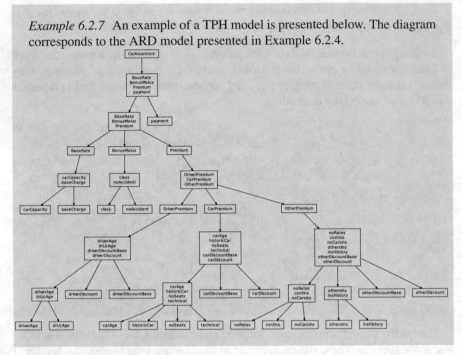

Originally [4], the ARD method was proposed as a simple method for generating a knowledge base structure. Thus, using the ARD dependencies, it is possible to generate structures (schemes) of decision tables [8]. Then, such schemes can be filled in with rules either manually by a domain expert or automatically mined from some additional data. Figure 6.1 presents an exemplary schema of a decision table generated from the identified ARD dependencies (specifically from the ARD dependency between two physical attributes carCapacity and baseCharge, see: Example 6.2.6) and the same table filled in with rules.

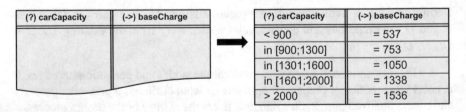

(?) carCapacity	(->) baseCharge

(?) carCapacity	(->) baseCharge
< 900	= 537
in [900;1300]	= 753
in [1301;1600]	= 1050
in [1601;2000]	= 1338
> 2000	= 1536

Fig. 6.1 Decision table schema (*left*) and table filled in with rules (*right*), after [11]

Example 6.2.8 The whole design process for the PLI case study, starting from the very general conceptual property `CarInsurance` and ending at the lowest ARD level, is depicted below.

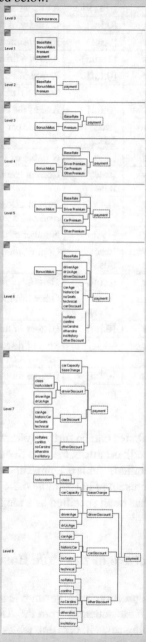

Table 6.1   Attributes for the PLI case study

Attribute name	Type	Range	Description of attribute
noAccident	Integer	[0;inf]	Number of accidents in last 12 months
class	Integer	[−1; 9]	A customer class
carCapacity	Integer	[0;inf]	Capacity of the car engine [cm^3]
baseCharge	Integer	[0;inf]	Base insurance charge [PLN]
driverAge	Integer	[16;inf]	Age of a driver (owner of the car)
drLicAge	Integer	[0;inf]	Period of holding a driving license
driverDiscount	Integer	—	Sum of driverAge and drLicAge
carAge	Integer	[0;inf]	Age of the car
historiCar	Boolean	[true; false]	Historic car
noSeats	Integer	[2;9]	Number of seats in the car
technical	Boolean	[true; false]	Current technical examination of the car
carDiscount	Integer	—	Sum of discounts: carAge, historiCar, noSeats and technical
noRates	Integer	[1;2]	Number of instalments
contIns	Boolean	[true; false]	Continuation of insurance
noCarsIns	Integer	[0;inf]	Number of insured cars
otherIns	Boolean	[true; false]	Other insurances
insHistory	Integer	[0;inf]	History of the driver insurance
otherDiscount	Integer	—	Sum of discounts: noRates, contIns, noCarsIns, otherIns, insHistory
payment	Float	[0;inf]	Charge for the car insurance

All the physical ARD attributes can be further refined and defined in terms of the ALSV(FD) simple and generalized attributes (see Sect. 4.1). Thus, the domains of the attributes have to be specified. Table 6.1 presents the attributes that have been identified for the case study presented in Example 6.2.4 with their short description. For each attribute a domain has been specified as a range of the specific type.

6.3   Prototyping Structure of the Rule Base

The goal of the table schema prototyping algorithm is to automatically build prototypes for rules described by table schemas from the ARD+ diagrams. The targeted rule base is structured, grouping rules in decision tables with explicit inference control. It is especially suitable for the XTT2 representation.

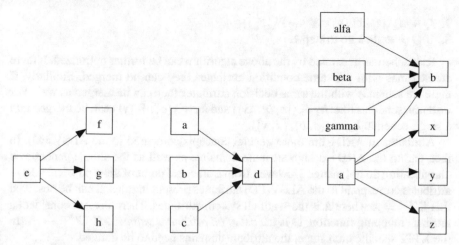

Fig. 6.2 Possible subgraphs in the ARD+ structure

In order to formulate the algorithm some basic subgraphs in the ARD+ structure can be considered. Examples of these are presented in Fig. 6.2. Now, considering the functional dependencies among the properties the corresponding rule prototypes are as follows:

- for the first case (left in Fig. 6.2): $h = (\{e\}, \{f, g, h\})$,
- for the second case (middle in Fig. 6.2): $h = (\{a, b, c\}, \{d\})$.

Let us consider a sub-graph presented on the right in Fig. 6.2. It corresponds to the following rule prototypes: $h' = (\{alpha, beta, gamma, a\}, \{b\})$ and $h'' = (\{a\}, \{x, y, z\})$. Using these cases a general prototyping algorithm was formulated in [7] and is discussed in the following section.

Let the set of schemas have the following structure:

$$\mathbb{H} = \{h_i : s_i = (\mathbb{C}^{cond}, \mathbb{C}^{dec}), h_i \in 2^{\mathbb{C}_{phys}} \times 2^{\mathbb{C}_{phys}}\}$$

Table Schema Prototyping Algorithm

INPUT: the most detailed ARD diagram G_n containing only simple properties that contain only physical attributes.
OUTPUT: a set of rule schemas $\mathbb{H}$.

1. Select a dependency $Y = (Q_i, Q_d)$, from all dependencies in the diagram G_n.
2. Find all properties Q_j that Q_d depends on: let $\mathbb{Q}_I = \{Q_j : \exists (Q_j, Q_d), Q_j \neq Q_i\}$.
3. Find all properties Q_k that depend on Q_i alone:
   let $\mathbb{Q}_D = \{Q_k : \exists (Q_i, Q_k), Q_k \neq Q_d, \nexists Q_x : (Q_x, Q_k), Q_x \neq Q_i)\}$.
4. If $\mathbb{Q}_I \neq \varnothing$ then generate schema: $h' = (Q_I, \{Q_d\}), \mathbb{H} \leftarrow \mathbb{H} \cup h'$
5. If $\mathbb{Q}_D \neq \varnothing$ then generate schema: $h'' = (\{Q_i\}, Q_D), \mathbb{H} \leftarrow \mathbb{H} \cup h''$
6. If $\mathbb{Q}_I = \varnothing$ and $\mathbb{Q}_D = \varnothing$ then generate schema: $h''' = (\{Q_i\}, \{Q_d\}), \mathbb{H} \leftarrow \mathbb{H} \cup h'''$

7. $\mathbb{Q} \leftarrow \mathbb{Q}\backslash(\mathbb{Q}_I \cup \mathbb{Q}_D)$, $\mathbb{Y} \leftarrow \mathbb{Y}\backslash\{Y_{i,d}\}$.

8. If $\mathbb{Q} \neq \varnothing$ then go to step 1.

Rule schemas generated by the above algorithm can be further optimized. If there are schemas with the same condition attributes they can be merged. Similarly, if there are schemas with the same decision attributes they can be merged as well. For instance, schemas like: $h_1 = (\{a, b\}, \{x\})$ and $h_2 = (\{a, b\}, \{y\})$ can be merged into a single schema: $h_3 = (\{a, b\}, \{x, y\})$.

Attributes in ARD+ are more general concepts compared to the ALSV(FD). In fact, during the ARD+ design attribute domains, as well as the simple/generalized classification do not matter. However, from a practical point of view physical ARD+ attributes correspond to the ALSV(FD) attributes present in rules. It can be observed that $\mathbb{C}^{phys} = \mathbb{A}$ where $\mathbb{A}$ is the set of all system attributes. Therefore, we consider an attribute mapping function: $\mathbb{D}$ is the *set of all attributes values*: $M: \mathbb{C}^{phys} \rightarrow \mathbb{A}$. In the XTT2 specification stage, the attribute domains need to be defined.

Another prototyping approach concerns the integration of business processes with XTT2 rules. ARD+ can be used to support the design of such integrated models as described in the following section.

6.4 Design of Business Processes with Business Rules

Business Processes and Business Rules are mostly designed manually. Although a simplified process model can be generated using process mining tools [12] or from natural language text using NLP techniques [13], they are not directly suitable for execution. In the approach presented here, an algorithm for generating a BPMN executable model is introduced. Using the algorithm, a BPMN model is generated along with decision table schemas for business rule tasks and from attributes for user tasks from ARD+ Diagrams. It can be seen as the extension and implementation of the ideas introduced in [14, 15]. The overview of the approach can be observed in Fig. 6.3.

Outline of the Approach

Models of processes and rules are based upon pieces of information acquired through structured interviews or documentation provided by the company (such as system description, requirement specification or some documents describing products and services). In the modeling task, the analysts take advantage of their knowledge and experience. As it is not a clearly defined mapping, it can be seen as it would involve the famous ATAMO[2] procedure to obtain Business Processes and Business Rules. Depending on the representation languages, such models often have to be significantly refined or implemented in order to be executed.

---

[2]"And then a miracle occurs" – the phrase, popularized by the Sidney Harris cartoon, is often used in BPM papers to describe procedures which take place but are hard to describe or algorithmize, e.g. [16, 17].

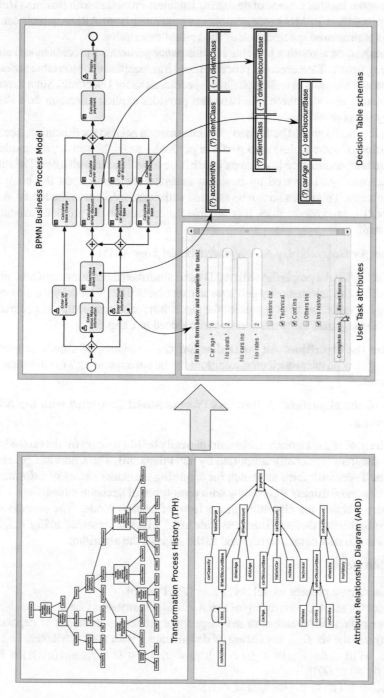

Fig. 6.3 Prototyping process model with rules

We propose another method of designing Business Processes with Business Rules. In our approach, a BPMN process model is generated from ARD+, which can be treated as a structured specification of the system description.

The translation algorithm presented in this paper generates a process model and a rule base structure. The generated process model has specified decision table schemas for Business Rule tasks as well as attribute specifications for User tasks. Such a model can be treated as a structured rule base that provides explicit inference flow determined by the process control flow.

Next, the model should be refined with information needed for execution, such as specification of users (roles) who perform particular tasks. Moreover, the model is integrated with decision table schemas which have to be filled in with rules. The rules, which are mostly discovered by business analysts, are based upon the company documentation. This task also can be assisted with some software in the future. After refining of the integrated model, it is suitable for enacting in the provided execution environment.

Algorithm for Generation of an Integrated Model from ARD+

Having identified the properties with ARD+ and described them in terms of attributes, the algorithm can automatically generate an executable BPMN model with the corresponding BR tasks. Let us now present the algorithm for integrated model generation from ARD diagrams, using the notation introduced in Chap. 5

Input for the algorithm: An ARD diagram G_A consisting of simple properties containing only physical attributes, additionally the corresponding TPH diagram G_T can be used.

Output of the algorithm: A Business Process Model Integrated with the XTT2 Rules $\mathcal{M}_{SKE}$.

Goal: The goal of the algorithm is to automatically build a process model on the basis of ARD diagram (optionally supported by TPH diagram). The algorithm generates both User Tasks with form attributes for acquiring particular pieces of information from a user and Business Rule Tasks with prototypes of decision tables.

Figure 6.4 shows the algorithm in the form of process model. The process uses the call subprocess "Develop Business Rule task" which is presented in Fig. 6.5. The numbers in the task names correspond to the steps of the algorithm.

Algorithm steps:

1. Create a new process model $\mathcal{M}_{SKE} = (\mathcal{P}, \mathcal{X}, map)$.
2. Select the set A_{tmp} consisting of the ARD input attributes (the attributes which occur only as independent or self-dependent properties in the set of dependencies) and the set D_{tmp} consisting of dependencies with these attributes, i.e.:
   $A_{tmp} = \{a \in A: \quad (a \in f_i) \land ((\exists_{t_j \in P} d(f_i, t_j) \in D \land \nexists_{t_k \in P} d(t_k, f_i) \in D) \lor (d(f_i, f_i) \in D))\}$,
   $D_{tmp} = \{d(f_i, t) \in D: \exists_{a \in A} a \in f_i\}$.
3. Loop for each dependency $d \in D_{tmp}: d(f, t), f \neq t$.

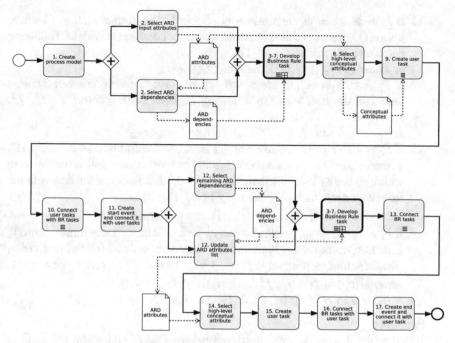

Fig. 6.4 Prototyping steps presented in a process model

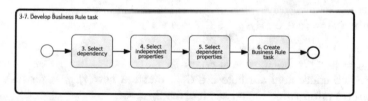

Fig. 6.5 The "Develop Business Rule task" subprocess

4. Select all independent properties (other than f) that t depends on.
   Let $F_t = \{f_t^i \in P : d(f_t^i, t) \wedge f_t^i \neq f\}$.
   Remove the considered dependencies from the set: $D_{tmp} := D_{tmp} \setminus F_t$.
5. Select all dependent properties (other than t) that depend only on f.
   Let $T_f = \{t_f^i \in P : d(f, t_f^i), t_f^i \neq t, \nexists_{f_x} (d(f_x, t_f^i), f_x \neq f)\}$.
   Remove the considered dependencies from the set: $D_{tmp} := D_{tmp} \setminus T_f$.
6. Based on F_t and T_f create Business Rule tasks and add them to the $\mathcal{T}_{BR}$ set.[3]
   A BR task ($\tau_{BR} \in \mathcal{T}_{BusinessRule}$) with the corresponding decision table ($t_X \in T_X$)
   schema can be developed in the following way:

---

[3]For simplicity, $\mathcal{T}_{BusinessRule}$ will be denoted as $\mathcal{T}_{BR}$, and its exemplary elements as $\tau_{BR}^1, \tau_{BR}^2 \in \mathcal{T}_{BusinessRule}$.

6.1 If $F_t = \varnothing, T_f = \varnothing$, create a new τ_{BR} BR task "Determine[4] $name(t)$", where $name(t)$ is a name of the t attribute, and associate the task with the following decision table schema: $f \mid t$, i.e.:
$schema(t_\chi) = (\{f\}, \{t\})$, and $map(\tau_{BR}) = t_\chi$.

6.2 If $F_t \neq \varnothing, T_f = \varnothing$, create a new τ_{BR} BR task "Determine $name(t)$" and associate the task with the following decision table schema: f, f_t^1, f_t^2, ... $\mid t$, i.e.:
$schema(t_\chi) = (\{f, f_t^1, f_t^2, \ldots\}, \{t\})$, and $map(\tau_{BR}) = t_\chi$.

6.3 If $F_t = \varnothing, T_f \neq \varnothing$, create a new τ_{BR} BR task "Determine $name(T_f \cup \{t\})$)", where $name(T_f)$ is a name of the lower-level conceptual attribute from which all the $T_f \cup \{t\}$ attributes are derived,[5] and associate the task with the following decision table schema: $f \mid t, t_f^1, t_f^2, \ldots$, i.e.:
$schema(t_\chi) = (\{f\}, \{t, t_f^1, t_f^2, \ldots\})$, and $map(\tau_{BR}) = t_\chi$.

6.4 If $F_t \neq \varnothing, T_f \neq \varnothing$, create new two τ_{BR}^1, τ_{BR}^2 BR tasks "Determine $name(t)$" and "Determine $name(T_f)$", and associate them with the following decision table schemas respectively: $f, f_t^1, f_t^2, \ldots \mid t$ and $f \mid t_f^1, t_f^2, \ldots$, i.e.:
$schema(t_\chi^1) = (\{f, f_t^1, f_t^2, \ldots\}, \{t\}), map(\tau_{BR}^1) = t_\chi^1$,
$schema(t_\chi^2) = (\{f\}, \{t_f^1, t_f^2, \ldots\})$, and $map(\tau_{BR}^2) = t_\chi^2$.

7. End loop.

8. Based on the A_{tmp} set of input attributes and the TPH model, select the set C_{tmp} of high-level conceptual attributes from which these input attributes are derived, i.e.:
$$derive(a) = \{c \in C : \exists_{\substack{q \in Q \\ q=(f,t) \\ Q \in G_T}} c \in f \wedge a \in t\},$$

$$C_{tmp} = \{c \in C : \exists_{a \in A_{tmp}} c \in derive(a)\}.$$

9. For each conceptual attribute $c \subset C_{tmp}$ create a new τ_{User}^c User task "Enter $name(c)$ information",[6] and add it to the $\mathcal{T}_{User}$ set.

10. Connect each User task from the $\mathcal{T}_{User}$ set using control flow with the proper BR tasks that require the input attributes related to the User task (with g_+ parallel gateway if necessary[7]), i.e.:
$$\mathcal{T}_{BR}^a = \{\tau \in \mathcal{T}_{BR} : \exists_{a \in A_{t_\chi}^{cond}} map(\tau) = t_\chi\},$$
$$\forall_{a \in A_{tmp}} \quad |\mathcal{T}_{BR}^a| = 1 \Rightarrow ((\tau_{User}^c, \tau_{BR}^a, \Lambda_{\tau_{User}^c, \tau_{BR}^a}) \in \mathcal{F} \wedge \tau_{BR}^a \in \mathcal{T}_{BR}^a \wedge c \in der-ive(a)),$$

---

[4]For the user-friendliness of task names, if the attribute t is of the symbolic type or a derived one, the word "Determine" should be used in the task name. In other cases (i.e. numeric types), one can use the word "Calculate" instead.

[5]The conceptual attribute name can be found in the corresponding TPH model, if it is available for the algorithm. In the other case, in the task name the names of all the attributes from the T_f set can be used.

[6]If a particular conceptual attribute covers a single input attribute, create a User task "Enter $name(a)$" instead.

[7]The g_+ parallel gateway is necessary if there are more than one BR tasks to be connected.

$$\forall_{a \in A_{tmp}} \quad |\mathcal{T}_{BR}^a| > 1 \Rightarrow (g_+ \in \mathcal{G} \wedge (\tau_{User}^c, g_+, \Lambda_{\tau_{User}^c, g_+}) \in \mathcal{F} \wedge c \in derive(a)$$
$$\wedge \; \forall_{\tau_{BR}^a \in \mathcal{T}_{BR}^a} (g_+, \tau_{BR}^a, \Lambda_{g_+, \tau_{BR}^a}) \in \mathcal{F})$$

11. Create the Start event e_{Start} and connect it with all User tasks from the $\mathcal{T}_{User}$ set using control flow (with g_+ parallel gateway if necessary), i.e.

$$\forall_{\mathcal{T}_{User} \in \mathcal{T}_{User}} |\mathcal{T}_{User}| = 1 \Rightarrow (e_{Start}, \tau_{User}, \Lambda_{e_{Start}, \tau_{User}}) \in \mathcal{F},$$
$$\forall_{\mathcal{T}_{User} \in \mathcal{T}_{User}} |\mathcal{T}_{User}| > 1 \Rightarrow (g_+ \in \mathcal{G} \wedge (e_{Start}, g_+, \Lambda_{e_{Start}, g_+}) \in \mathcal{F} \wedge (g_+, \tau_{User},$$
$$\Lambda_{g_+, \tau_{User}}) \in \mathcal{F}).$$

12. Select the set D_{tmp} consisting of all dependencies that have no input attributes in their properties, i.e.

$$D_{tmp} = \{d(f_i, t_i) \in D : \forall_{a \in A_{tmp}} \; a \notin f_i \wedge a \notin t_i\},$$

and the set A_{tmp} consisting of all the attributes occurring in these dependencies, i.e.

$$A_{tmp} = \{a \in A : \exists_{\substack{f_i \in P \\ t_i \in P \\ d(f_i, t_i) \in D}} \; a \in f_i \vee a \in t_i\},$$

and repeat steps 3–7 based on this set.

13. Using control flow, connect the BR tasks from the $\mathcal{T}_{BR}$ set one another (with g_+ parallel gateway if necessary) according to the following rule: two BR tasks $\tau_{BR}^1, \tau_{BR}^2 \in \mathcal{T}_{BR}$ should be connected if a decision table schema of τ_{BR}^2 contains at least one attribute a as an input attribute which is an output attribute of the τ_{BR}^1 decision table schema.

For formal description of this step, it is useful to define a temporary $\mathcal{F}_{tmp}$ set consisting of potential connections between BR tasks as well as two auxiliary sets $\mathcal{T}_{tmp}^{\tau, in}$ and $\mathcal{T}_{tmp}^{\tau, out}$ consisting of BR tasks from the temporary set:

$$\mathcal{F}_{tmp} = \{(\tau_1, \tau_2, \Lambda_{\tau_1, \tau_2}) \in \mathcal{F} : \exists_{a \in A} \quad a \in A_{t_1}^{dec} \wedge a \in A_{t_2}^{cond} \wedge map(\tau_1) = t_\chi^1 \wedge$$
$$map(\tau_2) = t_\chi^2)\},$$
$$\mathcal{T}_{tmp}^{\tau, in} = \{\tau_{in} \in \mathcal{T} : (\tau_{in}, \tau, \Lambda_{\tau_{in}, \tau}) \in \mathcal{F}_{tmp}\},$$
$$\mathcal{T}_{tmp}^{\tau, out} = \{\tau_{out} \in \mathcal{T} : (\tau, \tau_{out}, \Lambda_{\tau, \tau_{out}}) \in \mathcal{F}_{tmp}\},$$

Using the above sets, the control flows that have to be added to the process model can be formally defined as follows:

$$\forall_{(\tau_1, \tau_2, \Lambda_{\tau_1, \tau_2}) \in \mathcal{F}_{tmp}} (|\mathcal{T}_{tmp}^{\tau_1, in}| = 1 \wedge |\mathcal{T}_{tmp}^{\tau_2, out}| = 1) \Rightarrow (\tau_1, \tau_2, \Lambda_{\tau_1, \tau_2}) \in \mathcal{F},$$
$$\forall_{(\tau_1, \tau_2, \Lambda_{\tau_1, \tau_2}) \in \mathcal{F}_{tmp}} (|\mathcal{T}_{tmp}^{\tau_1, in}| > 1 \vee |\mathcal{T}_{tmp}^{\tau_2, out}| > 1) \Rightarrow$$
$$(g_+ \in \mathcal{G} \wedge \forall_{\tau_{in} \in \mathcal{T}_{tmp}^{\tau_1, in}} ((\tau_{in}, g_+, \Lambda_{\tau_{in}, g_+}) \in \mathcal{F}) \wedge \forall_{\tau_{out} \in \mathcal{T}_{tmp}^{\tau_2, out}} ((g_+, \tau_{out},$$
$$\Lambda_{g_+, \tau_{out}}) \in \mathcal{F}).$$

14. Select a subset $\mathcal{T}_{BR}^{out}$ of $\mathcal{T}_{BR}$, consisting of BR tasks that have no out coming control flows,[8] i.e.

$$\mathcal{T}_{BR}^{out} = \{\tau \in \mathcal{T}_{BR} : \nexists_{o \in \mathcal{O}} (\tau, o, \Lambda_{\tau, o}) \in \mathcal{F}\}.$$

Select the high-level conceptual attribute c from which the output attributes of task from $\mathcal{T}_{BR}^{out}$ are derived, i.e. $c \in derive(A_{t_\chi}^{dec})$, where $map(\tau) = t_\chi \wedge \tau \in \mathcal{T}_{BR}^{out}$.

15. Add a User task τ_{User}^{end} "Display $name(c)$ result".[9]

16. Connect the selected tasks from $\mathcal{T}_{BR}^{out}$ with it.

---

[8]This subset of output BR tasks should not be empty.

[9]If there is only one output attribute, its name should be used instead of $name(c)$.

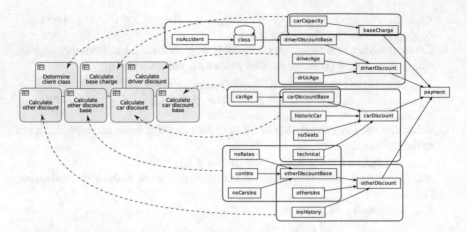

Fig. 6.6 First phase: Generating Business Rule tasks

17. Create the End event e_{End} and connect the User task τ_{User}^{end} with it, i.e. (τ_{User}^{end}, e_{End}, $\Lambda_{\tau_{User}^{end}, e_{End}}$) $\in \mathcal{F}$.
18. Return $\mathcal{M}_{SKE}$.

In the following section we will use the selected design example – PLI – to illustrate how this algorithm works in practice.

Design Example – Algorithm Applied to the PLI *Case*

Let us present the algorithm applied to the PLI case study. For simplicity, the algorithm steps have been grouped into five phases. Figure 6.6 illustrates the first phase of the algorithm.

1. At the beginning, a set of dependencies with input attributes are selected. Based on these dependencies, in the 3–7 steps of algorithm, a set of Business Rule tasks and the appropriate decision table schema for each task are generated. In the PLI case study, 7 Business Rule tasks are generated with the corresponding decision table schemas.
2. In the 8th step of the algorithm, in order to get information (required by the BR tasks) from the user, corresponding User tasks are generated. Additionally, the corresponding input attributes are added to these tasks in their XML-specification of the model in order to provide forms during execution for acquiring information from a user.
3. In the 11th step of the algorithm, a Start event is added at the beginning of the model; and in the 12th step all remaining dependencies are explored, in order to generate new BR tasks. In the case of the PLI example, two additional Business Rule tasks "Calculate payment" and "Calculate driver discount base" are generated.

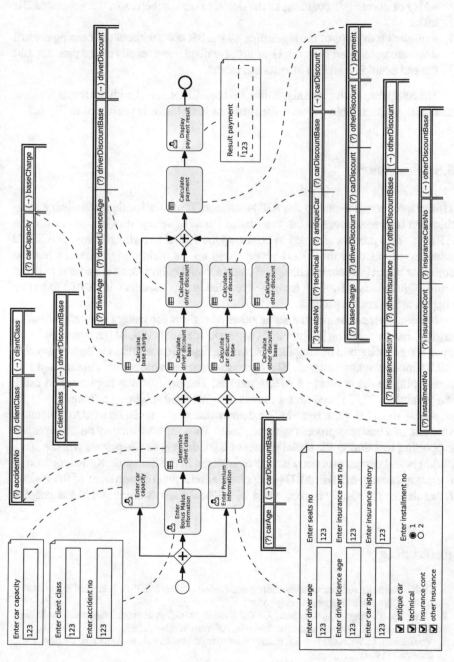

Fig. 6.7 The BPMN model for the PLI case study with forms and rules

4. In the 13th step of the algorithm, the appropriate control flow connections are added or corrected according to the decision table schemas in the generated BR tasks.
5. In the last four steps of the algorithm, all the BR tasks without out coming control flows are connected with the User task for displaying results of the process, and the end event is added to the process model.

The complete result of the algorithm, i.e. the BPMN model with the forms for User tasks and decision table schemas for Business Rule tasks, is presented in Fig. 6.7.

6.5 Summary

This chapter was devoted to the ARD+ method. It allows for the preliminary identification of the structure of the knowledge base. Table schemas obtained from the ARD+ design process are used to bootstrap the XTT2 knowledge base. The XTT2 schemas provided by the ARD+ phase speed up the main design phase, when the complete logical structure is designed using a visual editor. Once the structure of the knowledge base is identified, it can be specified in more details with XTT2 links that support the inference process.

It is worth mentioning that some other directions for using knowledge discovery [18] and data mining [19] methods were considered. In [20, 21] an approach based upon textual subgroup mining for discovering dependencies that are then mapped to ARD+ models was presented. The approach enables rapid model capture and rule prototyping in the context of ARD+ models. The process was implemented using the Vikamine [22, 23] system for knowledge-intensive subgroup mining.

ARD+ model is quite generic. We demonstrated how it can be used to generate a structure of a business process with rule tasks. Thus the method may be an important supporting tool during the initial phases of KBS design. In Chap. 9 we will discuss a multi-phased design process which is considered in the Semantic Knowledge Engineering approach. In SKE, ARD+ supports the conceptual design stage of the system. It can also be practically implemented with CASE tools as discussed in that chapter.

References

1. OMG: Semantics of Business Vocabulary and Business Rules (SBVR). Technical report dtc/06-03-02, Object Management Group (2006)
2. Connolly, T., Begg, C., Strechan, A.: Database Systems, A Practical Approach to Design, Implementation, and Management. 2nd edn. Addison-Wesley (1999)
3. Vanthienen, J., Wets, G.: From decision tables to expert system shells. Data Knowl. Eng. **13**(3), 265–282 (1994)
4. Nalepa, G.J., Ligęza, A.: Conceptual modelling and automated implementation of rule-based systems. In: Software Engineering: Evolution and Emerging Technologies, vol. 130. Frontiers in Artificial Intelligence and Applications, pp. 330–340. IOS Press, Amsterdam (2005)

5. Ligęza, A.: Logical Foundations for Rule-Based Systems. Springer, Berlin (2006)
6. Nalepa, G.J., Wojnicki, I.: Towards formalization of ARD+ conceptual design and refinement method. In: Wilson, D.C., Lane, H.C. (eds.) FLAIRS-21: Proceedings of the Twenty-First International Florida Artificial Intelligence Research Society Conference, 15–17 May 2008, Coconut Grove, Florida, USA, Menlo Park, California, pp. 353–358. AAAI Press (2008)
7. Nalepa, G.J., Wojnicki, I.: Hierarchical rule design with HaDEs the HeKatE toolchain. In: Ganzha, M., Paprzycki, M., Pelech-Pilichowski, T. (eds.) Proceedings of the International Multiconference on Computer Science and Information Technology, vol. 3, pp. 207–214. Polish Information Processing Society (2008)
8. Nalepa, G.J., Wojnicki, I.: ARD+ a prototyping method for decision rules. method overview, tools, and the thermostat case study. Technical report CSLTR 01/2009, AGH University of Science and Technology, June 2009
9. Kluza, K., Nalepa, G.J.: Towards rule-oriented business process model generation. In: Ganzha, M., Maciaszek, L.A., Paprzycki, M. (eds.) Proceedings of the Federated Conference on Computer Science and Information Systems – FedCSIS 2013, Krakow, Poland, 8-11 September 2013, pp. 959–966. IEEE (2013)
10. Nalepa, G.J.: Semantic Knowledge Engineering. A Rule-Based Approach. Wydawnictwa AGH, Kraków (2011)
11. Kluza, K., Nalepa, G.J.: Generation of hierarchical business process models from attribute relationship diagrams. In: Advances in ICT for Business, Industry and Public Sector, pp. 57–76. Springer (2015)
12. van der Aalst, W.M.P.: Process Mining - Discovery, Conformance and Enhancement of Business Processes. Springer, Berlin (2011)
13. Friedrich, F., Mendling, J., Puhlmann, F.: Process model generation from natural language text. In: Mouratidis, H., Rolland, C. (eds.) Advanced Information Systems Engineering. Lecture Notes in Computer Science, vol. 6741, pp. 482–496. Springer, Berlin (2011)
14. Nalepa, G.J., Mach, M.A.: Conceptual modeling of business rules and processes with the XTT method. In: Tadeusiewicz, R., Ligęza, A., Szymkat, M. (eds.) CMS'07: Computer Methods and Systems 21–23 November 2007, pp. 65–70. Poland, AGH University of Science and Technology, Cracow, Oprogramowanie Naukowo-Techniczne (september, Kraków, September 2007
15. Nalepa, G.J., Mach, M.A.: Business rules design method for business process management. In: Ganzha, M., Paprzycki, M. (eds.) Proceedings of the International Multiconference on Computer Science and Information Technology, vol. 4, pp. 165–170. Polish Information Processing Society. IEEE Computer Society Press (2009)
16. Dumas, M., La Rosa, M., Mendling, J., Reijers, H.A.: Fundamentals of Business Process Management. Springer, Berlin (2013)
17. Forster, F.: The idea behind business process improvement: toward a business process improvement pattern framework. BPTrends, pp. 1–13, April 2006
18. Klösgen, W., Żytkow, J.M. (eds.): Handbook of Data Mining and Knowledge Discovery. Oxford University Press, New York (2002)
19. Han, J., Kamber, M.: Data Mining: Concepts and Techniques. Morgan Kaufmann Publisher (2000)
20. Atzmueller, M., Nalepa, G.J.: A textual subgroup mining approach for rapid ARD+ model capture. In: Lane, H.C., Guesgen, H.W. (eds.) FLAIRS-22: Proceedings of the Twenty-Second International Florida Artificial Intelligence Research Society Conference, 19–21 May 2009, Sanibel Island, Florida, USA, Menlo Park, California, FLAIRS, pp. 414–415. AAAI Press (2009)
21. Atzmueller, M., Nalepa, G.J.: Towards rapid knowledge capture using textual subgroup mining for rule prototyping. Technical report Research Report Series no. 458, University of Würzburg, Institute of Computer (2009)

22. Atzmueller, M., Puppe, F.: Semi-automatic visual subgroup mining using VIKAMINE. J. Univ. Comput. Sci. **11**(11), 1752–1765 (2005)
23. Atzmueller, M., Puppe, F.: A knowledge-intensive approach for semi-automatic causal subgroup discovery. In: Proceedings of the Workshop on Prior Conceptual Knowledge in Machine Learning and Knowledge Discovery (PriCKL'07), at the 18th European Conference on Machine Learning (ECML'07), 11th European Conference on Principles and Practice of Knowledge Discovery in Databases (PKDD'07), pp. 1–6. University of Warsaw, Poland (2007)

Chapter 7
Handling Uncertainty in Rules

In this chapter we present extensions to the XTT2 model aimed at handling uncertain knowledge. The primary motivation for this research were studies in the area of the context-aware systems, introduced in Sect. 3.4. In fact we implemented such systems on mobile platforms, including smartphones or tablets. Such an environment poses a number of challenges addressed by our work. The solutions to the problems only briefly described here were delivered in the [1]. Some of them were published in [2, 3].

Contextual data can be delivered to the mobile context-aware system in several different ways, e.g. directly from the device sensors [4], from other devices sensors, over peer-to-peer communication channels [5, 6], from external data sources like contextual servers [7]. Moreover, it can be provided by reasoning engines that are based on the low-level context and a contextual-model that provides a higher-lever context [8]. In each of these cases, the system may experience problems caused by the uncertainty of contextual information.

Among many proposals of uncertainty handling mechanisms [9] only some were considered successful in the area of context-awareness. The first are probabilistic approaches, mostly based on Bayes theorem, that allows for describing uncertainty caused by the lack of machine precision and lack of knowledge [10, 11]. Fuzzy logic provides a mechanism for handling uncertainty caused by the lack of human precision [12, 13]. It allows describing imprecise, ambiguous and vague knowledge. Certainty factors (CF) describe both uncertainties due to lack of knowledge and lack of precision [14, 15]. They are mostly used in expert systems that rely on the rule-based knowledge representation. Machine learning approaches use a data driven rather than model driven approach for reasoning [16]. They allow for handling both uncertainties due to lack of knowledge and lack of precision.

We identified the use of a rule-based solution for context-based modeling and reasoning about uncertainty to be the most profitable approach [2]. Therefore, the main motivation for the research presented here was finding the best uncertainty han-

© Springer International Publishing AG 2018
G.J. Nalepa, *Modeling with Rules Using Semantic Knowledge Engineering*,
Intelligent Systems Reference Library 130,
https://doi.org/10.1007/978-3-319-66655-6_7

dling mechanism that will support rule-based knowledge representation in solving the most common uncertainties that are present in mobile context-aware systems. This mechanism should additionally meet the (4R) requirements that every mobile context-aware system should fulfill [2], including: intelligibility, robustness, privacy and efficiency. After the analysis of the modeling methods with respect to these requirements [1, 8], we decided to provide a hybrid approach that is a combination of rule-based solutions, and machine learning methods. Additionally, the aforementioned methods were enhanced with the time-parametrised operators that help coping with noisy data by averaging uncertainty over specified time spans.

In this chapter we present the classification of most common uncertainty sources present in mobile context-aware systems in Sect. 7.1. Later in Sect. 7.2, we provide a short survey of methods that aim at modeling and handling these uncertainties. In Sect. 7.3 we present the approach developed for XTT2 to cover uncertainties caused by the imprecise data based on modified certainty factors algebra. Furthermore, in Sect. 7.4 we discuss its probabilistic extensions. Then in Sect. 7.5 the time-parametrised operators for handling noisy batches of data are provided. Finally, Sect. 7.6 gives an insight into a probabilistic interpretation of rule-based models for handling uncertainties caused by the missing data, alongside with the hybrid reasoning algorithm. The chapter ends with a summary in Sect. 7.7.

7.1   Uncertainty in Mobile Systems

Sources of Uncertainty in Mobile Context-Aware Systems

The mobile environment is highly dynamic which requires from the uncertainty handling mechanism to adjust to rapidly changing conditions. What is more, despite the existence of various modeling approaches, there is arguably no method that is able to deal with two very different sources of uncertainty: aleatoric uncertainty, and epistemic uncertainty [17].

Aleatoric uncertainty is caused by the inherent randomness of data, or statistical variability of data, so, it can be handled and processed but it cannot be reduced. In the area of mobile context-aware systems this can be reflected as an uncertain sensor readings which cannot be reduced due to the low quality of sensors, or external environmental conditions. Such imprecise sensor readings are depicted in a form of horizontal lines in Fig. 7.1. Aleatoric uncertainty is depicted as imprecise readings in a form of horizontal lines. Epistemic uncertainty is depicted as ambiguity in choosing an appropriate model without any additional readings or background knowledge on the nature of data.

Epistemic uncertainty is caused by the missing data, or lack of background knowledge about the nature of this data. Such uncertainty may cause ambiguities in selecting appropriate model that best describes the data. This case was depicted in Fig. 7.1, where two different models (linear and quadratic functions) can be considered correct for a set of imprecise sensor readings. Only additional knowledge in the form of

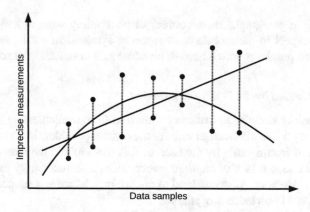

Fig. 7.1 Visualization of epistemic and aleatoric sources of uncertainty [1]

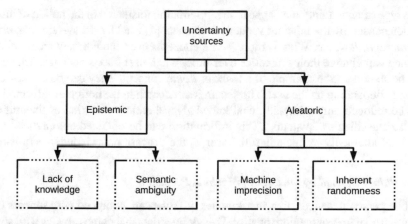

Fig. 7.2 Classification of sources of uncertainty in mobile context-aware systems [1]

additional readings, or the information whether the data is linear or polynomial, can reduce the uncertainty. Although it is not possible to cope efficiently with aleatoric uncertainty, as it is not possible to derive certain conclusions from uncertain data, there is a way to compensate these problems by reducing epistemic uncertainty.

These two general sources of uncertainty can be further unfolded into four more precise concepts [18] as depicted in Fig. 7.2, which includes uncertainty resulting from:

1. lack of knowledge – that comes from incomplete models or missing data,
2. lack of semantic precision – that may appear due to semantic mismatch in the notion of the information,
3. lack of machine precision – which covers machine sensors imprecision and ambiguity, and
4. the inherent randomness of measured phenomena.

In the following paragraphs these sources of uncertainty were described in more detail, with respect to the mobile context-aware systems. In the figure the dotted squares are the problems which have to be addressed in mobile context-aware systems.

Uncertainty Caused by the Lack of Knowledge

Here, we consider knowledge that can be inferred from available data. In this case, uncertainty may occur when one or more of the context providers is not available (i.e. it was turned of intentionally by the user, or it is broken) and the reasoning cannot be performed due to a lack of required information. The uncertainty caused by the lack of knowledge is an epistemic kind of uncertainty, because it cannot be reduced unless additional knowledge is available.

Uncertainty Caused by the Lack of Semantic Precision

This type of uncertainty may appear due to semantic mismatch in the notion of information caused by the inherent vagueness of concepts used by the system, like *cold, warm, high, low*, etc. What is more, some concepts even though they are precisely defined can change their semantics over time, when user habits changes. This case can be shown on the example of a *work* or *home* concept. They can be defined differently depending on the user. The semantic mismatch in the notion of information, can be reduced only with additional knowledge, therefore this kind of uncertainty is also classified as epistemic. This information can be obtained automatically, or semi-automatically by including the user in the process of knowledge acquisition [19].

Uncertainty Caused by the Lack of Machine Precision

Mobile personal devices, like smartphones or tablets are equipped with sensors that are usually of low, or medium quality. They deliver data that comes always with some degree of uncertainty. The example of this can be the GPS sensor, which always estimates location with some degree of accuracy. Such uncertainty can be handled, but cannot be reduced without changes to the hardware, therefore is classified as aleatoric. Another important aspect of the sensor readings is that they outdate over time, increasing uncertainty associated with them. What is more, they outdate with different rates, depending on the time of a day, sensor type, or other kind of context.

Uncertainty Caused by the Inherent Randomness of Data

Another type of aleatoric uncertainty is the one caused by the inherent randomness of processes that are modeled. Some information, even though it is delivered without any loss of certainty caused by the sensor readings, have the uncertainty inherently assigned to it by the definition of what they describe. The example of such data can be: weather forecasts, future location estimates, and other information of a stochastic nature.

The following section discusses selected state-of-the-art methods of handling uncertain data caused by the factors defined in the above classification.

7.2 Improving Uncertainty Handling

This section presents a comparison of most common uncertainty handling mechanisms in context-aware systems, omitting other formalisms like Hartley Theory, Shannon Theory and Dempster–Shafer Theory that are not widely used in such systems. Table 7.1 provides a summary of this comparison according to the type of uncertainty they handle and effort required to incorporate them into existing rule-based knowledge representation. Full circles represent full support, whereas empty circles represent low or no support.

Rules assure an efficient, and human readable form of knowledge representation that can be understood by the user and even modified by him if necessary. The inference in rule-based models is traceable, so any decision of the system can be justified to the user, improving the intelligibility of the system. Therefore, the main goal of this section is to survey possible uncertainty handling mechanisms that can be combined with rules to assure efficient, intelligible and robust modeling language for mobile context-aware systems.

Probabilistic Approaches

These approaches in the area of context-aware systems are mostly based on the probabilistic graphical models (PGM) [20] sometimes referred also as belief networks [21]. They are mainly used to deal with uncertainty caused by the lack of machine precision and lack of knowledge. However, the exact inference in complex probabilistic models is an NP-hard task and is not always tractable, which violates intelligibility and efficiency requirements for mobile context-aware systems. Therefore, to provide both intelligibility and effective uncertainty handling mechanism, probabilistic approaches have to be combined with rules. There were several attempts made to bind rules and probabilistic reasoning. ProbLog is a probabilistic extension of Prolog programming language [22]. A similar approach was implemented in AILog2 [23] (formerly CILog). It is an open-source purely declarative representation and reasoning system, that includes pure Prolog (including negation as failure) and allows for probabilistic reasoning. Although the idea of incorporating probability into rules

Table 7.1 Comparison of uncertainty handling mechanisms [1]

	Uncertainty source				Rules integration
	Lack of knowledge	Semantic ambiguity	Machine imprecision	Inherent randomness	
Probabilistic	●	○	●	●	Medium
Fuzzy logic	○	◗	◗	○	Easy
Certainty factors	◗	○	●	◗	Easy
Machine learning	◗	○	●	◗	Difficult

is not new, all of the aforementioned approaches use an unstructured knowledge model, and assumes that the reasoning will be done in a pure probabilistic manner. This makes the system difficult to modify and understand by the user, reducing its intelligibility.

Fuzzy Logic Approaches

Fuzzy logic is useful in capturing and representing imprecise notions. Fuzzy OWL and FiRE fuzzy inference engine was used in [24] to express and infer the user's dynamic context, in distributed heterogeneous computing environments. The fuzzy logic approach was also used in the pervasive health care system CARA [13], where the authors proposed a scalable and flexible infrastructure for the delivery, management and deployment of context-aware pervasive health care services based on imperfect or ambiguous contextual data. Another approach for introducing fuzzy reasoning into a health care context-aware system was presented in [12], where an integrated environment aimed at providing personalized health care services was proposed. Fuzzy Logic approaches are usually used to handle uncertainties caused by imprecision in defining concepts and by the semantic ambiguities between them. The uncertainty caused by semantic imprecision can be resolved with a combination of ontological and rule-based approach in the form of modules that are responsible for mediating ambiguous concepts with the user [19]. Fuzzy approaches can be integrated with a rule-based knowledge representation, as both formalisms have similar foundations. The example of that can be FuzzyCLIPS [25], which is a plugin introducing fuzzy rules to a well known expert system shell CLIPS [26].

Machine Learning Approaches

Machine learning solutions are used in situations where the contextual knowledge cannot be easily captured by an expert, or the number of features that have to be modeled is relatively big. The example of such a system can be found in [16] where authors used K-Means clustering algorithm and probabilistic Markov models to determine users location and physical activity. An overview of machine learning methods used in the domain of context-aware systems can be found in [27]. Although very popular nowadays, machine learning methods have several important drawbacks in regard to mobile context-aware systems. In this approach a model of the system is not given, but rather learned from data. This provides adaptability, but on the other hand, the model is usually sub-symbolic, which makes it difficult to adjust and understand by the user, decreasing the intelligibility property of the system. What is more, machine learning approaches similarly to probabilistic graphical models need time and data to adjust to new situation. On the other hand, the machine learning approaches are solutions that can very efficiently cope with uncertainties caused by the lack of precision, but also lack of knowledge and inherent randomness of data. In our work, machine-learning methods are used only as a mechanism that supports rule-based representation. The incremental rule-learning algorithms are used for supporting long term adaptability of the system.

Certainty Factors Approaches

Certainty factors [28] have an important advantage over other uncertainty handling mechanisms. They can be easily incorporated into an existing rule-based system without the necessity of redesigning or remodeling the knowledge base. This makes the certainty factor approach take full advantage of rule-based knowledge representation which provides: fast and traceable reasoning and a human readable and modifiable knowledge base. What is more, the certainty factors approach does not require from uncertain contextual data to be defined in a probabilistic manner, which is useful in situations, where there is no information about the probabilistic distribution of this data.

Certainty factors approaches are not very popular among context-aware solutions. This is caused by the fact, that there are very few context-aware frameworks that use rule-based knowledge representation [29, 30]. In [14] a hierarchical rule-based context modeling approach was proposed with certainty factors used for uncertainty handling. The certainty factors were obtained from an expert and did not change over time. AMBIONT [15] is an ontology for ambient context, that was designed to handle uncertain data in smart spaces. The system uses semantic representation of contextual data based on the OWL language, enriched with certainty factors for describing vagueness and ambiguity. In [31] the authors used another heuristic approach of a form of utilities functions that allow inexact matching between compared elements. The research was part of a MUSIC [32] middleware for building pervasive mobile context-aware systems. However, certainty factors are able to describe and handle uncertainty related to lack of machine precision, which is the most common uncertainty in context-aware systems. Although they do not provide mechanisms to cope with the uncertainty caused by the lack of knowledge, they still provide a method to model it and resolve using other formalism or mediation techniques.

Certainty factors are one of the most popular methods for handling uncertainty in RBS. However, for a long time they were under strong criticism regarding a lack of theoretical background and the assumption of independence of conditions for rules of the same conclusion which not always hold [28]. As a response to these, the Stanford Modified Certainty Factors Algebra was proposed [33]. It accommodated two types of rules with the same conclusion: cumulative rules (with an independent list of conditions) and disjunctive rules (with a dependent list of conditions). As it will be shown, this makes the certainty factors fit ALSV(FD) logic *generalized* and *simple* attributes, which are the principle components of XTT2 rules.

Rule in CF algebra is represented according to formula:

$$condition_1 \wedge condition_2 \wedge \ldots \wedge condition_k \rightarrow conclusion \qquad (7.1)$$

Each of the elements of the formula from Eq. (7.1) can have assigned a certainty factor $cf(element) \in [-1; 1]$ where 1 means that the element is absolutely true; 0 denotes the element about which nothing can be said with any degree of certainty; -1 denotes an element, which is absolutely false. The CF of the conditional part of a rule is determined by the formulae:

$$cf(condition_1 \wedge \ldots \wedge condition_k) = \min_{i \in 1 \ldots k} cf(condition_i)$$

The CF of conclusion C of a single i-th rule is calculated as follows:

$$cf_i(C) = cf(condition_1 \wedge \ldots \wedge condition_k) * cf(rule) \quad (7.2)$$

The $cf(rule)$ defines a certainty of a rule which is a measure of the extent, to which the rule is considered to be true. It is instantiated by the rule designer, or it comes from a machine learning algorithm (like for instance an association rule mining algorithms). A major departure from the traditional Stanford Certainty Factor Algebra [34] is an attempt to remove the major objection raised against it concerning conditional dependency of rules with the same conclusions. To address this issue, rules with the same conclusions were divided into two groups: *cumulative* ans *disjunctive*. Cumulative rules have the same conclusions and have independent conditions (i.e. value of any of the conditions does not determine values of other rules conditions). The formula for calculating the certainty factor of the combination of two cumulative rules is given in (7.3).

$$cf(C) = \begin{cases} cf_i(C) + cf_j(C) - cf_i(C) * cf_j(C) & \text{if } cf_i(C) \geq 0, cf_j(C) \geq 0 \\ cf_i(C) + cf_j(C) + cf_i(C) * cf_j(C) & \text{if } cf_i(C) \leq 0, cf_j(C) \leq 0 \\ \frac{cf_i(C) + cf_j(C))}{1 - \min\{|cf_i(C)|, |cf_j(C)|\}} & \text{if } cf_i(C)cf_j(C) \notin \{-1, 0\} \end{cases}$$

$$(7.3)$$

Disjunctive rules have the same conclusions but are conditionally dependent (i.e. the value of any of the conditions determine the values of other rules conditions).

The equation for calculating the certainty factor of a disjunctive rule is presented in (7.4).

$$cf(C) = \max_{i \in 1 \ldots k} \{cf_i(C)\} \quad (7.4)$$

The calculation of the CF for the rules are performed incrementally. This means that for instance for a pair of rules i-th and i-th $+ 1$, there is a calculated certainty factor $cf_k(C)$ that later is taken into the Eq. (7.3) or (7.4) together with rule i-th $+ 2$ to calculate $cf_{k+1}(C)$.

7.3 Modified Certainty Factors Algebra for XTT2 Rules

We discuss mechanisms for modeling and handling uncertainty caused by the lack of machine precision with an usage of modified certainty factors algebra and XTT2 knowledge representation. Because of the tabular structure of the XTT2 models, this mechanism has to be considered on two different levels. The first level concerns uncertainty handling in a single rule. This problem can be reduced to an issue of evaluation of a rule conditions with uncertain attributes values. The second level involves designing an algorithm that will allow for the uncertain reasoning in XTT2 models. This issue is related to the problem of handling uncertainty on the XTT2 table level and propagating uncertain conclusions during the inference process.

Conditional parts of rules in XTT2 notation consists of a conjunction of ALSV(FD) formulae (see Sect. 4.3). ALSV(FD) formulae are basic parts of the XTT2 rule, which can be represented as:

$$(A_i \propto d_i) \wedge \ldots \wedge (A_n \propto d_n) \longrightarrow (A_k := d_k) \wedge \ldots \wedge (A_z := d_z) \quad (7.5)$$

Every ALSV(FD) formula is a logical expression that can be either true or false according to a value of an attribute in consideration. One can therefore translate every formula to a conjunction or alternative of equality formulae. In particular the formula $A_i \in V_i$ can be translated into a form:

$$(A_i = V_i^0) \vee (A_i = V_i^1) \vee \ldots \vee (A_i = V_i^k) \quad (7.6)$$

where the V_i^k is a k-th element from a subset V_i of domain $\mathbb{D}_i$, and A_i is a simple attribute. On the other hand, for the general attributes A_i, the formulae of a form $A_i \sim V_i$ can be translated into:

$$(A_i^0 \in V_i) \vee (A_i^1 \in V_i) \vee \ldots \vee (A_i^k \in V_i) \quad (7.7)$$

where A_i^k is a k-th element of a set represented by the general attribute A_i. This formula can be further recursively rewritten as a conjunction of formulae from Eq. (7.6). Similarly it can be continued for every formula in the ALSV(FD) logic and reduce it into the form of a conjunction or an alternative of simple equality statements. Such a notation allows us to use certainty factors algebra for evaluating the formulae for uncertain attribute values, treating these formulae as a set of simple cumulative or disjunctive rules. In particular, one can represent the formula from Eq. (7.6), as a set of disjunctive logical rules of a form:

$$(A_i = V_i^0) \rightarrow \textit{rule satisfied}$$
$$(A_i = V_i^1) \rightarrow \textit{rule satisfied}$$
$$\ldots \quad (7.8)$$
$$(A_i = V_i^k) \rightarrow \textit{rule satisfied}$$

The notation from the example above means that the entire formula from Eq. (7.6) is true, when at least one of the rules from the Eq. (7.8) is true. Therefore, calculating CF for this rule, one can obtain a certainty factor of an ALSV(FD) formula that is considered. Every rule CF is assigned a value of 1 for clarity of this example, so the certainty of a formula is determined by the certainty of conditional expressions on the left hand side. The rules are disjunctive, as the value of A_i can be the only one (as it is a simple attribute), hence the Eq. (7.4) applies to this. On the other hand, rule interpretation of formulae (7.7) generates a set of cumulative rules, as the attribute A_i can take multiple values that do not depend on each other, and hence the Eq. (7.3) applies to this case. The rest of this section provides a complete set of transformations for all the operators in ALSV(FD) logic.

Certainty Factors for Simple Attributes

There are four relations for simple attributes that has to be covered by the uncertainty handling mechanism. For the sake of the further discussion, let us define *Val* operator that takes as a parameter an attribute name, and returns attributes value, without certainty factor. Additionally, we will consider an expression $cf(A_i)$ as the certainty associated with a value of an attribute A_i. Similarly, when we write $cf(A_i \propto d_i)$, we consider a certainty of a formula $A_i \propto d_i$ being true.

1. **Equality operator** $(A_i = d_i)$ – There are two cases that have to be considered while evaluating uncertain formula that contains equality operator. The first case is when the $Val(A_i) = d_i$ regardless the certainty, and the other is when the $Val(A_i) \neq d_i$ regardless the certainty. Therefore the $cf(A_i = d_i)$ can be defined as follows:

$$cf(A_i = d_i) = \begin{cases} cf(A_i) & \text{when } Val(A_i) = d_i \\ -cf(A_i) & \text{when } Val(A_i) \neq d_i \end{cases}$$

2. **Negation of equality operator** $(A_i \neq d_i)$ – This operator is inverse to equality operator, hence it can be defined as:

$$cf(A_i \neq d_i) = -cf(A_i = d_i)$$

3. **Membership operator** $(A_i \in V_i)$ – This operator can be split into an alternative of equality statements, as the formula $A_i \in V_i$ is true, when A_i equals to at least one of the elements from V_i. Therefore we can say that $cf(A_i \in V_i) = cf((A_i = V_i^0) \vee (A_i = V_i^1) \vee \ldots \vee (A_i = V_i^k))$, where the V_i^k is a k-th element from a subset V_i of domain $\mathbb{D}_i$. Finally, we can define the certainty factor of the membership operator as:

$$cf(A_i = d_i) = \max_{j=1\ldots k} cf(A_i = V_i^j)$$

4. **Negation of the membership operator** $(A_i \notin d_i)$ – This operator is inverse to the membership operator, hence it can be defined as follows:

$$cf(A_i \notin V_i) = -cf(A_i \in V_i)$$

Certainty Factors for Generalized Attributes

There are six relations for generalized attributes that have to be covered by the uncertainty handling mechanism. For the sake of further discussion lets assume that A_i is a generalized attribute, and $Val(A_i)$ is an n-*element* set of a form: $\{a_i^1, a_i^2 \ldots a_i^n\}$. Let also assume that the V_i is a value that stands on the right hand side of the ALSV(FD) formulae, and it is an m-*element* set of a form: $\{v_i^1, v_i^2 \ldots v_i^m\}$.

1. **Nonempty intersection operator** $(A_i \sim V_i)$ – This operator is listed first, as it is used by other operators to evaluate their certainty factors. $A_i \sim V_i$ is true, iff $\exists a_i^j \in A_i : a_i^j \in V_i$. Therefore, it can be shown that:

$$cf(A_i \sim V_i) = cf(a_i^1 \in V_i \vee a_i^2 \in V_i \vee \ldots \vee a_i^n \in V_i)$$

   Because the values of the elements in A_i are independent, Eq. (7.3) for the cumulative rules can be applied to the right hand side of the formula above. This results in the fact that the more of the elements of A_i that are also members of V_i, the stronger the similarity of these two sets, and also the higher the certainty of an *nonempty intersection* operator.

2. **Empty intersection operator** $(A_i \nsim V_i)$ – This operator is inverse to the non-empty interaction operator, hence it can be defined as:

$$cf(A_i \nsim V_i) = -cf(A_i \sim V_i)$$

3. **Equality operator** $(A_i = V_i)$ – This operator, similarly to the equality operator for simple attributes, can be defined differently, depending on the logical value of the expression $Val(A_i) = V_i$ regardless certainty of a value of A_i. In the first case, when $Val(A_i) = V_i$ is true, the certainty of the entire formula can be evaluated by finding the *weakest link* that keeps the formula true, that is the element of A_i that has the smallest certainty factor. On the other hand, when $Val(A_i) \neq V_i$, it is crucial to find how much $Val(A_i)$ differs from V_i. To do this, *non empty intersection* operator is used to calculate how certain it is that the A_i has a non empty intersection with $(A_i \backslash V_i)$. The compact evaluation formula for the equality operator is presented below:

$$cf(A_i = V_i) = \begin{cases} \min_{j=1\ldots n} cf(a_i^j) & \text{when } Val(A_i) = V_i \\ -cf(A_i \sim (A_i \backslash V_i)) & \text{when } Val(A_i) \neq V_i \end{cases}$$

4. **Negation of the equality operator** $(A_i \neq V_i)$ – This operator is inverse to the equality operator, hence it can be defined as:

$$cf(A_i \neq V_i) = -cf(A_i = V_i)$$

5. **Subset operator** $(A_i \subset V_i)$ – Evaluation of the certainty of this operator is equivalent to the evaluation of a certainty of an equality operator, as an equality is a special case of subset relation.

   If the logical value of $Val(A_i) \subset V_i$ is true, the certainty of the entire formula can be evaluated by finding the *weakest link* that keeps the formula true, that is the element of A_i that has the smallest certainty factor. Otherwise, when the logical value of $Val(A_i) \subset V_i$ is false, the certainty of the formula is determined by how much A_i differs form V_i. Therefore:

$$cf(A_i \subset V_i) = \begin{cases} \min_{j=1...n} cf(a_i^j) & \text{when } Val(A_i) \subset V_i \\ -cf(A_i \sim (A_i \backslash V_i)) & \text{when } Val(A_i) \not\subset V_i \end{cases}$$

6. **Superset operator** $(A_i \supset V_i)$ – This operator is complement to the subset operator, hence it can be defined as:

$$cf(A_i \supset V_i) = cf(V_i \subset A_i)$$

Making Use of Negative Certainty Factors

When dealing with logic that operates on finite domains, such as ALSV(FD), the negative certainty factors may be as valuable as the positive ones. Let us consider the example from Eq. (7.8). Assuming that $V_i' = \mathbb{D}_i \backslash V_i$, we can add an additional rule to the equation, that will cover the *false* cases of the ALSV(FD) formula $A_i \in V_i$:

$$(A_i \neq V_i'^0) \wedge (A_i \neq V_i'^1) \wedge \ldots \wedge (A_i \neq V_i'^l) \rightarrow \textit{rule satisfied} \qquad (7.9)$$

Supposing that we have no positive certainty on the value of attribute A_i, but we know which of the values the attribute does not take for sure, we can notice the dependence below:

$$\left(cf(A_i = V_i'^l) = -1\right) \Rightarrow \left(cf(A_i \neq V_i'^l) = 1\right)$$

The formula above can now be applied together with rule from Eq. (7.9) to infer the certainty factor of the ALSV(FD) formulae in consideration.

7.4 Certainty Factors on Table Level

The uncertainty handling mechanism based on certainty factors algebra described in Sect. 7.3 operates on the level of ALSV(FD) formulae. To take full advantage of this mechanism, it has to be applied also to rules, and included into the reasoning mechanism. This section describes the approach for reasoning in uncertain XTT2 models with the use of certainty factors algebra. The evaluation of a certainty of a single rule is straightforward, according to Eq. (7.2). However, as it was stated in Sect. 7.3, in certainty factors algebra rules are divided into cumulative and disjunctive. This distinction has to be also provided for the XTT2 models.

In the XTT2 models, the basic decision component is a table, which groups rules using the same attributes. By convention, rules within the same table are considered disjoint, as there is no state that is covered by more than one rule [35]. This makes all the rules within the single table disjunctive in the understanding of Certainty Factors Algebra. Therefore, Eq. (7.4) can be applied for the evaluation of the certainty of the rules within a single table. On the other hand, in order to model cumulative rules, it is necessary to split such rules into separate tables. Rules that are located in separate tables, but have the same attribute in their decision parts are considered cumulative

Data: E – the set of all known attributes values

A – the set of attributes which values are to be found

Result: V – values for attributes from the set A

1 Create a stack of tables T that needs to be processed to obtain V;
2 **while** *not empty* T **do**
3     $C = \varnothing$;
4     $t = pop(T)$;
5   Select rule $r \in t$ using Eq. (7.4) and add it to C;
6   Identify schema $(\mathbb{COND}, \mathbb{DEC})$ of table t;
7   **while** *not empty* T **do**
8   Identify schema $(\mathbb{COND}', \mathbb{DEC}')$ of table $t' = peek(T)$;
9   **if** $\mathbb{DEC}' \cap \mathbb{DEC} \neq \varnothing$ **then**
10             $pop(T)$;
11             $\mathbb{DEC} = \mathbb{DEC} \cup \mathbb{DEC}'$;
12   Select rule $r \in t$ using Eq. (7.4) and add it to C;
13   **else**
14   **break**;
15   **end**
16   **end**
17   Execute rules from C and store all assertions in E';
18   **if** $size(C) > 1$ **then**
19   Calculate $cf(C)$ using Eq. (7.3);
20   Assign $cf(C)$ to appropriate values in E';
21   **end**
22     $E = E \cup E'$;
23     $V = V \cup (E' \cap A)$;
24 **end**
25 **return** V;

Algorithm 7.1: Algorithm for inference in XTT2 models with certainty factors algebra [1]

with respect to certainty factors algebra, and hence Eq. (7.3) applies to them. This inference strategy was presented in a more formal way in the Algorithm 7.1.

The algorithm builds a stack of XTT2 tables that need to be processed in order to obtain values of the given attributes. When there are at least two tables that have shared attributes in their decision part, the rules that were selected inside them to be fired are added to a set of cumulative rules C (see lines 7–16). Finally, when there are at least two cumulative rules in set C, the final certainty of their conclusions is calculated according to the cumulative formula (see lines 18–21). In all the other cases, rules selected to be executed are treated as disjunctive.

7.5   Time-Parametrised Operators for XTT2

Certainty factors approach provides a mechanism for handling values which are explicitly defined as uncertain at some degree. However, it does not cope well with noisy data and is not well prepared for reacting on long- and short-term changes in the environment. Detection of such changes is possible by creating a model of system dynamics, which is based on the analysis of historical data. Although there exist methods that allow for modeling dynamics of the system, like Markov models, or dynamic Bayesian networks, they do not provide intelligibility features, which can be easily achieved with rules.

Therefore, the primary motivation for this work was to develop methods that would allow for an analysis of historical data in rule-based systems. Such methods would allow for better modeling of the dynamics of the context-aware systems that are immersed in a mobile environment. What is more, a rule-based model will provide intelligibility capabilities, which can be further used to obtain user feedback and improve overall system performance.

Every ALSV(FD) formula can be represented as $A_i \propto value_i$, where A_i is an attribute, $value_i$ is a value (or set of values) from its domain, and $\propto$ is any of the valid operators. Such a notation allows for testing only current values of attributes, but does not allow for referring into the past states in any way. Time-based operators, allow for referring to past states in an aggregated manner. However, they are not based on any statistical measure, but rather are extended ALSV(FD) formulae operators. Such operators allow for checking if the formula is satisfied on the selected portion of the historical data. Defining rules conditions on series of historical data provides a declarative way for encoding simple patterns and temporal dependencies between contexts. This improves short-term adaptability, as it can be used to build meta-models for monitoring and adjusting sampling rates of sensors, but also for resolving temporal ambiguities between user contexts.

To extend standard ALSV(FD) operators with time-based conditions, the formula $A_i \propto value_i$, was changed into the following form:

$$A_i \propto \{Q(v), T\}\ value_i \tag{7.10}$$

where $\{Q(v), T\}$ is a parameter which makes the entire formula true, if and only if the operator $\propto$ holds for $Q(v)$ times in a time range defined by T. The $Q(v)$ is a quantitative relationship that defines how is the amount of the states satisfying the condition related to the specified number of states. There are three possible relationships: min (at least as many as ...), max (at most as many as ...), and exact (exactly as many as ...).

The parameter v in operator $Q(v)$ is defined as an *amount* of data that is required for the quantitative relationship to be true. It can be specified in two ways:

1. As a number (for example min 5 means *at least 5 states*).
2. As a fraction of the total number of states (for example exact 50% means *exactly the half of the states*).

The last element of the parametrised operator is vector t of past states, which also can be written twofold:

1. As a range of non–positive numbers which should be regarded as indices of time samples relative to the current state designated by 0. For example, -15 to 0 means last sixteen states including the current state.
2. Using time units and Matlab/Octave–like notation.[1] For example, -2h:1min:0 means the last two hours of states sampled once per minute (121 samples including the current one).

Using this notation it is possible to create rules which are fired only when the a certain amount of past states satisfies the specified condition. It is a simple generalization allowing for the capturing of the basic dynamic features of user context.

Evaluating Uncertain Parametrised Operators

The evaluation of a time-parametrised operator with uncertain attributes values, requires additional computations. However, the interpretation is straightforward. Let us assume that we need to determine the certainty factor of a conditional formula which uses a parametrised operator, as shown in the Eq. (7.10). For the sake of simplicity, let us refer to this formula as F. Therefore, the certainty of the formula is represented as an weighted average of certainty factors of conditional formulas at specific points in time, as it was shown in the following equation:

$$cf(F) = \frac{1}{N} \sum_{t_i \in T} cf\, [Val(A, t_i) \propto value_i] \tag{7.11}$$

where N is a total number of values against which the formula had to be evaluated in a specified time range T. This equation has no impact on the evaluation result of the time-parametrised operator, when all the values of an attribute were certain. However, it decreases the certainty of the entire formula, when the uncertainty appears.

---

[1]GNU Octave is a high-level interpreted language, primarily intended for numerical computations. See http://www.gnu.org/software/octave/.

7.6 Probabilistic Interpretation of XTT2 Models

So far we presented a mechanism to handle uncertainty caused by the lack of machine precision. One of the biggest problems of this approach was lack of proper handling for uncertainty caused by the lack of knowledge. In a case when the value of an attribute from the conditional part of the rule was unknown, the certainty of the entire rule was defined as zero. In such a case every rule within a table was considered as completely uncertain and no further reasoning was possible. This was caused by the fact that certainty factors algebra does not take into consideration historical data while evaluating. In contrary, probabilistic approaches are strictly based on statistical analysis of historical data. The idea of exploiting this strength, for the purpose of providing an efficient uncertainty handling mechanism in rule based mobile context-aware systems, was the primary motivation for the research presented in this work. The XTT2 decision tables allow for the building of structured probabilistic models in a human-readable way. What is more, the idea of dividing rules into separate tables allows building hybrid rule-based models, that uses probabilistic reasoning only when needed.

Transforming XTT2 Models into a Bayesian Network

Although the XTT2 representation is based on rules, the structure of the XTT2 formalism allows for its probabilistic interpretation. In such interpretation every attribute can be considered a random variable, and every XTT2 table a deterministic conditional distribution table. The connections between tables can be further interpreted as dependencies between random variables, and the XTT2 model can be easily transformed to a Bayesian network.

Let us consider a basic example of the XTT2 table presented in Fig. 7.3. It describes a fragment of a mobile context-aware recommendation system, that based on the user activity, location and time, suggests applications for the user and switches profiles in the user mobile phone. Figure 7.4 represents a Bayesian interpretation of the XTT2 model presented in Fig. 7.3.

In such an interpretation, every rule schema (COND, DEC) can be represented as a conditional probability of a form:

$$P(DEC \mid COND) \tag{7.12}$$

Therefore, in the probabilistic interpretation of schema (COND, DEC), every rule is represented by a pair $\langle r, p \rangle$, where r is an XTT2 rule defined in the Eq. (7.5) and $p \in [0; 1]$ is the conditional probability assigned to it. Probability p defines a certainty that the particular rule should be fired given the evidence.[2] In the following discussion p will be referred as the certainty, not the probability. Therefore, in the case where all the attributes from the COND part of the rule are known, the conditional

---

[2]In the probabilistic inference, evidence is not only limited to the preconditions, but can include all the available attribute values.

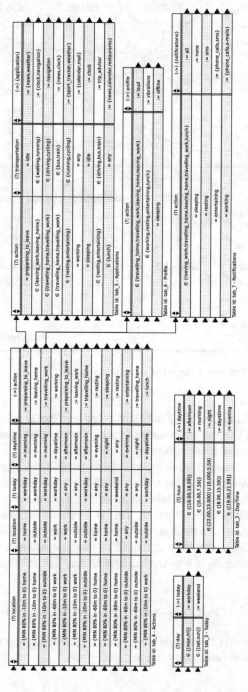

Fig. 7.3 Fragment of an XTT2 model for context-aware system for applications recommendation and phone profile switching [1]

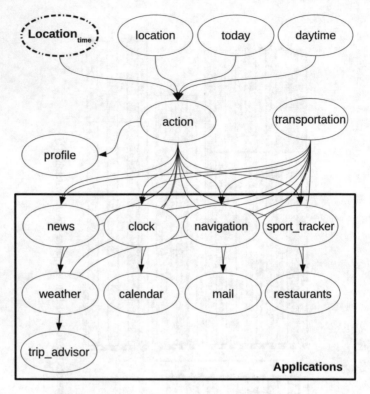

Fig. 7.4 Bayesian network representation of a considered model [1]

probability distribution (CPD) is deterministic and traditional rule-based reasoning can be performed. In the case when some of the attributes values from the conditional part are unknown, probabilistic reasoning is triggered.

The XTT2 representation allows generalized attributes to be present in both COND and DEC. This may lead to problems in the probabilistic interpretation of the rules, as the generalized attributes have to be treated as random variables with multiple independent binary values allowed. This is a serious departure from standard Bayesian understanding of a random variable. Therefore, the following interpretation was proposed. Let us assume that the XTT2 model contains a schema of a form $(\{A_i, A_j\}, \{A_g\})$, where A_g is a generalized attributes. The rule having this schema is given as follows:

$$r : (A_i \propto d_i) \wedge (A_j \propto d_j) \longrightarrow A_g = \{v_1, v_2, \ldots v_n\}$$

Following the Eq. (7.12), the rule from the above formula can be represented in the form of conditional probability, defined as follows:

$$P\left(A_g = \{v_1, v_2, \ldots v_n\} \mid A_i, A_j\right)$$

And further, assuming that the values of a random variable A_g are independent, the conditional probability can be rewritten as:

$$P(A_g = \{v_1, v_2, \dots v_n\} \mid A_i, A_j) = \\ P(v_1 \mid A_i, A_j) \cdot P(v_2 \mid A_i, A_j) \cdot \dots \cdot P(v_n \mid A_i, A_j) \quad\quad (7.13)$$

The interpretation of the generalized attributes as a set of independent random variables are extremely important in the inference process in special cases when attributes from the decision part of the rules are treated as evidence.

The other consequence of the fact that XTT2 knowledge representation is based on attributive logic, is that it has some advantages over the traditional table distribution approaches in terms of notation compactness. One of the most important advantages is that the ALSV(FD) logic introduces operators like $=, \neq, \sim, \approx, \subset, \supset$. This allows for the representation of a probability distribution in a more efficient way. For instance to encode the conditional probability distribution presented in table *Applications* from Fig. 7.3 using traditional conditional probability tables (CPT), one will need 50 rows to cover all the combinations of attribute values presented in the *Applications* table. The complexity of the representation is highly dependent on the nature of the problem, and in the worst case even for XTT2 it can be the same as for the standard CPT representation. However, in most cases there will be an advantage of usage of the XTT2 notation over the standard CPT as it presents probability distributions in a human readable rule-based form. What is more, the XTT2 representation allows explaining probabilistic reasoning by exploiting a rule-based system capabilities of intelligibility.

Probabilistic Interpretation of Time-Parametrised Operators

Our previous research presented in [36] did not take into account the time-parametrised operators in the probabilistic interpretation of the XTT2 models. This exposed a serious drawback of the solution, as the operators appeared to be extremely useful in modeling the dynamics of the processes that occur in the mobile environment. Due to the fact, that the time-based operators perform the evaluation of the ALSV(FD) in a specified time span by performing N separate evaluations of the formula, we decided to use Hidden Markov Models (HMM) to represent this type of construct. The example interpretation of the time-parametrised operator from table *Actions* presented in Fig. 7.3 was given in Fig. 7.5. The length of a HMM chain is determined by the number of states that the time-parametrised operator takes into consideration. In the case of 10 min time range, with a sampling set to one minute, the chain will consists of 10 nodes.

The probabilistic interpretation of the XTT2 models presented in this work assumes that the XTT2 tables and rules are given. They can be provided by an expert, or mined with data mining algorithms. Although the learning structure of the model and automatic discovery of the rules is a very important task in terms of adaptability of the system, it is beyond the scope of this work. In [2] we discussed learning the distribution of the random variables (attributes) for a given set of rules and XTT2 schemas.

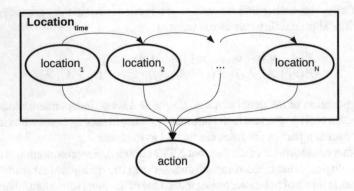

Fig. 7.5 Bayesian network representation of a time-parametrised operator from the previous model
[1]

Inference in the Probabilistic XTT2 Models

There are three possibilities of reasoning in the probabilistic XTT2 models:

1. Purely deterministic inference – in such an inference only tables that have all
   values of attributes from their conditional parts known can be processed. This
   may therefore end up in interrupted inference when some values are missing.
2. Purely probabilistic inference – in such an inference the XTT2 model is queried
   as if it was Bayesian network. No deterministic reasoning is performed.
3. Hybrid inference – in such an inference tables are processed in a deterministic
   way when possible, and the probabilistic reasoning is triggered only in the other
   cases.

To exploit fast and efficient reassigning provided by the rule-based approach,
with probabilistic uncertainty handling, a hybrid inference model was proposed. The
procedure for processing the XTT2 tables in such an approach was presented in
Algorithm 7.2.

The first step of the algorithm is the identification of a list of the XTT2 tables T
that have to be processed to obtain the values of a given set of attributes A. This is
done according to one of the inference modes available [37]. For every table $t \in T$
popped from the list, a deterministic inference is performed if possible, and the
values of the attributes from the conclusion part of the executed rule are added to
evidence set E. When it is impossible to run deterministic inference (e.g. some values
of the attributes are missing), the probabilistic inference is triggered. It uses all the
evidences E to calculate probability of the attributes values from the current schema.
After that, a rule with the highest certainty (or probability in this case) is selected and
triggered, and the reasoning returns to be deterministic. In cases when the probability
of a rule is very low, say less than some value ϵ, no rule is executed. However, if the
conclusion part of the schema for a currently processing table contains an attribute
that belongs to a set A, the most probable estimation of this attribute value is added
to the result.

Data: E – the set of all known attributes values
       A – the set of attributes which values are to be found
Result: V – values for attributes from the set A

1   Create a stack of tables T that needs to be processed to obtain V;
2   **while** *not empty T* **do**
3      $t = pop(T)$;
4   Identify schema $(\mathbb{COND}, \mathbb{DEC})$ of table t;
5   **if** $\forall c \in \mathbb{COND}, Val(c) \in E$ **then**
6   Execute table t using Algorithm 7.1;
7         $\forall a \in \mathbb{DEC} \cap A$: add $Val(a)$ to E and V;
8   **else**
9   Run probabilistic reasoning to obtain $P(a) \forall a \in \mathbb{DEC}$;
10   Select rule $\langle r_{max}, p_{max} \rangle$ such that: $\forall \langle r, p \rangle \in t : p \leq p_{max}$;
11   **if** $p_{max} \geq \epsilon$ **then**
12   execute rule r;
13          $\forall a \in \mathbb{DEC} \cap A$: add $Val(a)$ to E and V;
14   **else**
15          $\forall a \in \mathbb{DEC} \cap A$: add $P(a)$ to E and V;
16          $t = pop(T)$;
17   Identify schema $(\mathbb{COND}, \mathbb{DEC})$ of table t;
18   **goto 9**
19   **end**
20   **end**
21 **end**
22 **return** V;

Algorithm 7.2: Algorithm for probabilistic inference in XTT2 models [1]

The evidences for the set E are obtained from two types of sources:

- from the working memory component of the inference engine that stores all the attributes values, and
- from the reasoning process, when new values are inferred.

The XTT2 notation allows three types of attributes: *comm, in,* and *out*. Attributes that are marked as *out* cannot be treated as evidences, even though their value is known. For example if the value of the attribute *Profile* from the model presented in Fig. 7.4 was known, and marked as *in* or *comm*, it will be included as evidence in the reasoning process. In other cases it will not be used in the inference process. This is important for the probabilistic reasoning strategy, where every evidence can have an impact on the reasoning results.

7.7 Summary

In this chapter, several uncertainty handling extensions for the XTT2 formalism were discussed. The classification of sources of uncertainties was given, and the two most common causes of uncertainty in mobile environments were distinguished. Following this classification three complementary methods were proposed for handling these types of uncertainties. These methods are inherent parts of XTT2 rule-based knowledge modeling language. First, based on modified certainty factors algebra, handles aleatoric uncertainty caused by the lack of machine precision. It is supported by the time-parametrised operators in rules to handle noisy data over specified time periods. Moreover, a probabilistic interpretation of the XTT2 knowledge base in a form of Bayesian network was given. These methods are applicable as long, as there are readings that come from the context providers. In the case, when the information is missing, the probabilistic interpretation of XTT2 model can be used to reduce the epistemic uncertainty caused by the lack of knowledge.

We developed models presented in this chapter with specific applications in mind, concerning mobile context-aware systems. Such systems run on personal mobile devices, which have probably most advanced and user-friendly human-computer interfaces that are available on the market. This feature can be used to obtain missing information directly from the user, by asking him appropriate questions in an context-dependent manner. Such an approach helps to reduce uncertainty and to improve the adaptability of the system. The preliminary results that are aimed at providing such methods were given by us in [19]. We will present the application part of this work in Chap. 17 where we introduce the KNOWME architecture. It integrates all of these ideas into a complete software framework. It includes the HEARTDROID inference engine which implements the models presented in this chapter. The engine runs on Android mobile devices, and extends the original HEART inference engine for XTT2.

References

1. Bobek, S.: Methods for modeling self-adaptive mobile context-aware systems. Ph.D. thesis, AGH University of Science and Technology (April 2016) Supervisor: Grzegorz J. Nalepa
2. Bobek, S., Nalepa, G.J.: Uncertain context data management in dynamic mobile environments. Future Gener. Comput. Syst. **66**, 110–124 (2017)
3. Bobek, S., Nalepa, G.J.: Uncertainty handling in rule-based mobile context-aware systems. Pervasive and Mobile Computing (2016)
4. Kjaer, K.E.: A survey of context-aware middleware. In: Proceedings of the 25th Conference on IASTED International Multi-Conference: Software Engineering, SE'07, pp. 148–155. ACTA Press (2007)
5. Benerecetti, M., Bouquet, P., Bonifacio, M.: Distributed context-aware systems. Hum. Comput. Interact. **16**(2), 213–228 (2001)
6. Hu, H., of Hong Kong, U.: ContextTorrent: A Context Provisioning Framework for Pervasive Applications. University of Hong Kong (2011)
7. Chen, H., Finin, T.W., Joshi, A.: Semantic web in the context broker architecture. In: PerCom, IEEE Computer Society, pp. 277–286 (2004)

8. Nalepa, G.J., Bobek, S.: Rule-based solution for context-aware reasoning on mobile devices. Comput. Sci. Inf. Syst. **11**(1), 171–193 (2014)
9. Parsons, S., Hunter, A.: A review of uncertainty handling formalisms. In: Hunter, A., Parsons, S. (eds.) Applications of Uncertainty Formalisms. Lecture Notes in Computer Science, vol. 1455, pp. 8–37. Springer, Berlin (1998)
10. van Kasteren, T., Kröse, B.: Bayesian activity recognition in residence for elders. In: 3rd IET International Conference on Intelligent Environments, IE 07, pp. 209–212 (2007)
11. Bui, H.H., Venkatesh, S., West, G.: Tracking and surveillance in wide-area spatial environments using the abstract hidden Markov model. Int. J. Pattern Recognit. Artif. Intell. **15** (2001)
12. Fenza, G., Furno, D., Loia, V.: Hybrid approach for context-aware service discovery in healthcare domain. J. Comput. Syst. Sci. **78**(4), 1232–1247 (2012)
13. Yuan, B., Herbert, J.: Fuzzy cara - a fuzzy-based context reasoning system for pervasive healthcare. Procedia Comput. Sci. **10**, 357–365 (2012)
14. Hao, Q., Lu, T.: Context modeling and reasoning based on certainty factor. In: PACIIA 2009 Asia-Pacific Conference on Computational Intelligence and Industrial Applications, November 2009, vol. 2, pp. 38–41 (2009)
15. Almeida, A., Lopez-de Ipina, D.: Assessing ambiguity of context data in intelligent environments: towards a more reliable context managing systems. Sensors **12**(4), 4934–4951 (2012)
16. Krause, A., Smailagic, A., Siewiorek, D.P.: Context-aware mobile computing: learning context-dependent personal preferences from a wearable sensor array. IEEE Trans. Mob. Comput. **5**(2), 113–127 (2006)
17. Senge, R., Bösner, S., Dembczyński, K., Haasenritter, J., Hirsch, O., Donner-Banzhoff, N., Hüllermeier, E.: Reliable classification: learning classifiers that distinguish aleatoric and epistemic uncertainty. Inf. Sci. **255**, 16–29 (2014)
18. Niederliński, A.: RMES, Rule- and Model-Based Expert Systems. Jacek Skalmierski Computer Studio (2008)
19. Köping, L., Grzegorzek, M., Deinzer, F., Bobek, S., Ślażyński, M., Nalepa, G.J.: Improving indoor localization by user feedback. In: 2015 18th International Conference on Information Fusion (Fusion), July 2015, pp. 1053–1060 (2015)
20. Koller, D., Friedman, N.: Probabilistic Graphical Models: Principles and Techniques. MIT Press, Cambridge (2009)
21. Korver, M., Lucas, P.J.F.: Converting a rule-based expert system into a belief network. Med. Inform. **18**, 219–241 (1993)
22. De Raedt, L., Kimmig, A., Toivonen, H.: ProbLog: a probabilistic prolog and its application in link discovery. In: Proceedings of the 20th International Joint Conference on Artificial Intelligence, IJCAI'07, San Francisco, CA, USA, pp. 2468–2473. Morgan Kaufmann Publishers Inc. (2007)
23. Poole, D., Mackworth, A.K.: Artificial Intelligence – Foundations of Computational Agents. Cambridge University Press, Cambridge (2010)
24. Kang, D., Sohn, J., Kwon, K., Joo, B.G., Chung, I.J.: An intelligent dynamic context-aware system using fuzzy semantic language. In: Park, J.J., Adeli, H., Park, N., Woungang, I. (eds.) MUSIC. Lecture Notes in Electrical Engineering, vol. 274, pp. 143–149. Springer, Berlin (2013)
25. Orchard, R.A.: FuzzyCLIPS Version 6.04A. User's Guide, Integrated Reasoning Institute for Information Technology National Research Council Canada (October 1998)
26. Giarratano, J.C.: CLIPS User's Guide (December 2007)
27. Khan, W.Z., Xiang, Y., Aalsalem, M.Y., Arshad, Q.: Mobile phone sensing systems: a survey. IEEE Commun. Surv. Tut. **15**(1), 402–427 (2013)
28. Heckerman, D.: Probabilistic interpretations for MYCIN's certainty factors. In: Proceedings of the First Conference Annual Conference on Uncertainty in Artificial Intelligence (UAI-85), Corvallis, Oregon, pp. 9–20. AUAI Press (1985)
29. Salber, D., Dey, A.K., Abowd, G.D.: The context toolkit: Aiding the development of context-enabled applications. In: Proceedings of the SIGCHI Conference on Human Factors in Computing Systems, CHI '99, New York, NY, USA, pp. 434–441. ACM (1999)

30. Etter, R., Costa, P.D., Broens, T.: A rule-based approach towards context-aware user notification services. In: 2006 ACS/IEEE International Conference on Pervasive Services, June 2006, pp. 281–284 (2006)
31. Vanrompay, Y., Kirsch-Pinheiro, M., Berbers, Y.: Context-aware service selection with uncertain context information. ECEASST **19** (2009)
32. Floch, J., Fra, C., Fricke, R., Geihs, K., Wagner, M., Lorenzo, J., Soladana, E., Mehlhase, S., Paspallis, N., Rahnama, H., Ruiz, P.A., Scholz, U.: Playing music – building context-aware and self-adaptive mobile applications. Softw. Pract. Exp. **43**(3), 359–388 (2013)
33. Parsaye, K., Chignell, M.: Expert Systems for Experts / Kamran Parsaye, Mark Chignell. Wiley, New York (1988)
34. Buchanan, B.G., Shortliffe, E.H.: Rule Based Expert Systems: The Mycin Experiments of the Stanford Heuristic Programming Project (The Addison-Wesley Series in Artificial Intelligence). Addison-Wesley Longman Publishing Co., Inc., Boston (1984)
35. Nalepa, G., Bobek, S., Ligęza, A., Kaczor, K.: HalVA – rule analysis framework for XTT2 rules. In: Bassiliades, N., Governatori, G., Paschke, A. (eds.) Rule-Based Reasoning, Programming, and Applications. Lecture Notes in Computer Science, vol. 6826, pp. 337–344. Springer, Berlin (2011)
36. Bobek, S., Nalepa, G.: Compact representation of conditional probability for rule-based mobile context-aware systems. In: Bikakis, A., Fodor, P., Roman, D. (eds.) Rules on the Web: From Theory to Applications. Lecture Notes in Computer Science. Springer International Publishing, Berlin (2015)
37. Nalepa, G., Bobek, S., Ligęza, A., Kaczor, K.: Algorithms for rule inference in modularized rule bases. In: Bassiliades, N., Governatori, G., Paschke, A. (eds.) Rule-Based Reasoning, Programming, and Applications. Lecture Notes in Computer Science, vol. 6826, pp. 305–312. Springer, Berlin (2011)

Chapter 8
Formalizing Interoperability in Rule Bases

With the increasing number of rules application areas, the number of different rule representations is also growing. As a result, rule-based knowledge cannot be easily shared among different rule bases. The goal of translation methods is to facilitate the process of interoperability between representations by providing an intermediate and formalized format for knowledge translation (see Sect. 2.5). An efficient interoperability method cannot be limited to translation of rule language syntax. It also has to take into account the semantics of the complete knowledge base. In the research presented in this chapter, it is assumed that two rule bases, expressed in different representations, have the same semantics[1] if for a given initial state both production systems infer the same conclusion. Such a definition is simultaneously the most important requirement that, in our opinion, must be satisfied by a rule interoperability method. Of course, this definition cannot be applied for any type of rules as they may be processed in different ways e.g. derivation rules can be processed in both a forward and backward direction. In such a case, the inferred conclusion may also depend on the applied inference direction. Therefore, we focus on production rules, where forward chaining is the native inference mode whereas the backward chaining is mostly not applicable. The next sections provide a definition of the formalized model for production rule representation. The proposed model is intended to be used as the intermediate format for rule interoperability between rule languages like CLIPS [2], JESS [3], DROOLS [4] or XTT2 [5].

The discussed model is based on ALSV(FD) logic and significantly extends a formal model of XTT2 towards production rule systems. The most important extensions are discussed in subsequent sections, where different parts of the formalization are introduced. Section 8.1 provides an overview of the multilevel approach to rule interoperability that is considered in this research. In Sect. 8.2 definitions of data types, objects, and attributes used in the model are given. The formulae and operators con-

---

[1]The *semantics* of a knowledge base corresponds in fact to the so-called *operational semantics* as presented in [1] and describes the changes of a fact base after rules application is considered from a user perspective.

© Springer International Publishing AG 2018
G.J. Nalepa, *Modeling with Rules Using Semantic Knowledge Engineering*,
Intelligent Systems Reference Library 130,
https://doi.org/10.1007/978-3-319-66655-6_8

sidered in the model are discussed in Sect. 8.3. Rule formalization is presented in Sect. 8.4. As we consider structured rule bases module formalization is introduced in Sect. 8.5. The knowledge base definition for interoperability is given in Sect. 8.6. The chapter is summarized in Sect. 8.7.

8.1 Formalized Approach to Rule Interoperability

The focus of model presented here is on production rules. RBS using such rules can be considered as a dynamic system having an input, an output and a certain internal state (see Sect. 8.2). The state is defined by a set of values of facts that are stored within a fact base that, in turn, can be changed by the rules stored within a rule base (see Sect. 8.4). Considering the nature of the production rules, the fact base (and hence state of the system) can be changed by invoking actions allowing for asserting new facts into a fact base or removing or modifying the existing ones. Such actions are defined within the RHS of a rule and is invoked only when such a rule is executed. Thus, the set of rules defines a set of possible transitions between states and thereby the dynamics of RBS.

In the proposed perspective, the rule language, provided by the specific RBS, is used only for describing a rule base. The expressiveness and semantics of a rule language determines how this description can be made and what can be described. In turn, the expressiveness and semantics of the knowledge have a significant impact on the way how the system can process this knowledge and what can be inferred based on it. This perspective is based on a simple assumption in which the unequivocal and consistent semantics of the knowledge is considered to be equivalent in various representations only when, the systems that use these representations, are able to infer the same conclusions (systems reaches the same state). This assumption is very important in the context of providing an interoperability method that allows for preserving knowledge semantics during translation, as it defines the main effect of the interoperability. It is assumed that the proposed method does not take methods of knowledge processing and inference issues itself into account but focuses on the knowledge semantics.

An interoperability method cannot be limited only to the semantics of rule language constructs, but it must provide a broader view on the rule base [6, 7]. This is why, the issues related to rule base structure and their impact on knowledge processing are also taken into account in this work. This chapter proposes a formalized model for rule interoperability that considers all these issues in three levels of abstraction of knowledge representation:

1. *The rule Base level* – takes the operational semantics of rule base into consideration.
2. *Individual Rule level* – involves semantics of all individual knowledge elements.
3. *The environment level* – provides the support for design environments.

The *rule base* level involves issues related to knowledge base structure. The majority of the existing tools provide mechanisms for knowledge modularization. The structured knowledge bases cannot be treated in the same way as unstructured (flat) ones. In a flat knowledge base, all rules are evaluated every time when the knowledge base is modified. In modularized rule bases, the knowledge structure may have an impact on i.a. inference control and the amount of the knowledge that is available for the engine in a specific point of reasoning. In JESS or CLIPS, the set of evaluated and fired rules is determined by both the inference algorithm and modularization mechanism. The Sect. 8.5 provides a formal definition of the rule base structure and operational semantics of defined structure elements.

The *individual rule* level considers semantics of the single knowledge elements. Thus, the rule language must provide an accurate definition of its meaning. This can be assured by providing underlying logic which allows for unequivocal interpretation of the knowledge elements. Nevertheless, the existing rule languages are merely programming solutions which provide only well defined syntax. They rarely provide an underlying logical interpretation with well defined semantics. Such languages cannot be properly used for knowledge interoperability because their semantics cannot be unequivocally translated. Therefore, the main goal of this research is to provide a formalized model of rule representation that will allow for expressing the semantics of the considered rule languages in terms of its semantics.

The *environment* level is important from a technical point of view. It concerns issues related to design environment such as support for rule language syntax, executing, tools integration, etc. This level involves many technical issues related to importing, sharing and mapping knowledge into appropriate formats. Nevertheless, this issue is out of the scope of this chapter.

The definition of the model follows and significantly extends the perspective introduced in [8]. The proposed model provides an object-based representation of the world and fact-based storing of the knowledge. Such an approach is also used by ALSV(FD) logic and supported by the considered rule languages. Thus, the formalization starts with the definition of the fundamental elements like system types and objects. Later, facts are introduced as well as the definition of the state of RBS in this model. Over these fundamental elements a complex expressions may be defined. Firstly, the formulation of syntax that uses supported operators is described. Furthermore, the precisely defined semantics of expressions as well as operators are provided. The definition of a rule, as the dynamic and most important element of the model, is based on the well-formed expressions. Due to the fact that the model is intended to take rule base structure into account, a module-like mechanism is proposed.

Thanks to the formalization, this model has a well-defined semantics. The semantics of a certain element is expressed with the help of a dedicated interpretation function I. Each interpretation function belongs to the set of all interpretation functions $\mathbb{I}$ and maps a certain element to the elements expressed in terms of the universum set Ω or other elements of the model. However, due to the limited space of this chapter, we omit the discussion of semantics. It was presented *in extenso* in [9].

In order to make the formalization more clear, the definition contains examples that show how the defined constructs can be used for defining the DROOLS model in Example 8.1.1. The DROOLS format is selected due to its transparency and the wide support for different rule language features.

Example 8.1.1

```
1. declare Car
2.    Capacity  : int          12.declare Result
3.    Age       : int          13.  Value      : double
4.    Historic  : boolean      14.end
5.    Seats     : int          15.rule "base-charge1"
6.    Technical : boolean      16.  agenda-group "base-charge"
7.    Accidents : int          17.when
8. end                         18.  Car($Capacity: Capacity,
                                          Capacity < 900)
9. declare Base                19.then
10.  Value      : double       20.  insert(new Base(537));
11.end                         21.  insert(new Result(537));
                               22.end
```

8.2 Data Types, Objects and Attributes

Data Types and Objects

The concept of an object is considered according to ALSV(FD). The proposed model extends the model by the introduction of data types. Data type is used to define a structure of all objects instantiating it as well as to restricting the set of possible values that an object can take. In the proposed model, every object is an instance of a certain existing type T, which belongs to the set of all types $\mathbb{T}$:

$$\mathbb{T} \stackrel{def}{=} \{T_i \mid i \in \mathbb{X}_\mathbb{T}\} \tag{8.1}$$

where $\mathbb{X}_\mathbb{T}$ is a finite set of type identifiers. Moreover, each object can be considered as an abstract portion of data belonging to the countably infinite set of all objects $\mathbb{O}$:

$$\mathbb{O} \stackrel{def}{=} \{o_i \mid i \in \mathbb{X}_\mathbb{O}\} \tag{8.2}$$

where $\mathbb{X}_\mathbb{O}$ is a finite set of object identifiers. An object o is an instance of certain type T:

$$o : T \tag{8.3}$$

Each object can be defined as an instance of only one type. Thus, the set of all objects can be divided into pairwise disjoint subsets containing objects of the same type: $\mathbb{O} = \bigcup_{i \in \mathbb{X}_\mathbb{T}} \mathbb{O}_{T_i}$, where: $\forall_{i \in \mathbb{X}_\mathbb{T}} \mathbb{O}_{T_i} = \{o \mid o : T_i\}$ and $\forall_{i \neq j} \mathbb{O}_{T_i} \cap \mathbb{O}_{T_j} = \varnothing$. A type T is defined as a pair:

$$T \stackrel{def}{=} (\mathbf{S}_T, \mathbf{A}_T) \tag{8.4}$$

where: $\mathbf{S}_T$ defines a set of *composite* types of T and determines the structure of objects, and $\mathbf{A}_T$ defines a set of attribute interpretation functions of objects.

The proposed model provides two kinds of predefined types: *Real* and *Smbl*. These types belong to the set of $\mathbb{T}$, but they do not provide any object structure and thus they are called *primitive* types:

$$Real \stackrel{def}{=} \left(\varnothing, (val)\right)$$
$$Smbl \stackrel{def}{=} \left(\varnothing, (val)\right) \tag{8.5}$$

where *val* is the name of an attribute of a primitive type.

In fact, the structure $\mathbf{S}_T$ is a tuple that is defined by the following Cartesian product:

$$\mathbf{S}_T \stackrel{def}{=} \left(T_{X_{\mathbb{T},1}}, T_{X_{\mathbb{T},2}}, \ldots, T_{X_{\mathbb{T},n}}\right) \tag{8.6}$$

where:

- $\{X_{\mathbb{T},1}, X_{\mathbb{T},2}, \ldots, X_{\mathbb{T},n}\}$ is a multiset over $\mathbb{X}_\mathbb{T}$.
- $T_{X_{\mathbb{T},i}} \equiv T_i$ and T_i denotes the ith element within the tuple.
- $\forall_{i \in \{1,2,\ldots,n\}}(T_i \in \mathbb{T} \wedge T_i \neq T)$ – each T_i is already defined type.

The semantics of a type $T \in \mathbb{T}$ can be defined as a tree providing type structure by means of other types and can be defined by the $I_\mathbb{T}$ interpretation function that belongs to the set $\mathbb{I}$:

$$I_\mathbb{T}(T) \stackrel{def}{=} (I_\mathbb{T}(\mathbf{S}_T), \mathbf{A}_T) \stackrel{def}{=} \begin{cases} ((I_\mathbb{T}(T_1), I_\mathbb{T}(T_2), \ldots, I_\mathbb{T}(T_n)), \mathbf{A}_T) & \text{if } T \in \mathbb{T} \backslash \{Real, Smbl\} \\ (\Omega, \mathbf{A}_T) & \text{if } T = Smbl \\ (\mathbb{R}, \mathbf{A}_T) & \text{if } T = Real \end{cases}$$
$$\tag{8.7}$$

According to Definitions (8.4) and (8.6), the proposed model allows for creating complex (nested) types of objects. This corresponds to the possibility of building classes or structures that are known from such programming languages like C++ or JAVA. The interpretation function provided by Definition (8.7) works recursively in order to provide an interpretation of all nested types. At the end, the type T is expressed as a complex structure consisting of only primitive types.

An object is defined as an instance of a certain type and is used for storing portions of data. This means that the data stored within an object corresponds to a certain existing type in terms of the structure $\mathbf{S}_T$ as well as in terms of the types of data. Assuming the object o is defined by Formula (8.3), the type T is defined by Formulae (8.4) and (8.6), the structure of an object can be written as follows:

$$o = (o_1, o_2, \ldots, o_n) \tag{8.8}$$

where $\forall_{i \in \{1,2,\ldots,n\}} o_i : T_i$ – each element of the structure of the object o is an another object that is an instance of the type which is provided by $\mathbf{S}_T$.

The semantics of an object, introduced by Formula (8.8), can be defined as an element of a certain set that has a structure corresponding to the structure of the object. The formal definition of object semantics is provided by the $I_\mathbb{O} \in \mathbb{I}$ interpretation function. It maps an object into a set, called object *domain*, that contains all possible values the object can take. Thus in general, the definition of the function $I_\mathbb{O}$ can be written as follows:

$$I_\mathbb{O} : \mathbb{O} \to \mathbb{D} \tag{8.9}$$

where $\mathbb{D}$ is the domain of the object o and thus contains elements (or sets of elements) having structure defined by Formula (8.8). It is worth to notice that in order to provide a complete interpretation of an object, the $I_\mathbb{O}$ function must be applied recursively:

$$I_\mathbb{O}(o) = \begin{cases} \{(I_\mathbb{O}(o_1), I_\mathbb{O}(o_2), \ldots, I_\mathbb{O}(o_n))\} & \text{if } o : T \in \mathbb{T} \backslash \{Real, Smbl\} \\ \omega \subset \Omega & \text{if } o : Smbl \\ r \subset \mathbb{R} & \text{if } o : Real \end{cases} \tag{8.10}$$

Expression (8.3) defines an object o to be of the type T. Formally, this expression determines the set of allowed values of the object o which, in turn, is determined by the type T. Hence, the semantics of the : operator can be defined by the following formula:

$$I_: (o : T) \stackrel{def}{=} I_\mathbb{O}(o) \in I_\mathbb{T}(2^T) \tag{8.11}$$

where the set $I_\mathbb{T}(2^T)$ restricts the domain $\mathbb{D}$ of the object o. In this way, the object interpretation function, defined by Formula (8.9), can be more precisely defined for subset of objects being of the same type T in the following way:

$$I_\mathbb{O} : \mathbb{O}_T \to I_\mathbb{T}(2^T) \tag{8.12}$$

Example 8.2.1 *Using the notation provided by Formula (8.3), it is possible to define an object o_{vw} that denotes the car of the client. This object is defined as instances of the T_{Car} type: $o_{vw} : T_{Car}$. Thanks to this definition, the $\mathbb{O}$ set of all objects contains one element: $\mathbb{O} = \{o_{vw}\}$. Similarly, the set of all object identifiers contains also one element: $\mathbb{X}_\mathbb{O} = \{vw\}$. In turn, considering the subsets of $\mathbb{O}$ containing objects of a certain type one can write: $\mathbb{O} = \mathbb{O}_{T_{Car}} = \{o_{vw}\}$. According to Definitions (8.8), (8.10) and (8.11) an object interpretation is a set that belongs to the power set of type. Thus, the following example shows possible interpretation (value) of the object in the form of a*

> *singleton containing tuple that has a structure consistent with the object type:*
> $$o_{vw} = \{(800, 10, 0, 5, 1, 0)\} \in I_{\mathbb{T}}(2^{T_{Car}}).$$

The interpretation of an object can be expressed using attributes functions. The set $\mathbb{A}$ contains all possible attribute functions $\mathbb{A} = \{A_i \mid i \in \mathbb{X}_\mathbb{A}\}$ where $\mathbb{A} \subset \mathbb{I}$ and $\mathbb{X}_\mathbb{A}$ is a finite set of attribute identifiers.

Within the proposed approach a concept of attribute is defined as an interpretation function of objects that maps a given object to a certain value. The possible attributes of an object are defined within the $\mathbf{A}_T$ tuple that is a complementary part of the structure of the object type T (see Definition (8.4)). This set contains n (the number of elements within $\mathbf{S}_T$ tuple) attribute functions and can be defined as follows:

$$\mathbf{A}_T \overset{def}{=} \left(A_{X_{\mathbb{A},1}}, A_{X_{\mathbb{A},2}}, \ldots, A_{X_{\mathbb{A},n}} \right) \tag{8.13}$$

where:

- $\{X_{\mathbb{A},1}, X_{\mathbb{A},2}, \ldots, X_{\mathbb{A},n}\}$ is a multiset over $\mathbb{X}_\mathbb{A}$.
- $A_{X_{\mathbb{A},1}} \equiv A_i$ and A_i denotes the ith element within the tuple.
- $\forall_{i \in \{1,2,\ldots,n\}} A_i \in \mathbb{A}$.

The ALSV(FD) logic and approach presented in [8] also define attributes as functions that map objects into values. In comparison to them, the concept of the attributes proposed here is more precisely defined as each attribute can be related to only one data type and therefore its domain can be more strictly defined.

Attribute interpretation function closely related to object interpretation function. Within this model an attribute is an object interpretation function that maps a given object to a value of its individual element of the structure which is determined by $\mathbf{S}_T$. Following this assumption, it is possible to provide an analogous definition of the attribute function to the definition of the object interpretation function specified by Formula (8.12). Assuming that A_i is an attribute function provided by a certain type T, which is defined by Formula (8.4), then A_i can be defined as follows:

$$A_i : \mathbb{O}_T \to D_{A_i} \tag{8.14}$$

where $D_{A_i} \in I_{\mathbb{T}}(2^{T_i})$ (according to Definition (8.11)) and it is called an *attribute domain* (in fact this is a codomain of the attribute interpretation function) and provides an additional restrictions concerning possible values of objects.

> **Example 8.2.2** *Having the definition of an attribute, it is possible to provide a complete specification of the Car, Base and Result types. All of the types are composed of fields that can be expressed using only Real and Smbl primitive types. Thus, the set of all types is defined as follows:* $\mathbb{T} = \{Real, Smbl, T_{Car}, T_{Base}, T_{Result}\}$. *In this way, the set of all types identifiers* $\mathbb{X}_\mathbb{T}$ *contains five elements:* $\mathbb{X}_\mathbb{T} = \{Real, Smbl, Car, Base, Result\}$.

The definition of Real and Smbl types is provided by Formula (8.5). In turn, according to Definitions (8.4) and (8.6) the types can be defined in the following way:

$$
\begin{aligned}
T_{Car} &= \big((Real, Real, Real, Real, Real, Real), (A_{Capacity}, \\
&\quad A_{Age}, A_{Historic}, A_{Seats}, A_{Technical}, A_{Accidents})\big) \\
T_{Base} &= \big((Real), (A_{Value})\big) \\
T_{Result} &= \big((Real), (A_{Value})\big)
\end{aligned}
\tag{8.15}
$$

In order to make the model consistent, the set of all attribute identifiers must contain the following elements:
$\mathbb{X}_{\mathbb{A}} = \{Capacity, Age, Historic, Seats, Technical, Value, Accidents\}$.
According to Definition (8.14) each attribute provided by a certain type maps an object of this type to an element of its domain. Thanks to that, each attribute provides an additional constraint for the possible values of an object element. In order to make the definition of the considered types more precise the following domains can be defined:

$$
A_{Capacity} : \mathbb{O}_{T_{Car}} \to \left\{ x \in \mathbb{R} \mid x > 0 \right\}
$$

$$
A_{Age} : \mathbb{O}_{T_{Car}} \to \left\{ x \in \mathbb{N} \mid x \in [0, 100] \right\}
$$

$$
A_{Historic} : \mathbb{O}_{T_{Car}} \to \left\{ x \in \mathbb{N} \mid x \in \{0, 1\} \right\}
$$

$$
A_{Seats} : \mathbb{O}_{T_{Car}} \to \left\{ x \in \mathbb{N} \mid x \in [1, 100] \right\}
$$

$$
A_{Technical} : \mathbb{O}_{T_{Car}} \to \left\{ x \in \mathbb{N} \mid x \in \{0, 1\} \right\}
\tag{8.16}
$$

$$
A_{Accidents} : \mathbb{O}_{T_{Car}} \to \left\{ x \in \mathbb{N} \mid x \in [0, 366] \right\}
$$

$$
A_{Value} : \mathbb{O}_{T_{Base}} \to \left\{ x \in \mathbb{N} \mid x > 0 \right\}
$$

$$
A_{Value} : \mathbb{O}_{T_{Result}} \to \left\{ x \in \mathbb{N} \mid x > 0 \right\}
$$

Due to the fact that a power set of certain type contains all possible values of object of this type, the sum of all power sets of all types contains all possible values that can be expressed with the help of the defined types. This set is the countably infinite set of all constant symbols and is referred as $\mathbb{C}$. It can be defined in the following way:

$$
\mathbb{C} \stackrel{def}{=} \{c_i \mid i \in \mathbb{N}\} \stackrel{def}{=} \left(\bigcup_{T \in \mathbb{T}} I_{\mathbb{T}}(2^T) \right) \cup \{false, true\}
\tag{8.17}
$$

Inheritance of Types

Proposed in [8] a definition of the model can be further extended by the concept of inheritance of types. This features is well known in OO languages. An inheritance supported by this model allows for defining types with the help of already exiting ones.[2] Thanks to that, it is not necessary to repeat definitions of a certain set of type elements that are already defined within other types. The objects of the derived type are considered to be special cases of the base type and thus such objects are treated as being of the base type as well.

Inheritance of the types can be expressed by the $::$ operator i.e. in order to express that a type T_D inherits a certain type T_B (or type T_B is a base type for T_D) one can write the following expression: $T_D :: T_B$ In general, the right hand side of the operator is a tuple containing a set of types that are inherited: $T_D :: (T_{B_1}, T_{B_2}, \ldots, T_{B_n})$.

Attributes of Objects

Assuming that an object can take any value that belongs to the power set of its type (see Definition (8.14)), in particular cases an object value can take the form of a set, an empty set or a singleton. In turn, the form of an object value is determined by the definition of the attribute interpretation function. In this context, the proposed model adopts two types of object attributes that are provided by the ALSV(FD)logic and supported by XTT2:

1. A *basic attribute* (A^b) that maps an object to a single value at a given point of time, and
2. A *generalized attribute* (A^g) that maps an object to a set of values i.e. such an attribute can take more than one value at the given point of time.

To be more precise, if an attribute is defined as basic then its domain must contain only singletons: if $A^b: \mathbb{O} \to D_A$ then $\forall_{d \in D_A} |d| = 1$. In turn, if an attribute is defined as generalized then its domain may contain any sets: if $A^g: \mathbb{O} \to D_A$ then $\forall_{d \in D_A} |d| \geqslant 0$. In this context, the definition of the attribute type plays a significant role, because it determines the allowed operators that can be used with attributes.

Facts

The original XTT2 approach uses objects and attributes for storing and representing both data and knowledge. Therefore, each object is a part of knowledge that can be processed by an inference engine. Such an approach does not fit to the production rule systems which distinguish concepts of data and knowledge. Therefore, the proposed approach introduces a concept of fact. The main difference between facts and objects, apart from their structure, is their purpose. Here, objects are intended to store a portion of data while facts are intended to store a portion of knowledge. It is assumed that the knowledge stored by facts is unconditionally true or it implicitly stems from the existing facts and rules. The proposed model defines a fact as a structure built over the object that provides additional elements strictly related to the knowledge that allows for more advanced processing.

---

[2]A new type T that inherits other types may provide its own structure which is further extended by the set of inherited types. Here, such a structure is called *initial structure of the type T* and is written as $T^0 = (\mathbf{S}_{T^0}, \mathbf{A}_{T^0})$.

The main motivation for separating objects and facts is twofold. First of all, such a distinction allows for separating data and knowledge elements. Secondly, in one hand an object-like representation of data is very intuitive especially in the context of object-oriented languages that usually have a strong impact on the rule language. On the other hand, the provided definition of facts allows for a more flexible and intuitive definition of semantics of rule representation.

$\mathbb{F}$ is called *fact base* and is defined in the following way:

$$\mathbb{F} \overset{def}{=} \bigcup_{i \in \mathbb{X}_\mathbb{F}} F_i \tag{8.18}$$

where $\mathbb{X}_\mathbb{F}$ is a finite set of fact identifiers.

Fact base contains all the existing facts (all the knowledge) i.e. everything that is currently known. In turn, a single fact F_o is a set of all triples that are related to one existing object o. This set can be defined in the following way:

$$F_o \overset{def}{=} \{f_{o,1}, f_{o,2}, \dots, f_{o,n_{F_o}}\} \tag{8.19}$$

where:

- F_o is not empty set that is called a single fact and $F_o \subset \mathbb{F}$.
- $\forall_{i \in \{1,2,\dots,n_{F_o}\}} f_{o,i} \in \mathbb{F}$.
- $\nexists f_{o_a,i}(f_{o_a,i} \in \mathbb{F} \wedge F_{o_b} \subset \mathbb{F} \wedge o_a \equiv o_b \wedge f_{o_a,i} \notin F_{o_b})$ i.e. each triple that refers to a certain object o must belong to the same fact F_o. Thus, there cannot be two different facts F_{o_a} and F_{o_b} that refers to the same object ($o_a \equiv o_b$).

The triple $f_{o,i}$ is the smallest portion of knowledge that can be provided by the model and refers to the value of a certain attribute for a given object. Assuming that an object o is of type T, the definition of $f_{o,i}$ is as follows:

$$f_{o,i} \overset{def}{=} (o, A_i, L_{f_{o,i}}) \in \mathbb{O} \times \mathbb{A} \times 2^\mathbb{L} \tag{8.20}$$

where:

- o refers to an existing object, of a certain type $T \in \mathbb{T}$, that belongs to the $\mathbb{O}_T \subset \mathbb{O}$.
- A_i is the ith attribute function from the $\mathbf{A}_T$ tuple that is provided by type T (see Definition (8.4)).
- $L_{f_{o,i}}$ is a set that is called *Logical Support* of the fact triple $f_{o,i}$.
- $\mathbb{L}$ is a set of all possible logical support sets.

Logical Support of the Facts

The proposed model supports *truth maintenance* mechanism. Such a mechanism is also provided by the majority of production systems. As in the case of these tools, this mechanism plays an analogous role within the proposed model and facilitates the maintenance of the consistency of fact base by automatically removing inconsistent facts. Nevertheless, the main difference between the logical support provided by the model and the mentioned tools lies in the level of abstraction to which this mechanism is related. In the case of CLIPS the logical support works at the level of facts while in case of DROOLS at the level of rules. Within this approach, the logical support works at the level of fact triples – the level below fact level. Thus, the model allows for more flexible and expressive management of logical support.

The logical support of a certain fact triple $f_{o,i}$ is defined as a group (set) of fact triples that determine the existence of $f_{o,i}$. Such a group is called a *supporting group* while the triples belonging to it are called *supporting triples*. Thus, the existence of $f_{o,i}$ is determined by the existence of all the supporting triples within group. This is why, if at least one of these triples is removed from the fact base then the supported triple must also be removed. In general, there can be more than one supporting group. In such a case, the existence of $f_{o,i}$ is determined by the existence of at least one of these groups. Within the proposed model, the set of all supporting groups for the triple $f_{o,i}$ is denoted as $L_{f_{o,i}}$ and can be defined as:

$$L_{f_{o,i}} \subset \mathbb{L} = 2^{\mathbb{F}} \tag{8.21}$$

Thanks to the provided connection between $\mathbb{L}$ and $\mathbb{F}$, the removal of a certain element from $\mathbb{F}$ implies the removal of all supporting groups containing this element, from the logical support sets of all fact triples in $\mathbb{F}$. This is because, according to Definition (8.21), a certain supporting group, containing the removed element, cannot belong to the $\mathbb{L}$ set.

The existence of a single fact is determined by the existence of at least one triple that is related to this fact. In turn, the existence of a triple is determined by the logical support set in the following way: $(o, A_i, L_{f_{o,i}}) = f_{o,i} \in \mathbb{F} \Leftrightarrow L_{f_{o,i}} \neq \varnothing$. Considering this definition, the existence of the fact is implicitly determined by the logical support sets provided by all the triples related to this fact.

Types of Facts

From Definitions (8.19) and (8.20) stems that each fact is related to one and only one object. On the other hand, each object is an instance of some existing type T. Taking this into account, it is possible to provide assignment of type to fact element. Similarly, as in case of objects, this can be done by using : operator. The following formula states that a fact F_o is of type T: $F_o : T$. The semantics of such an expression is defined by the $I_.$ function and corresponds to the semantics provided by Formula (8.3): $I_.(F_o : T) \overset{def}{=} I_.(o : T)$, where $T \in \mathbb{T}$ and $I_.(o : T)$ is defined by Formula (8.11). Going further, the set $\mathbb{F}$ of all facts can be divided into pairwise disjoint subsets (possibly empty) containing facts of a certain type:

$$\mathbb{F} = \bigcup_{i \in \mathbb{X}_\mathbb{T}} \mathbb{F}_{T_i} \tag{8.22}$$

where $\mathbb{F}_{T_i} = \bigcup_{F_o : T_i} F_o$.

In general, a concept of fact is very close to the concept of object.

Example 8.2.3 *Example 8.2.1 provides a definition of one object. In terms of rule languages, this object cannot be treated as a portion of knowledge because it is not defined as fact. Currently it can be considered as a portion of some data and treated as a variable. In order to make the information stored within this object available for inference, an appropriate fact must be defined. Using a provided notation, this can be done as follows:*

$$F_{o_{vw}} : T_{Car} \tag{8.23}$$

According to Definitions (8.19) and (8.20), the $F_{o_{vw}}$ fact can be in the form:

$$
\begin{aligned}
F_{o_{vw}} &= \left\{ f_{o_{vw},1}, f_{o_{vw},2}, f_{o_{vw},3}, f_{o_{vw},4}, f_{o_{vw},5}, f_{o_{vw},6} \right\} \\
&= \Big\{ \left(o_{vw}, A_1, \{\{f_{o_{vw},1}\}\}\right), \left(o_{vw}, A_2, \{\{f_{o_{vw},2}\}\}\right), \\
&\quad \left(o_{vw}, A_3, \{\{f_{o_{vw},3}\}\}\right), \left(o_{vw}, A_4, \{\{f_{o_{vw},4}\}\}\right), \\
&\quad \left(o_{vw}, A_5, \{\{f_{o_{vw},5}\}\}\right), \left(o_{vw}, A_6, \{\{f_{o_{vw},6}\}\}\right) \Big\}
\end{aligned}
\tag{8.24}
$$

Having this definition and according to Formula (8.22), the set of all facts contains the following elements: $\mathbb{F} = \mathbb{F}_{T_{Driver}}$.

Definition (8.23) allows information stored within o_{vw} object to be treated as knowledge. Nevertheless, the structure of a set of all facts allows for storing each information related to the values of the object's elements separately. Therefore, it is possible to remove single portion of information provided by a certain fact without removing the remaining part. Such an approach allows for storing only necessary information as knowledge, and thus increases efficiency of inference.

System State and Trajectory

Production rules usually allow the current fact base to be modified through adding new knowledge and removing or changing the already existing knowledge. This makes the inference process non monotonic. Thus, the changes of fact base can be considered as a dynamic aspect of the system and the system itself as a dynamic one. Going further with this analogy, for such a dynamic rule-based system, a *system state* [10] can be defined as well. Within the control theory, the system state is usually defined by the possibly smallest subset of internal system variables that can represent the entire state of the system at single point of time. In the context of a rule-based system, the system state can be defined using current knowledge that is stored within the fact base. Thus, the current system state is defined as a set of all fact triples belonging to $\mathbb{F}$:

$$S_c \stackrel{def}{=} \mathbb{F} \tag{8.25}$$

where S_c denotes the current system state.

During the inference process some of the facts (triples) can be removed, new facts (triples) can appear or some attribute functions can be changed within the fact base $\mathbb{F}$. Because of modifications of $\mathbb{F}$, the current state of the system also changes (according to Definition (8.25)). The difference between the current and the previous state of the system is defined as *state transition* and is determined by the *transition function* expressed, in this model, as a rule. Thus, it can be said that rules define the dynamics of the system.

Considering all the state changes that were made from the initial state of the system, a concept of *system trajectory* can be introduced. Within this model, a system trajectory $\mathbb{S}$ is defined as a sequence of states from the initial until the current one:

$$\mathbb{S} \stackrel{def}{=} S_0, S_1, \ldots, S_{c-1}, S_c \tag{8.26}$$

where S_i denotes the system state after i transitions. In particular S_0 is the initial state of the system while S_c is the current system state. According to Definitions (8.25) and (8.26), a system trajectory reflects the inference path within the system. This may correspond to the explanation mechanism that is implemented in the majority of RBS shells. Additionally, the formal definition of the system state and trajectory gives an ability to provide an advanced mechanism that allows for defining the set of allowed transitions between system states called *dynamic constraints*.

The system history set, denoted as $\mathbb{H}$, is defined as follows:

$$\mathbb{H} = \bigcup_{i=0}^{c} S_i \tag{8.27}$$

The $\mathbb{H}$ set stores all the facts that appear within the all system states. It includes the current facts, from the $\mathbb{F}$ set, as well as all the facts that belong to any of the previous system states.

Variables

In comparison to the XTT2, variable is a new concept supported by the proposed model. Within this model, variables play a twofold role. First of all, a variable can be used to denote an *unknown but specific element* i.e. the same variable can refer to a certain object, fact or even constant value that may be, in particular case, unknown. However, it may be further specified by changing the interpretation function (semantics of the variable) that determines the mapping of variables to their values. The case where mapping to an unknown value is changed to mapping to a specific one is called *variable instantiation*. Secondly, variables play the role of *element co-reference and data carriers* – each occurrence of the same variable denotes the same element that stores some portion of data. This stems directly from the definition of variable semantics through the interpretation function. Its value is determined by taking variable name as only argument without taking any wider context of variable occurrence into consideration.

Each variable belongs to the countably infinite set of all variables that is denoted as $\mathbb{V}$. It is defined as follows:

$$\mathbb{V} \stackrel{def}{=} \{v_i \mid i \in \mathbb{X}_V\} \tag{8.28}$$

where v_i is a single variable, and $\mathbb{X}_V$ is a finite set of variable identifiers.

Example 8.2.4 *Considering the example provided at the beginning of this section, the complete definition of the T_{Car} type (see Example 8.2.2) and the $F_{o_{vw}}$ fact (see Example 8.2.3) into account, a definition of the $v_{Capacity}$ variable can be introduced. In this way, the set of all variables contains one element: $\mathbb{V} = \{v_{Capacity}\}$. Simultaneously, the set of all variable identifiers also contains one element: $\mathbb{X}_V = \{Capacity\}$. The provided example shows that the $\$Capacity$ variable refers to the $Capacity$ field of a certain fact being of Car type. The following formula defines reference of the $v_{Capacity}$ variable to value of the $A_{Capacity}$ attribute of the $F_{o_{vw}}$ fact can be written in an analogous way*[2]: $v_{Capacity} \ominus I_{\mathbb{F}}\left(f_{o_{vw}},1\right) \equiv v_{Capacity} \ominus I_{\mathbb{F}}\left(\left(o_{vw}, A_{Capacity}, \{\{f_{o_{vw}},1\}\}\right)\right) \equiv v_{Capacity} \ominus A_{Capacity}(F_{o_{vw}})$. *It is worth to notice that this example involves the case where the defined variable refers to a constant value that is a result of attribute interpretation function. This is consistent with rule languages in which change of a variable that stores value of some field does not cause change of this field.*

8.3   Taxonomy of Formulae and Operators

The abstract syntax of the formulae that can be defined within the proposed model. Its syntax is based on the set of supported operators and different types of formulae. This is why, within this section a detailed definition of the set of supported operators, their syntax and semantics as well as definition of different types of formulae are provided.

The set of all possible formulae which can be expressed, is denoted by Φ. In the presented approach, formulae are constructed in a way that is analogous to propositional logic. In comparison to this logic, a concept of *objects*, *variables* and *quantifiers* are introduced. Each created formula is based on four primitive elements: *constant* elements, *object* elements, *fact* elements and *variable* elements. All elements of these four types are grouped within four separate sets that are further referred as:

- $\mathbb{C}$ – the set of all constants (for definition see Formula (8.17)).
- $\mathbb{O}$ – the set of all objects (for definition see Formula (8.2)).
- $\mathbb{F}$ – the set of all facts (fact triples) (for definition see Formula (8.18)).
- $\mathbb{V}$ – the set of all variables (for definition see Formula (8.28)).

If we have in mind a specific constant value $c \in \mathbb{C}$, object $o \in \mathbb{O}$, fact $F_o \in \mathbb{F}$ or variable $v \in \mathbb{V}$ occurring in a formula $\phi \in \Phi$, we shall write $\phi(c), \phi(o), \phi(F_o)$, and $\phi(v)$, respectively.

Quantum Formulae

Within the proposed model, one can distinguish two major types of formulae: *quantum* formulae and *complex* formulae. The main difference between these two types lies in the fact that quantum formulae do not provide any deeper structure, that is, do not contain connectives, or equivalently, have no strict subformulae. In turn, the complex formula is composed of the finite number of quantum formulae.

Quantum formulae constitute the most basic type of formulae that can be constructed. They consist of single element that belongs to one of the sets $\mathbb{C}$, $\mathbb{O}$, $\mathbb{F}$ or $\mathbb{V}$. The set of all possible quantum formulae is the subset of Φ and is referred as Φ_q. It can be divided into four subsets corresponding to the four sets of primitive elements:

$$\Phi_q = \Phi_{\mathbb{C}} \cup \Phi_{\mathbb{O}} \cup \Phi_{\mathbb{F}} \cup \Phi_{\mathbb{V}} \tag{8.29}$$

where:

- $\Phi_{\mathbb{C}}$ – is the set of all *constant* formulae, where each is constructed of a single element from the set $\mathbb{C}$.
- $\Phi_{\mathbb{O}}$ – is the set of all *object* formulae, where each consists of a single element belonging to the set $\mathbb{O}$.
- $\Phi_{\mathbb{F}}$ is the set of all *fact* formulae, where each formula consists of a single element from the set $\mathbb{F}$.
- $\Phi_{\mathbb{V}}$ – is the set of all *variable* formulae, in which each is composed of a single element from the set $\mathbb{V}$.

Example 8.3.1 *Considering the already defined elements, it is possible to distinguish the following quantum formulae:*
$\phi_{q_,} \equiv (800, 10, 0, 5, 1, 0) \in \mathbb{C}$, $\quad \phi_{q_,} \equiv o_{vw} \in \mathbb{O}$, $\quad \phi_{q_,} \equiv F_{o_{vw}} \in \mathbb{F}$, $\quad \phi_{q_,} \equiv v_{Capacity} \in \mathbb{V}$

The distinction between different types of quantum formulae is important in the context of precise abstract syntax and a semantics definition of the model.

Complex Formulae

Complex formula is a formula that can be divided into subformulae. Within the proposed model, the components of a complex formula are connected by operators. The approach adopts, inter alia, operators that are used within SKE [5] and extends this set by several new operators. The set of all supported operators $\mathbb{P}$ is defined as follows:

$$\mathbb{P} = \{\oplus, \ominus, \oslash, \otimes, \oslash, \cap, \cup, \oslash, \triangle, \oslash, \otimes, \ominus,$$
$$\oslash, \oslash, \oslash, \oslash, \oslash, \oslash, \oslash, \oslash, \oslash, \oslash, \oslash, \oslash, \oslash, \oslash, \oslash, \oslash,$$
$$\mathcal{P}, \oslash, \oslash, \oslash, \mathcal{A}, \mathcal{R}, \oslash\} \tag{8.30}$$

These operators are used to build complex formulae that allow for expressing functional dependencies between already known portions of data or knowledge. Similarly to quantum formulae, the set of complex formulae Φ_x can also be divided into several subsets containing elements being of the same type.

These operators allow for building different types of formulae. Thus, the set can be divided into pairwise disjoint subsets containing operators that can be used for building formulae of the same type: $\mathbb{P} = \mathbb{P}_a \cup \mathbb{P}_s \cup \mathbb{P}_r \cup \mathbb{P}_l \cup \mathbb{P}_p \cup \mathbb{P}_c \cup \mathbb{P}_k$, where:

- $\mathbb{P}_a = \{\oplus, \ominus, \oslash, \odot, \oslash\}$ is the set of *algebraic* operators.
- $\mathbb{P}_s = \{\cap, \cup, \oslash, \triangle, \oslash, \otimes\}$ is the set of *set theoretic* operators.
- $\mathbb{P}_r = \{\ominus, \neq, \oslash, \lessgtr, \ominus, \gtrless, \in, \notin, \subseteq, \oslash, \ominus, \oplus, \ominus, \approx\}$ is the set of *relational* operators.
- $\mathbb{P}_l = \{\wedge, \vee, \ominus\}$ is the set of *logical* operators.
- $\mathbb{P}_p = \{\mathcal{P}\}$ is the set of *pattern* operators.
- $\mathbb{P}_c = \{\ni, \not\ni, \vee\}$ is the set of *constraint* operators.
- $\mathbb{P}_k = \{\mathcal{A}, \mathcal{R}, \ominus\}$ is the set of *knowledge* operators.

Some of the provided subsets of $\mathbb{P}$ can be further divided into smaller subsets according to the different nature of operators domains and codomains.

Basic Formulae

Within this approach, a value (of an attribute, a fact, etc.) is, in general, considered as a set. Thus, it can be said that a certain fact has more than one value in the same time i.e. it takes a set of values as its value. In turn, value is obtained by the evaluation of a certain formula which, in a particular case, may be a single element e.g. evaluation of a formula consisting of basic attribute function (see Sect. 8.2). In this case, such a formula is called *basic formula* and the set of all basic formulae, and we later refer to it as Φ_b.

Algebraic Formulae

Within this approach, an *algebraic formula* is a basic formula that can be evaluated to a numeric value i.e. the interpretation of such a formula is a single element that belongs to the set of real numbers $\mathbb{R}$. Thus, in a particular case, a certain algebraic formula can be in the form of a quantum formula e.g. o is an algebraic formula if $o \in \mathbb{O}$ and $o : Real$. In turn, more complex algebraic formulae can be constructed by using algebraic operators (from the $\mathbb{P}_a$ set). The set of algebraic formulae Φ_a is defined in the following way:

1. $\Phi_a \subseteq \Phi_b$;
2. $\phi_{\mathbb{C},}(c) \in \Phi_{\mathbb{C}} \subset \Phi_q$ and $c \in \mathbb{R}$, then $\phi_{\mathbb{C},} \in \Phi_a$;
3. $\phi_{\mathbb{O},}(o) \in \Phi_{\mathbb{O}} \subset \Phi_q$ and $o : Real$, then $\phi_{\mathbb{O},} \in \Phi_a$;
4. $\phi_{\mathbb{F},}(F_o) \in \Phi_{\mathbb{F}} \subset \Phi_q$ and $F_o : Real$, then $\phi_{\mathbb{F},} \in \Phi_a$;
5. $\phi_{\mathbb{V},}(v) \in \Phi_{\mathbb{V}} \subset \Phi_q$ and v refers to constant value, object or fact of the *Real* type, then $\phi_{\mathbb{V},} \in \Phi_a$;
6. If $\phi_{a,1}$ and $\phi_{a,2}$ are algebraic formulae, i.e. $\phi_{a,1}, \phi_{a,2} \in \Phi_a$, then
   $(\phi_{a,1} \oplus \phi_{a,2}), (\phi_{a,1} \ominus \phi_{a,2}), (\phi_{a,1} \oslash \phi_{a,2}), (\phi_{a,1} \odot \phi_{a,2}), (\oslash(\phi_{a,1}, \phi_{a,2})) \in \Phi_a$.

Where:

- $\oplus, \ominus, \odot, \ominus$ are the *addition*, *subtraction*, *multiplication* and *division* operators in the model.
- $\oslash$ is the *root* operator in the model, where the first argument corresponds to the *index* and the second to the *radiand*.

The elements of Φ_a are called well-formed algebraic formulae in the model.

Set Formulae

The proposed model allows for defining set theory operations like sum, intersection, complement, etc. A set is a crucial element of these operations and it can contain numbers, symbols or even mixed types of elements. Each statement that can be interpreted as a set is called a *set formula*. In the context of the previously defined types of formulae, each basic formula can be considered as a special case of set formula. This stems from the assumption in which a single element is treated as a singleton. This is why, each algebraic formula can also be considered as a special case of a set formula. In turn, the quantum formulae constitute a simple set formulae while more complex ones can be formed using set theoretic operators (belonging to $\mathbb{P}_s$). Thus, the formal definition of all well-formed set formulae Φ_s can be as follows:

1. $\Phi_a \subseteq \Phi_s$;
2. If ϕ_s is a set formula, then $(\phi_{s,})^{\oslash} \in \Phi_s$;
3. If $\phi_{s,1}$ and $\phi_{s,2}$ are set formulae, then $(\phi_{s,1} \cup \phi_{s,2})$, $(\phi_{s,1} \cap \phi_{s,2})$, $(\phi_{s,1} \setminus \phi_{s,2})$, $(\phi_{s,1} \triangle \phi_{s,2})$, $(\phi_{s,1} \otimes \phi_{s,2}) \subset \Phi_s$.

Where:

- $\oslash$ is the *complement* operator in the model.
- $\cup, \cap, \setminus$ are the *union*, *intersection* and *set difference* operators in the model.
- $\triangle$ is the *symmetric difference* operator in the model.
- $\otimes$ is *Cartesian Product* operator in the model.

The elements of Φ_s are called well-formed set formulae in the model.

Relation Formulae

Relation formulae allow for comparing values of the evaluation of algebraic and set formulae. In comparison to the previously discussed types of formulae, relation formula is always complex. It must be defined with the help of a certain operator (belonging to $\mathbb{P}_r$) which determines the type of relation (comparison). What is more, each relation formula can be evaluated to *true* or *false*. Considering the possible types of arguments for relational operators, among the set of all relation formulae Φ_r, two subsets of special types of relations formulae can be distinguished. First of them contains algebraic relation formulae Φ_{ra} ($\Phi_{ra} \subseteq \Phi_r$) that provide comparisons of algebraic formulae evaluations. In turn, the second subset, referred as Φ_{rb}, contains relation formulae consisting of only basic formulae. Moreover, the set of relational operators $\mathbb{P}_r$, can also be divided into subsets of operators that can be used with given types of relation formulae. $\mathbb{P}_{ra}$ is the first subset of $\mathbb{P}_r$ and contains operators that can be used for defining formulae belonging to the Φ_{ra} set. The $\mathbb{P}_{ra}$ set contains the following operators $\mathbb{P}_{ra} = \{\ominus, \oslash, \odot, \oslash, \odot, \oslash\}$. In turn, the basic relation formulae can be defined using relational operators belonging to the $\mathbb{P}_{rb}$ set and $\mathbb{P}_{rb} = \{\ominus, \oslash\}$. The remaining part of the $\mathbb{P}_r$ set $\{\in, \oslash, \odot, \oplus, \ominus, \oplus, \ominus, \otimes\}$ can be used in every case. In case when an argument is a single element then it is treated as singleton. In order to summarize, the definition of well-formed relation formulae can be expressed as follows:

1. $\Phi_r \subseteq \Phi_b$;
2. If $\phi_{a,1}$ and $\phi_{a,2}$ are algebraic formulae i.e. $\phi_{a,1}, \phi_{a,2} \in \Phi_a$, then $(\phi_{a,1} \odot \phi_{a,2}), (\phi_{a,1} \otimes \phi_{a,2}), (\phi_{a,1} \odot \phi_{a,2}), (\phi_{a,1} \otimes \phi_{a,2}) \in \Phi_{ra}$;
3. If $\phi_{b,1}$ and $\phi_{b,2}$ are basic formulae i.e. $\phi_{b,1}, \phi_{b,2} \in \Phi_b$, then $(\phi_{b,1} \ominus \phi_{b,2})$, $(\phi_{b,1} \otimes \phi_{b,2}) \in \Phi_{rb}$;
4. $\Phi_{ra} \subseteq \Phi_r$;
5. $\Phi_{rb} \subseteq \Phi_r$;
6. If ϕ_b is a basic formula such that $\phi_b \in \Phi_b$, ϕ_s is a set formulae such that $\phi_s \in \Phi_s$, then $(\phi_b \odot \phi_s), (\phi_b \odot \phi_s) \in \Phi_r$;
7. If $\phi_{s,1}, \phi_{s,2} \in \Phi_s$ are set formulae, then $(\phi_{s,1} \ominus \phi_{s,2}), (\phi_{s,1} \otimes \phi_{s,2}), (\phi_{s,1} \odot \phi_{s,2})$, $(\phi_{s,1} \oplus \phi_{s,2}), (\phi_{s,1} \ominus \phi_{s,2})$, $(\phi_{s,1} \oplus \phi_{s,2}), (\phi_{s,1} \ominus \phi_{s,2}), (\phi_{s,1} \otimes \phi_{s,2}) \in \Phi_r$.

Where:

- $\odot, \oslash, \odot, \oslash$ are the *lower than, lower than or equal, greater than, greater than or equal* operators.
- $\ominus, \oslash$ are the *equality* and *inequality* operators in the model.
- $\in, \oslash$ are the *membership* and *non-membership* operators in the model.
- $\odot, \oplus, \ominus, \oplus$ are the *subset, not subset, superset, not superset* operators in the model.
- $\ominus, \otimes$ are the *similarity* and *non-similarity* operators that check if two sets have an empty intersection or not.

On the first sight it may seem that a relation formula cannot contain other relation formulae. However, it is important to notice that according to the points 1 and 3 of the provided definition, a relation formula can contain basic formula while each relation formula belongs to the Φ_b set. Thus, assuming that $\phi_{b,1}, \phi_{b,2}, \phi_{b,3}, \phi_{b,4}$ are basic formulae, the following formula is considered as well-formed: $(\phi_{b,1} \otimes \phi_{b,2})$ $\ominus (\phi_{b,3} \ominus \phi_{b,4})$.

Logical Formulae

The main goal of the logical operators $\mathbb{P}_l$ is to allow for combining relation formulae and other logical formulae. These operators play a role of connectives between these types of formulae. The formulae built with the help of logical operators are called *logical formulae*. In comparison to relation formulae, each logical formula also belongs to the set of basic formulae Φ_b as well as is also evaluated to *true* or *false*. Nevertheless, in contrast to relation formulae, a certain logical formula can be composed of only those types of formulae whose codomain is equal to $\{true, false\}$ set. In this approach a set of well-formed logical formulae is defined as follows:

1. $\Phi_l \subseteq \Phi_b$;
2. If ϕ is relation or logical formula i.e. $\phi \in \Phi_r \cup \Phi_l$, then $(\ominus \phi) \in \Phi_l$;
3. If ϕ_1, and ϕ_2 are relation or logical formulae i.e. $\phi_1, \phi_2 \in \Phi_r \cup \Phi_l$, then $(\phi_1 \otimes \phi_2)$, $(\phi_1 \oslash \phi_2) \in \Phi_l$.

Where:

- $\ominus$ is the *negation* operator in the model.
- $\otimes, \oslash$ are the *and* and *or* operators.

The elements of Φ_l are called well-formed logical formulae in the model.

Pattern Formulae

The main goal of *pattern formulae* is to define a set of facts that satisfy a given condition. In comparison to XTT2, this type of formulae is a new element. The conditional part of the XTT2 rule uses a specific instance of an attribute of a given type and thus the provided condition concerns only this single instance. In turn, the pattern formulae allows for expressing conditions related to all instances of a given type. Therefore, each rule can be executed several times depending on the number of facts that satisfy the condition.

This type of formula can be expressed by using pattern operators that belong to the $\mathbb{P}_p$ set. It is important to emphasize that the facts that satisfy the provided condition do not have to belong to the $\mathbb{F}$ set, but they can belong to a subset of $\mathbb{F}$ or even can be obtained from different sources e.g. from the previous states of the system. This is an important feature of these formulae because it has a crucial meaning in the context of expressing dynamic constraints within the model. This formula is composed of logical formula that specifies the condition that must be satisfied by all the facts belonging to the set defined by the pattern formula. The set of all pattern formulae Φ_p can be defined in the following way: If S is a certain set of facts, T is an existing type from the $\mathbb{T}$ set and $\phi_{l,}$ is a logical formula that belongs to the Φ_l set, then $(\mathcal{P}(S, T, \phi_{l,})) \in \Phi_p$. Where $\mathcal{P}$ is the *pattern* operator in the model that defines a set of facts that belong to the S set, are of type T and satisfy the $\phi_{l,}$ formula. The elements of Φ_p are called well-formed pattern formulae in the model.

Example 8.3.2 *In the considered example, one can distinguish an expression that can be treated as pattern formulae in terms of the proposed model. This expression compose* LHS*of the exemplary rule:* `Car($Capacity : Capacity, Capacity < 900)`. *This expressions defines a subset of facts satisfying the provided condition – this is consistent with the semantics of pattern formulae and thus each of these expressions can be written as pattern formulae in the following way*

$$Car(Capacity < 900) \text{ corresponds to } \mathcal{P}\left(\mathbb{F}, T_{Car}, A_{Capacity}(F) \ominus 900\right)$$

$$(8.31)$$

It is worth to notice that this expression refers to the $\mathbb{F}$ set as the source of elements from which the facts satisfying provided logical formula are selected. This is because, the discussed rule languages do not provide any mechanism making other sources of facts available for reasoning. In this context, the proposed model provides a significant extension and allows for the definition of the fact source within the pattern formula. Therefore, it is possible to define pattern formula that checks facts from the previous state of the system against the provided condition: $\mathcal{P}\left(S_{c-1}, T_{Car}, A_{Capacity}(F_o) \ominus 900\right)$.
One can also notice the following equivalence:

$$\mathcal{P}\left(S_c, T_{Car}, A_{Capacity}(F_o) \ominus 900\right) \equiv \mathcal{P}\left(\mathbb{F}, T_{Car}, A_{Capacity}(F_o) \ominus 900\right).$$

Constraint Formulae

The pattern formulae allow for defining a set of facts that satisfy the provided condition. Thus, it can be said, that this type of formula works on the fact level. In comparison to pattern formulae, *constraint formulae* work on the set of facts level and allow for looking at a set of facts as a whole. In this way, this type of formulae allows for specifying constraints with the help of constraint operators ($\mathbb{P}_c$) which corresponds to existential or universal quantifiers that are known from FOL. These constraints allow for checking if a given set of facts contains a specific element or all elements of a given set satisfy a certain condition. This type of formula can be evaluated to *true* or *false* values and it is built over a single pattern formula. The set of all constraint formulae is denoted as Φ_c and is defined as follows: If $\phi_{p,}$ is a pattern formula i.e. $\phi_p, \in \Phi_p$, then $(\exists(\phi_{p,})), (\not\exists(\phi_{p,})), (\vee(\phi_{p,})) \in \Phi_c$. Where:

- $\exists$ is the *existence* operator that checks if the set defined by the pattern formula contains at least one element.
- $\not\exists$ is the *non existence* operator in the model that checks if the set defined by the pattern formula is empty.
- $\vee$ is the *universal* operator in the model that checks if the set defined by the pattern formula contains all the facts of type T that belong to the S set.

The elements of Φ_c are called the well-formed constraint formulae in the model.

Knowledge Formulae

According to the source of origin of the facts, the $\mathbb{F}$ set can be divided into two disjoint subsets. This first subset contains all the knowledge that is assumed to be true, i.e. this subset is composed of axioms. This subset can change during the inference process, in particular some of the axioms can be removed from $\mathbb{F}$. Of course the inference process can also add a new fact to the $\mathbb{F}$ set. Nevertheless, all the newly added facts belong to the second subset that contains facts that are inferred to be true basing on the already existing knowledge and rules. These modifications of the set of facts $\mathbb{F}$ can be done by using knowledge operators belonging to the $\mathbb{P}_k$ set. The formulae build with the help of these operators are called *Knowledge Formulae*. The set of all knowledge formulae is denoted as Φ_k, and defined as follows:

1. If T is a certain type such that $T \in \mathbb{T}$, c is a constant value such that $c \in I_{\mathbb{T}}(2^T)$ and $|c| = 1$, $L \subset \mathbb{L}$ (see Definition (8.21)), then $(\mathcal{A}(T, c, L)) \in \Phi_{k,}$;
2. If F is a set of elements belonging to $\mathbb{F}$ i.e. $F \subset \mathbb{F}$, then $(\mathcal{R}(F)) \in \Phi_{k,}$;
3. If o is an object such that $o \in \mathbb{O}$ and it is of a certain type T such that $T \in \mathbb{T}$ and $\phi_{s,}$ is a well-formed set formula that can be evaluated to element of $I_{\mathbb{T}}(2^T)$, then $(o \ominus \phi_{s,}) \in \Phi_{k,}$;
4. If $f_{o,i}$ is a fact triple such that $f_{o,i} \in F_o \subset \mathbb{F}$ and $f_{o,i} = (o, A_i, L_{f_{o,i}})$ and $\phi_{s,}$ is a set formula such that $\phi_{s,} \in \Phi_s$ and $\phi_{s,}$ can be evaluated to element of attribute domain D_{A_i}, then $(A_i(F_o) \ominus \phi_{s,}) \in \Phi_{k,}$;
5. If v is a variable such that $v \in \mathbb{V}$ and $\phi_{s,}$ is a well-formed set formula, then $(v \ominus \phi_{s,}) \in \Phi_{k,}$.

Where:

- $\mathcal{A}$ is the *assert* operator in the model that allows for adding a new fact to knowledge base.
- $\mathcal{R}$ is the *retract* operator in the model that allows for removing fact(s) from the knowledge base.
- $\ominus$ is the assignment operator in the model.

The elements of $\Phi_{k,}$ are called well-formed knowledge formulae in the model.

This section introduced the most important types of formulae supported by the model. It also has provided syntax of supported operators. The following section provides formal definition of operators and formulae semantics.

8.4 Rule Level

Rules are the most important elements within the provided model of the rule-based system because they allow for manipulating existing knowledge. Considering such a system as a dynamic one, rules are the element that define the dynamics of this system by specifying possible transitions between system states. The proposed model considers a rule as a transition function that is defined with the help of knowledge formulae that determine a new system state according to the previous ones. Each single rule defines a single transition that can be made always when a certain condition is satisfied. This condition is defined as the domain of a rule function.

Rule Definition

This model provides a set of all rules Γ that is defined as:

$$\Gamma = \{r_i \mid i \in \mathbb{X}_\Gamma\} \tag{8.32}$$

where r_i is a single rule and $\mathbb{X}_\Gamma$ is a set of all rule identifiers.

This model defines a rule as a transition function that maps a combination of facts into a new system state. To be more precise, assuming that r is a single rule such that $r \in \Gamma$, the definition of rule function can be written as follows:

$$r : \Gamma \times \mathbb{H}^n \to \Gamma \times \mathbb{F}^1 \times \mathbb{I}^{1m} \tag{8.33}$$

where:

- $\mathbb{H}$ is a system history set (see Definition (8.27)).
- n is a value of exponent of Cartesian power such that $n \geqslant 0$ and $\mathbb{H}^n = \underbrace{\mathbb{H} \times \mathbb{H} \times \cdots \times \mathbb{H}}_{n}$.
- $\mathbb{F}^1$ is the set of all facts modified by rule r.
- $\mathbb{I}^{1m}$ is the set of all interpretation functions modified by rule r, where $m \geqslant 0$ and is a number of assignment operations performed within rule.

The arity of a rule depends on the value of n. In a particular case, n can be equal to zero. Then, such a rule is called an *unconditional rule* and it can be used only once.

Assuming that $\Phi_{k,\mathcal{A}}$, $\Phi_{k,\mathcal{R}}$ and $\Phi_{k,\mathbb{I}}$ are subset of well-formed knowledge formula Φ_k, containing formulae expressed by $\mathcal{A}$, $\mathcal{R}$ and $\ominus$ operator, respectively, M and M^1 are subsets of Γ and $H = \{H_1, H_2, \ldots, H_n\}$ is a set of some facts belonging to $\mathbb{H}$, the rule r can be defined in the following way:

$$r(M, H_1, H_2, \ldots, H_n) = (M^1, \mathbb{F}^1, \mathbb{I}^m) \overset{def}{=} \begin{cases} M^1 \\ \mathbb{F}^1 = \mathbb{F} \cup \left(\bigcup_{\phi_k \in \Phi_{k,\mathcal{A}}} \left\{ I_{\Phi_{k,\mathbb{F}}}(\phi_k(H^{i_{\phi_k}})) \right\} \right) \\ \setminus \left(\bigcup_{\phi_k \in \Phi_{k,\mathcal{R}}} \left\{ I_{\Phi_{k,\mathbb{F}}}(\phi_k(H^{i_{\phi_k}})) \right\} \right) \\ \mathbb{I}^m = \mathbb{I} \setminus \mathbb{I}_{\ominus} \cup \left(\bigcup_{\phi_k \in \Phi_{k,\mathbb{I}}} \left\{ I_{\Phi_{k,\mathbb{I}}}(\phi_k(H^{i_{\phi_k}})) \right\} \right) \end{cases} \tag{8.34}$$

where:

- M^1 defines a subset of rules that are taken into account during the next iteration of an inference algorithm.
- $\phi_k(H^{i_{\phi_k}})$ means that within each knowledge formula ϕ_k any combination of elements of the H set may occur. i_{ϕ_k} is the value of the exponent of Cartesian power of the H set which defines a number of elements from the H set that occurs in the formula ϕ_k and $0 \leqslant i_{\phi_k} \leqslant n$.
- m is the number of assignment formulae within the $\Phi_{k,\mathbb{I}}$ set i.e. $m = |\Phi_{k,\mathbb{I}}|$.
- The definition of $\mathbb{I}_{\ominus}$ is related to a general knowledge interpretation function (see [9]),
- $I_{\Phi_{k,\mathbb{I}}} \in \mathbb{I}$.

Considering this definition one of the tasks of inference mechanism is to instantiate each rule according to its definition by finding all the tuples that belong to the $\mathbb{H}^n$. In this step, an inference engine may produce a great number of rule instantiations e.g. having 3 states containing 10 facts each and a rule taking two elements, an inference engine should produce about 900 instantiations of only this single rule. On the other side, it is not necessary that each rule is instantiated every time and for each fact permutation. Thus, it is assumed that a rule can be instantiated only in a specific situation i.e. if some conditions are satisfied. This is why, each rule is related to a set of conditions that must be satisfied in order to permit its instantiation. This model allows for specifying conditions as the domain and constraints of the rule arguments.

The domain for a single rule argument can be specified using pattern formulae while constraints using constraint formulae. Thus, for the rule defined by Formula (8.34), the domain can be written as follows:

$$\left((H_1 \in \phi_{p,1}) \otimes (H_2 \in \phi_{p,2}) \otimes \ldots \otimes (H_n \in \phi_{p,n}) \otimes (\phi_{c,1}) \otimes (\phi_{c,2}) \otimes \ldots \otimes (\phi_{c,r})\right)$$

$$(8.35)$$

where:

- $\forall_{i \in \{1,2,\ldots,n\}} \phi_{p,i} \in \Phi_p$.
- n is the arity of a rule.
- $\forall_{i \in \{1,2,\ldots,r\}} \phi_{c,i} \in \Phi_c$.
- r is a number of additional (besides domains) constraints of rule arguments and $r \geqslant 0$.
- Within each pattern and constraint formulae any subset of facts being rule arguments may occur.

A complete rule definition must provide all of the following elements: specification of domains and constraints for arguments, specification of formulae that modify system state and formula that defines a subset of rules for the next evaluation. Definitions of all these elements are provided by Definitions (8.34) and (8.35). Thus, in order to improve their readability, the following notation combines both of them:

$$r(M, H_1, \ldots, H_n) = \text{if} \begin{cases} H_1 \in \phi_{p,1} \\ \ldots \\ H_n \in \phi_{p,n} \\ \phi_{c,1}(H^{i_1}) \\ \ldots \\ \phi_{c,r}(H^{i_r}) \end{cases} \text{then} \begin{cases} \phi_{k1}(II^{j_1}) \\ \ldots \\ \phi_{ks}(H^{j_s}) \\ \{ M^1 \end{cases}$$

$$(8.36)$$

where $\forall_{i \in \{1,2,\ldots,n\}} \phi_{p,i} \in \Phi_p$ and $\forall_{i \in \{1,2,\ldots,i_r\}} \phi_{c,i} \in \Phi_c$ and $\forall_{i \in \{1,2,\ldots,j_s\}} \phi_{ki} \in \Phi_{k,}$. Within each pattern, constraint and knowledge formulae any combination of facts being rule arguments may occur.

Example 8.1 In order to make the rule definition more clear, this example provides specification of the rule from the case study, that is considered at the beginning of this section, expressed in terms of the model. Keeping in mind all the exemplary pattern and considered knowledge formulae it is possible to write the following rule:

$$r_{base-charge1}(M, H_1) = \text{if} \left\{ H_1 \in \mathcal{P}\left(\mathbb{F}, T_{Car}, A_{Capacity}(H_1) \otimes 900\right) \text{ then } \begin{cases} \mathcal{A}\left(T_{Base}, (537), \varnothing\right) \\ \mathcal{A}\left(T_{Result}, (537), \varnothing\right) \\ \{ M \end{cases} \right.$$

$$(8.37)$$

Having this rule, the set of all rules can be defined in the following way: $\Gamma = \{r_{base-charge1}\}$. Analogously, the set of all rule identifiers contains the following elements: $\mathbb{X}_r = \{base - charge1\}$. Another important issue is related to the domain of this rule. The domain is determined by the set specified by the pattern formulae.

According to the interpretation of this formula, the domain of the $r_{base-charge1}$ rule can be written as follows: $\left(H_1 \in \{F_{o_{vw}}\}\right)$. Therefore, the inference engine can create only one instance of this rule for $F_{o_{vw}}$ fact.

In the further part of this chapter, this notation for rules representation is used. In turn, the interpretation of a rule is provided together with the operational semantics of the modules (see Sect. 8.5).

Rule Priority

All of the considered rule languages use a concept of rule priority. In all of them, besides XTT2, this feature can be specified explicitly, while XTT2 uses it implicitly. A priority feature is mainly used by the inference engine in order to determine the order of rules within a conflict set. This is a significant feature and this is why this model provides a corresponding feature in the form of the function that maps a given rule into a real number. This function is denoted as π and is defined as follows: $\pi : \Gamma \rightarrow \mathbb{R}$. It is worth to emphasize that the definition of value of this function for a given rule must be provided directly e.g. $\pi(r) = a$. There is no way to calculate priority according only to rule definition. However, it is assumed that if there is no information about priority of a given rule, then zero value is used as default.

The interpretation of rule priority, provided by this model, assumes that the rule priority is higher the higher is the value of the π function. Thus considering two rules r_1 and r_2, it can be said that rule r_1 has higher priority that r_2 if and only if $\pi(r_1) > \pi(r_2)$.

8.5 Modules and Structure

Modules allow for building rule bases that have an internal structure. Such a structure increases the maintainability of a rule base as modules allow for grouping rules working together. Moreover, structure allows for more advanced inference control as the engine can process only one module at a time.

XTT2 and the majority of the rule-based tools support building structured rule bases. In XTT2 the concept of a module corresponds to a single decision table which groups rules having the same attributes in the conditional and decision parts (i.e. the same schema). Thanks to that, a single decision table contains all the rules that work in the same context and thus allows for providing logical verification of the rule base against such anomalies like redundancy, inconsistency, subsumption, etc.

In the case of XTT2 as well as such tools like DROOLS, JESS, OPENRULES, etc. rule base structure allows for advanced inference control. In comparison to flat rule bases (without structure) only rules belonging to a processed module can be activated and executed. Therefore, the order of the rule evaluation and execution can be partially determined by the order of module evaluation.

Apart from the concept of a module, the proposed model introduces the concept of a submodule. Thanks to submodules, the model allows for defining more advanced structures of a rule base that allow for expressing such features like `lock-on-active` or `activation-group`.[3]

Set of All Modules

The Γ set (all rules) can be divided into several non-empty pairwise disjoint subsets that are called modules: $\Gamma = M_1 \cup M_2 \cup \cdots \cup M_n$, where M_i is a non-empty set called *module*.

According to the provided definition a single module is a set of rules. In turn, M_i belongs to the ordered set of all modules $(\mathbb{M}, <_{\mathrm{M}})$, where the $<_{\mathrm{M}}$ relation defines the order of the modules. Using this relation the min_{M} and max_{M} functions that determine the minimal and maximal element of $(\mathbb{M}, <_{\mathrm{M}})$, respectively, can be defined: The min_{M} function is important in the context of modules evaluation order. Based on this function, the $next_{\mathrm{M}}$ function, that in terms of the provided ordering relation determines the next module with respect to a certain module M_i, can be defined in the following way: $next_{\mathrm{M}}(M_i) = M_j \Leftrightarrow M_j = min_{\mathrm{M}}\left((\mathbb{M}\backslash\{M_1, M_2, \ldots, M_k, M_i\}, <_{\mathrm{M}})\right)$, where $\{M_1, M_2, \ldots, M_k\}$ is the set of all modules belonging to $(\mathbb{M}, <_{\mathrm{M}})$ such that $\forall_{M \in \{M_1, M_2, \ldots, M_k\}} M <_{\mathrm{M}} M_i$.

In turn, the max_{M} function refers to the maximal, in terms of $<_{\mathrm{M}}$, element of the set of all modules that is always referred as M_a and is called an *autofocus* module. This is the only one module that can be empty and contains rules that simultaneously must belong to any of the remaining modules. This module has a special purpose that is related to the *autofocus* rule property supported by the considered rule languages.

Example 8.5.1 *According to the use case example provided at the beginning of this section, especially line 16th, the $M_{base-charge}$ module can be defined. Therefore, $\mathbb{X}_{\mathrm{M}} = \{base - charge\}$ and $(\mathbb{M}, <_{\mathrm{M}}) = \{M_{base-charge}, M_a\}$, where it is assumed that the order is defined in the following way: $M_{base-charge} <_{\mathrm{M}} M_a$. Considering the assignment of a rule to the module, that is specified in the mentioned line, it is possible to define the content of the module: $M_{base-charge} = \{r_{base-charge1}\}$ and $M_a = \varnothing$.*

Structure of a Single Module

In order to provide support for other features of rules like *lock-on-active* or *activation-group*, each single module has its own structure. Each module is divided into several

---

[3]These features are provided by the DROOLSrule language and allows for changing the default inference flow. In the proposed model they can be expressed with the help of `loc-on-active` and `xor` submodules.

submodules that correspond to these features of rules. This is why, a structure of a single module M_i is defined as follows: $M_i = M_i^o \cup M_i^l \cup M_i^{\oplus_1} \cup M_i^{\oplus_2} \cup \cdots \cup M_i^{\oplus_n}$, where:

- M_i^o is called *ordinary* submodule and contains only rules that can be processed in an ordinary way.
- M_i^l is called *lock-on-active* submodule[4] and contains rules that can be instantiated only by facts that are not modified by other rules from the module M_i.
- $M_i^{\oplus_j}$ is called *xor* submodule[5] and contains rules that can be activated only when there is no pending activation of rule from the submodule $M_i^{\oplus_j}$. In contrast to *lock-on-active* submodule, there can be more than one *xor* submodule within single module M_i.
- n is the number of *xor* submodules.
- It is assumed that an *ordinary* submodule is disjoint with all the remaining submodules, while all the submodules, besides *ordinary* one, can contain the same rules i.e. they are allowed to have a non-empty intersection.

Example 8.5.2 *Considering the provided case study, one can notice that the $r_{bonus-malus0}$ rule provides* `lock-on-active` *property (see line 23rd) while the $r_{base-charge1}$ rule does not provide any structure-specific features (besides module assignment). Therefore in both cases, structures of the already defined modules (see Example (8.5.1)) are composed of only a single submodule containing one rule (the remaining submodules are empty):*

$$M_{base-charge} = M_{base-charge}^o = \{r_{base-charge1}\} \qquad (8.38)$$

The next part of this section shows how the modules and submodules should be interpreted by an inference algorithm. This is done by providing *operational semantics* of the modules.

Operational Semantics of Modules and Submodules

Modules provide an additional level of abstraction for the knowledge stored within the knowledge base. This level defines the structure of knowledge base that cannot be omitted by efficient interoperability methods. In turn, all of the considered rule languages use some variation of modules. Nevertheless, in every case, a small change in the structure of the knowledge base may cause a different response of the inference. Therefore, the operational semantics of rules belonging to modules and submodules supported by the model must be provided. In order to do that, the following notation is introduced:

---

[4]This submodule corresponds to the *lock-on-active* feature supported by DROOLS.
[5]This submodule corresponds to the *activation-group* feature supported by DROOLS.

- $E(\{r_1, r_2, \ldots, r_n\})$ – is the function that returns the set of rules that were executed. The returned set must be a subset of the set provided as an argument.
- $E(\Gamma_i) \triangleright E(\Gamma_{i+1}) \triangleright \ldots \triangleright E(\Gamma_{i+m})$ – indicates the order of rule executions. It means that rules belonging to the Γ_i set must be executed before rules from the Γ_{i+1} set. In turn, rules from the Γ_{i+1} set must be executed before rules from the Γ_{i+2} set, etc. It is important to notice that $\triangleright$ operator defines execution order only in terms of the rules belonging to the $\Gamma_i \cup \Gamma_{i+1} \cup \cdots \cup \Gamma_{i+m}$ set. Therefore, it is possible that between execution of these rules, other rules, that do not belong to sum of these sets, may be executed as well.
- $\frac{r}{E(M_i) \triangleright E(M_{i+1}) \triangleright \ldots \triangleright E(M_{i+m})}$ – means that the execution of the r rule causes execution of rules from the M_i module (set) at first, later executions from the M_{i+j} modules (sets), where $j \in \{1, 2, \ldots, m - 1\}$, and execution from the M_{i+m} module (set) at the end. The order of rules evaluation (and execution) provided by the operational semantics of rules, has priority over order provided by the $<_M$ operator. This is because this operator defines only the default order that can be modified by rules.

Using this notation the operational semantics of rules belonging to a particular submodule can be introduced.

An *ordinary* submodule does not impose any additional restrictions on the rule execution. Therefore, after a rule execution all the rules belonging to the M_c module, besides those in a *lock-on-active* submodule, can be activated and executed. In this case, the operational semantics of rules belonging to an *ordinary* submodule can be defined in the following way:

$$\frac{r \in M_c^o \subset M_c}{E(M_c \setminus M_c^l) \triangleright E(next_M(M_c)) \triangleright \ldots \triangleright E(next_M(\ldots next_M(M_c) \ldots))} \quad (8.39)$$

Moreover, each submodule can contain rules having different priorities. In such a case, rules with the highest value of priority must be executed at first. Assuming that the highest value of rule priority within module M_c is equal to n, while the lowest is equal to $n - m$, the module can be divided in terms of rules priorities in the following way: $M_c = M_{cn} \cup M_{cn-1} \cup \cdots \cup M_{cn-m}$, where each M_{cn-i} submodule contains rules having the same value of priority that is equal to $n - i$. In this case, the operational semantics of rule belonging to M_c can be defined as follows:

$$\frac{r \in M_c}{E(M_{cn} \setminus M_c^l) \triangleright E(M_{cn-1} \setminus M_c^l) \triangleright \ldots \triangleright E(M_{cn-m} \setminus M_c^l) \triangleright E(next_M(M_c)) \triangleright \ldots} \quad (8.40)$$

The issue related to rule priority can be considered for every submodule and their combinations. However, such discussion is analogous and for the sake of transparency it is omitted in the remaining considerations.

Rules that belong to a *lock-on-active* submodule can be activated and executed only against state of the system at the moment when their module becomes current. Therefore, any further modifications made by rules belonging to a current module cannot activate these rules. This scenario can already be observed in the definition of operational semantics of rules from an *ordinary* submodule. Considering Definition (8.39), it can be noticed that execution of a rule from an *ordinary* submodule cannot activate rules from *lock-on-active* submodule. However, rules belonging to an *lock-on-active* submodule of a next module may be activated and executed against modifications made by this rule. Operational semantics of rules from an *lock-on-active* submodule, can be defined in the same way as the operational semantics of rules belonging to an *ordinary* submodule:

$$\frac{r \in M_c^l \subset M_c}{E(M_c \backslash M_c^l) \rhd E(next_M(M_c)) \rhd \ldots \rhd E(next_M(\ldots next_M(M_c) \ldots))} \quad (8.41)$$

A certain module can provide more than one *xor* submodule. Rules belonging to a certain *xor* submodule can be activated only when there is no pending activation of other rule from the same *xor* submodule. Therefore, each instantiated and not executed rule belonging the *xor* submodule, is rejected from activation (i.e. is not added to a conflict set) if the conflict set $(CS, <_{CS})$ contains other rule from this submodule. Taking this fact into account, a definition of an operational semantics of rules belonging to a *xor* submodule must consider two cases. In the first case, where a certain rule from the same *xor* submodule is already in the conflict set, the operational semantics can be written in the following way:

$$\frac{\left(r \in M_c^{\oplus i} \subset M_c\right) \wedge \left(M_c^{\oplus i} \cap (CS, <_{CS}) \neq \varnothing\right)}{E(M_c \backslash (M_c^{\oplus i} \cup M_c^l)) \rhd E(next_M(M_c)) \rhd \ldots \rhd E(next_M(\ldots next_M(M_c) \ldots))} \quad (8.42)$$

In turn, the second case is related to situation where the conflict set does not contain any rule belonging to the same *xor* submodule as the execution rule. In this case, the operational semantics can be defined as follows:

$$\frac{\left(r \in M_c^{\oplus i} \subset M_c\right) \wedge \left(M_c^{\oplus i} \cap (CS, <_{CS}) = \varnothing\right)}{E(M_c \backslash M_c^l) \rhd E(next_M(M_c)) \rhd \ldots \rhd E(next_M(\ldots next_M(M_c) \ldots))} \quad (8.43)$$

The provided definitions of operational semantics of rules belonging to the supported submodules precisely define an impact of a certain rule execution on activations and executions of other rules as well as on the system state. Moreover, these definitions are later used to prove the correctness of semantically equivalent transformations of rule base structure.

8.6 Complete Knowledge Base

Knowledge base is the last element of this formalization. It groups all other already defined elements like: set of types, rules, interpretations, etc. Thus knowledge base $\mathbb{K}$ is defined as the following tuple:

$$\mathbb{K} \stackrel{def}{=} (\Omega, \mathbb{I}, \mathbb{X}, \mathbb{T}, \mathbb{O}, \mathbb{F}, \mathbb{S}, \mathbb{C}, \mathbb{V}, \Gamma, (\mathbb{M}, <_\mathbb{M}), \mathbb{P}, \Phi) \qquad (8.44)$$

where:

- Ω is the universum set.
- $\mathbb{I}$ is the set of all interpretation functions.
- $\mathbb{X}$ is the set of all identifiers.
- $\mathbb{T}$ is the set of all types, see Definition (8.1).
- $\mathbb{O}$ is the set of all objects, see Definition (8.2).
- $\mathbb{F}$ is the set of all facts, see Definition (8.18).
- $\mathbb{S}$ is the sequence of all system states, see Definition (8.26).
- $\mathbb{C}$ is the set of all constants, see Definition (8.17).
- $\mathbb{V}$ is the set of all variables, see Definition (8.28).
- Γ is the set of all rules, see Definition (8.32).
- $(\mathbb{M}, <_\mathbb{M})$ is the ordered set of all modules.
- $\mathbb{P}$ is the set of all operators, see Definition (8.30).
- Φ is the set of all formulae.

Example 8.6.1 *Considering all examples the provided in this section and the defined operational semantics, it is possible to check that during the inference the $r_{base-charge1}$ rule is executed. Therefore, the final state of the system differs from the initial one by two new facts asserted by this rule. Furthermore, the $\mathbb{S}$ tuple takes the following form: $\mathbb{S} - (S_0, S_1)$, where $S_c = S_1$ and:*

$$S_0 = \left\{ (o_{vw}, A_1, \{\{f_{o_{vw},1}\}\}), (o_{vw}, A_2, \{\{f_{o_{vw},2}\}\}), (o_{vw}, A_3, \{\{f_{o_{vw},3}\}\}), \right.$$
$$\left. (o_{vw}, A_4, \{\{f_{o_{vw},4}\}\}), (o_{vw}, A_5, \{\{f_{o_{vw},5}\}\}), (o_{vw}, A_6, \{\{f_{o_{vw},6}\}\}) \right\}$$
$$S_1 = \left\{ (o_{vw}, A_1, \{\{f_{o_{vw},1}\}\}), (o_{vw}, A_2, \{\{f_{o_{vw},2}\}\}), (o_{vw}, A_3, \{\{f_{o_{vw},3}\}\}), \right.$$
$$(o_{vw}, A_4, \{\{f_{o_{vw},4}\}\}), (o_{vw}, A_5, \{\{f_{o_{vw},5}\}\}),$$
$$(o_{vw}, A_6, \{\{f_{o_{vw},6}\}\}), (o_{base}, A_1, \{\{f_{o_{base},1}\}\}),$$
$$\left. (o_{result}, A_1, \{\{f_{o_{result},1}\}\}) \right\}$$

$$(8.45)$$

It is important to notice that the defined knowledge base does not provide any inference mechanisms, any explanation features, etc. It can be used to provide an unified description of the knowledge semantics. In this way, it can be treated as a principle element that allows for the definition of an efficient rule interoperability method preserving knowledge semantics.

8.7 Summary

In this chapter, a formalization of the production rule representation model was presented. Using it we demonstrate that the execution of each rule, and thereby final conclusion generated by the inference engine, is determined by the structure of a rule base. Therefore, the proposed model introduces a novel approach to rule interoperability that takes rules operational semantics into account. The model originates from XTT2 formalization and therefore it is compatible with this method and constitutes its super set. Hence, one of the important advantages of the proposed model is the possibility to use verification and validation (V&V) methods developed for XTT2 [11]. Of course, these methods can be used only for those models that can be completely translated to XTT2(e.g. DROOLSis more expressive than XTT2) or for only a part of the model in the case when its expressiveness goes beyond XTT2. Nevertheless, this approach opens opportunities for the formal verification of DROOLS, CLIPS or JESS models with the help of the XTT2 tools. This chapter focuses on the most important aspects of the formalization of the model. Its operationalization with supporting tools, as well as its use for knowledge interchange will be presented in Chap. 10.

References

1. Winskel, G.: The Formal Semantics of Programming Languages: An Introduction. MIT Press, Cambridge (1993)
2. Riley, G.: CLIPS - A Tool for Building Expert Systems (2008). http://clipsrules.sourceforge. net
3. Friedman-Hill, E.: Jess in Action. Rule Based Systems in Java. Manning, Greenwich (2003)
4. Browne, P.: JBoss Drools Business Rules. Packt Publishing, Birmingham (2009)
5. Nalepa, G.J.: Semantic Knowledge Engineering. A Rule-Based Approach. Wydawnictwa AGH, Kraków (2011)
6. Kaczor, K., Nalepa, G.J., Łysik, Ł., Kluza, K.: Visual design of Drools rule bases using the XTT2 method. In: Katarzyniak, R., Chiu, T.F., Hong, C.F., Nguyen, N. (eds.) Semantic Methods for Knowledge Management and Communication. Studies in Computational Intelligence, vol. 381, pp. 57–66. Springer, Berlin (2011). https://doi.org/10.1007/978-3-642-23418-7
7. Kaczor, K., Kluza, K., Nalepa, G.J.: Towards rule interoperability: design of drools rule bases using the XTT2 method. Trans. Comput. Collect. Intell. XI **8065**, 155–175 (2013)
8. Nalepa, G.J., Ligęza, A., Kaczor, K.: Formalization and modeling of rules using the XTT2 method. Int. J. Artif. Intell. Tools **20**(6), 1107–1125 (2011)
9. Kaczor, K.: Knowledge formalization methods for semantic interoperability in rule bases. PhD thesis, AGH University of Science and Technology (2015) (Supervisor: Grzegorz J. Nalepa)
10. Ligęza, A.: Logical Foundations for Rule-Based Systems. Springer, Berlin (2006)
11. Nalepa, G., Bobek, S., Ligęza, A., Kaczor, K.: HalVA – rule analysis framework for XTT2 rules. In: Bassiliades, N., Governatori, G., Paschke, A. (eds.) Rule-Based Reasoning, Programming, and Applications. Lecture Notes in Computer Science, vol. 6826, pp. 337–344. Springer, Berlin (2011)

Part III
Practical Studies in Semantic Knowledge Engineering

The first part of the book was devoted to the discussion of important issues in the state of the art in the domain of rule-based systems. The important aspects we emphasized included formalized representation, visual design, and knowledge interoperability, as well as opportunities for integration with other systems. In the second part, we introduced a number of formalized models for representing knowledge in these systems and supporting their design. Furthermore, the models were oriented on the integration of rule-based systems with other formalisms, as well as rule interoperability.

The third part of the book is devoted to the presentation of a number of practical case studies that use methods from the Semantic Knowledge Engineering approach (SKE). The approach itself is introduced in Chap. 9. It builds on the models from the second part of the book, especially ALSV(FD), XTT2, that were introduced in Chap. 4, as well as ARD+ introduced in Chap. 6. SKE also provides a number of important tools for these languages, including the HeaRT rule engine for XTT2 rules.

The first case demonstrates how SKE can be of substantial help in providing rule interoperability. Based on the model introduced in Chap. 8 we develop interoperability techniques with rule-based system shells in Chap. 10. Then in Chap. 11, we discuss how the XTT2 rule representation can be used together with UML to design software. We follow the MDA paradigm to use the SKE design concepts. Furthermore, in Chap. 12 we continue with the SE domain, showing the application of XTT2 to the automation of test cases in unit testing. We employ decision table-based testing and extend this concept with formalization based on XTT2. We remain in the area of business applications in Chap. 13 where we present practical integration of business processes with rules encoded with XTT2. To achieve this goal the model previously introduced in Chap. 5 was used.

The two next chapters regard knowledge representation on the Web. In Chap. 14, we discuss the DAAL language that allows for an integration of ALSV(FD) with Description Logics. Thus, it opens up opportunities for using XTT2 rules in the

Semantic Web applications. Chap. 15 extends some of these ideas with the use of practical Web-based knowledge engineering tools namely Semantic Wikis. The Loki platform is introduced which provides a Collaborative Knowledge Engineering environment.

The next chapters describe two other case studies of the practical use of SKE these are closer to specific hardware platforms. In Chap. 16, we discuss the use of the SKE design and application of XTT2 rules to the design of control logic for mobile robots. Then in the final chapter (Chap. 17), we demonstrate how XTT2 rules can be put in operation on mobile platforms to support reasoning in context-aware systems. We use the model previously introduced in Chap. 7 to handle uncertainty, which is omnipresent on such platforms.

Chapter 9
Semantic Knowledge Engineering Approach

In this chapter we introduce to *Semantic Knowledge Engineering* approach. It is a development approach for Knowledge-Based Systems that uses rule-based knowledge representation. The core of the approach is the formalized rule representation method XTT2 that was introduced in Chap. 4. Some supporting design methods, such as ARD+ introduced in Chap. 6, allowed the formulation of the SKE design process. The motivation for the approach, along with its distinctive features is given in Sect. 9.1. Then the SKE design process for rule-based systems is presented In Sect. 9.2. SKE was developed to support a heterogeneous architecture of rule-based applications as discussed in Sect. 9.3. The approach is well supported by a number of software tools discussed in Sect. 9.4. Three main and practical aspects are the knowledge base design support discussed in Sect. 9.5, generation of the executable rule format using the HMR notation in Sect. 9.6, and execution of the rule-based system discussed in Sect. 9.7. Furthermore, tools for rule analysis are discussed in Sect. 9.8. The chapter is concluded in Sect. 9.9.

This work is related to several research projects. The first project was MIRELLA, as it provided the first version of the XTT method. The primary motivation and results were first delivered in the HEKATE project[1] including ALSV(FD) and ARD+. The objectives were to investigate the possible bridging of knowledge and software engineering. The next important project was BIMLOQ (Business Models Optimization for Quality)[2] whose objectives were to build a declarative model for business processes, including business rules specification, with an emphasis on analysis and optimization. In that project preliminary work on BP integration with semantic wikis was carried out. The next project was PARNAS[3] which was aimed at developing tools for inference control and quality analysis in modularized rule bases, and provided the final formalized version of XTT2.

---

[1] See http://hekate.ia.agh.edu.pl.

[2] See http://geist.re/pub:projects:bimloq:start.

[3] See http://geist.re/pub:projects:parnas:start.

© Springer International Publishing AG 2018

G.J. Nalepa, *Modeling with Rules Using Semantic Knowledge Engineering*,
Intelligent Systems Reference Library 130,
https://doi.org/10.1007/978-3-319-66655-6_9

9.1 Introduction to the Approach

There are certain persistent limitations of the existing approaches to the design of Rule-Based Systems. They become apparent in the design and development of complex systems in new domains, making high quality design difficult. Therefore, the primary objective of Semantic Knowledge Engineering of Intelligent Systems is to overcome the limitations of existing methods and solutions. Specifically, the approach addresses the issues of:

1. *Informal description* – in existing systems rules are informally described and often have no formal semantics. Lack of formal relation of the knowledge representation language to classical logic leads to difficulties with understanding its expressive power. At the same time, the lack of standards for knowledge representation results in the lack of rule knowledge portability between different systems.
2. *Sparse knowledge representation* – single rules constitute items of low knowledge processing capabilities, while for practical applications a higher level of abstraction is needed. A flat rule base includes hundreds of different rules that are considered equally important, and equally unrelated. The rule-centric design, with no structure explicitly identified by the user, makes the hierarchization of the logic model very difficult.
3. *Blind inference algorithms* – common inference algorithms assume a flat rule base, where a brute force search for a solution is used, whereas a number of practical applications are goal-oriented. Common inference engines, especially forward-chaining ones, are highly inefficient with respect to the focus on the goal to be achieved. Moreover, classic massive inference mechanisms do not address the contextual nature of the rule base.
4. *The unstructured design approach* – no practical methodology for consecutive top-down design and development of RBS, acceptable by engineers and ensuring quality of rules, is widely available. Most of the rule-based tools are just shells providing rule interpreters and editing environments. Hence, the knowledge acquisition task constitutes a bottleneck and is arguably the weakest point in the design and development process. Typical rule shells are not equipped with consistent methodology and tools for an efficient development of the rule base.
5. *Limited quality analysis* – existing solutions to rule design consider system quality analysis only as a step in the system life-cycle (in a similar way to that of Software Engineering), not as an integral part of the design. This often makes the design of high quality[4] systems challenging. A well-defined systematic and complete design process that preserves the quality aspects of the rule model, and allows for gradual system design and automated implementation of rules. Moreover, with unformalized rule languages, rule quality analysis is mostly limited to the detection of syntax errors, not logical malformations.

---

[4]Here by "quality" we mostly mean the correctness of the system, providing its safety, reliability, etc. From this perspective, quality is mostly assured by system verification.

6. *Lack of integration on the semantic level* – the rule-based approach originates from the field of classic AI and most of the rule-based tools provide custom representations and languages that need to be aligned with another design and implementation approaches on the programming level. As a classic knowledge representation method, rules have a long history, and a number of custom rule languages and representations were developed. However, today software engineering practice uses methods and tools that are not conceptually compatible with the rule-based approach. Therefore, it is hard to benefit directly from rule technologies.

These challenges are addressed by the SKE approach. The most important solutions are introduced next.

The approach addresses the first issue by providing a *formalization of the rule language*. This is crucial for determining the expressive power, defining the inference mechanism and solving verification issues. The formalization uses the ALSV(FD) logic. It is more expressive than the classic (mostly propositional) rule languages, e.g. it allows for formal specification of non-atomic values in rule conditions. Considering the logical foundations of rule languages, such a strict formalization brings a number of practical benefits. Formalized rule language opens the possibility to partially formalize the design process which can, in turn, lead to better detection of design error in the early stage, as well as simplify the transitions between design phases. Moreover, formal methods can be used to identify logical errors in rule formulation. Finally, partially formalized translations to other knowledge representation formats are possible.

As of the second issue, in the SKE approach, a *custom rule representation formalism* is introduced, called XTT2. It ensures high density and transparency of visual knowledge representation. This representation combines decision trees and decision tables. Contrary to traditional, flat RBS, the representation is focused on groups of similar rules rather than single rules. There are two levels of abstraction in the XTT2 model:

- the lower level – where a single knowledge component defined by a set of rules working in the same context is represented in a single table, and
- the higher level – where the structure of the whole XTT2 knowledge base (consisting of XTT2 tables) is considered.

XTT2 forms a transparent and hierarchical visual representation composed of the decision tables linked into a network. This solution addresses the issue of low processing capabilities of single rules.

As of the third issue, an efficient inference in the XTT2 network is possible using *new inference methods for modularized rule bases*. It is ensured that only the rules necessary for achieving the goal (identified by context of inference and partial order among tables) are being fired by selecting the desired output tables and identifying the tables necessary to be fired first. Several inference modes are provided, including forward-chaining and goal-driven ones. This allows for reusing the same rule base in different applications. Furthermore, in our later research we demonstrated how the XTT2 decision network can be incorporated into a business process. Such a heterogeneous network can be run by a process engine that delegates the execution of rules grouped in the decision tables to the XTT2 inference engine. The whole network is described by a formalized model that allows for the integration of processes and rules.

The fourth issue is addressed by introducing a *systematic design procedure*. It is a complete, well-founded process that covers all of the main phases of the system life-cycle, from the initial conceptual design, through the logical formulation, to the physical implementation. This is in fact a top-down design methodology based on successive refinement of the RBS. It starts with the development of an ARD+, model which describes relationships among process variables represented by attributes. Based on the ARD+ model, schemas of particular tables and links between them are generated. This phase also includes the design of the BP integrated with XTT2 tables using the ARD+ design. The tables are filled with expert-provided definitions of constraints over the values of attributes forming the rule preconditions. The code for rule representation is generated and interpreted with the provided inference engine. The procedure involves verification of the system model with respect to important formal properties.

As of the fifth issue, the proposed design approach includes *quality analysis solutions*. What is important for RBS is their logical description. A rule base can be considered only a code in a programming language – this is in fact the case with most rule shells. Once an explicit formalization is introduced, it becomes an expression (or a set of expressions) in mathematical logic. This is in fact the case in the logic programming paradigm, including the Prolog language, and also the case of the XTT2 rule base. Its formal description in ALSV(FD) allows for the use of formal quality properties tests of the RBS. Thanks to the logical formulation, such an analysis can be an integral part of the logical design, when rules are built in an incremental way.

To address the sixth issue, the approach presented here is oriented towards a *practical integration of a rule-based approach* with important existing methods to software design. Rules are a classic knowledge representation method aimed at expressing the operational logic of the system. A number of design methods have been developed for this knowledge engineering solution. However, they are mostly focused on specific applications. Some recent implementations have introduced solutions for integrating rules with object-oriented Java-based applications. Unfortunately, this integration is conducted on the low-level mapping of the programming languages and runtimes. Not much has been done in the area of integrating the semantics of rule-based and object-oriented methods. The Semantic Web initiative had ambitious goals of developing rule base reasoning for an intelligent Web. However, the semantics of logical methods that are used (ontologies formalized with the use of the DL) are incompatible with rules. The approach discussed later in this monograph proposes integration solutions for both object-oriented as well as Semantic Web design methods on the semantic level. Moreover, we discuss the integration of BP systems with RBS using the above mentioned formalized model and practical tools. We also consider two important cases of integration in the area of Software Engineering, namely on the

UML design level and for supporting rule-based software testing. Finally, we also consider two other cases closer to specific hardware. This includes designing control logic for mobile robots, and reasoning in context-aware systems on mobile platforms.

After the introduction of the assumptions of SKE we discuss the design process it introduces and the software tools supporting it.

9.2 Design Process for Rule-Based Systems

Semantic Knowledge Engineering introduces a systematic design process, presented in Fig. 9.1 and described below. The process includes three main *phases* and several important *stages*. The phases are as follows:

1. **Conceptual modeling and rule prototyping** including: Linguistic Specification, Conceptualization, Conceptual Design and Structure Prototyping,
2. **Logical specification and analysis** including: Attribute Specification, Logical Design, and Formal Analysis,
3. **Physical design with prototype integration** including: Physical Implementation, and System Integration.

In the figure the phases can be observed as the horizontal swim-lanes including names of knowledge representation methods (on the left side of the figure) and design tools (on the right side). The detailed specific stages are depicted as rectangles. Arrows on the left side of the rectangles denote the output of a given stage, whereas arrows on the right side correspond to file formats. The details of these design stages are presented in the following paragraphs.

Linguistic Specification

At this first stage a general description of the system in the natural language is given. It is provided by a domain expert, customer, etc.[5] It is assumed that the description is prepared with the assistance of a knowledge engineer who takes care of its format. The description is stored in a textual form, with additional an media file also registered if available.[6] This stage is not explicitly formalized. It is supported by an on-line collaboration tool, as the LOKI system provides SBVR authoring for this, see Chap. 15.

Conceptualization

In this stage some primary concepts within the system model are identified. These can be attributes, properties, objects (in the general sense), or rules. The output of this stage is a preliminary list of possible attributes, or system properties. The list may be somehow structured. In a general case, the concepts are identified by the

---

[5]In fact, one can imagine that a number of documents related to the system can be used in this phase. This include norms, standards, and written policies.

[6]To simplify the transition from this stage to the next one, some basic natural language processing techniques might be considered. In fact, in [1] we explored a simple semi-formalization in Structured English, which is the part of the SBVR standard [2].

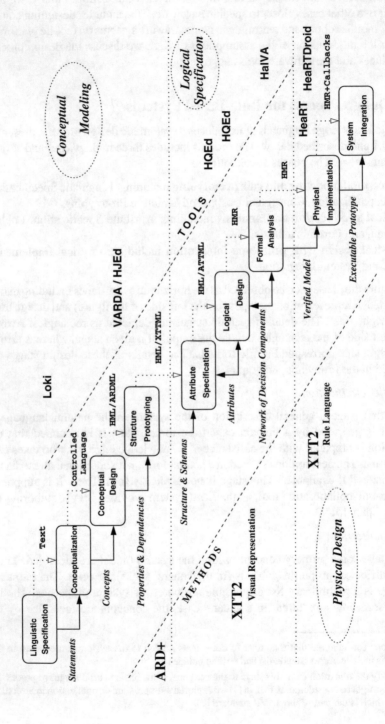

Fig. 9.1 Design process for the SEMANTIC KNOWLEDGE ENGINEERING

designer and captured using a simple representation. This is a preliminary attribute identification, which is not explicitly formalized, but some templates are provided. The stage is also supported by an on-line collaboration tool.

Conceptual Design

The conceptual design stage is the beginning of the formalized hierarchical process. This stage is fully supported by the ARD+ method described in Chap. 6, and supported by practical tools. Based on the concepts, or vocabulary, the ARD+ properties, attributes, and relations between them are identified. In a general case, where no explicit original rules are provided, the ARD+ diagram is built using the previously identified vocabulary. If the original rules are available, the physical attributes identified during the design should match them. The most specific ARD+ level should match the structure of the original rule base.

Structure Prototyping

Using the output of the conceptual design, in this stage an automated generation of the XTT2 table schemas (rule prototypes) is performed. Using the most specific ARD+ level XTT2 schemes are generated. The preliminary inference structure is also built, and serves as a guideline for XTT2 processing by the HEART inference engine.

Attribute Specification

In this stage the physical attributes from the ARD+ phase are mapped into XTT2 attributes in the ALSV(FD) logic (see Sect. 4.1). The formalized attribute specification can be shared with other systems, tools and approaches. At this stage it is possible to import attributes from other formalized sources.

Logical Design

In this stage the precise design of the XTT2 tables is formulated. Tables are filled with rules, and fine-grained inference links can be added to form the whole inference network. This is the primary stage of the rule base design. Thanks to the visual representation of XTT2 it is easier to design larger rule bases. This is also the stage where the rule base design can be optionally integrated with business processes.

Formal Analysis

In this stage the logical description of the XTT2 is used to conduct an analysis of selected formal properties. The results of the analysis are used during the design process, when the rule base is incrementally refined. For a complete discussion see Sect. 4.6, and for the tool support see Sect. 9.8.

Physical Implementation

In this stage an executable form of the system is generated. The designed and analyzed rule base is exported into a textual rule language, which then can be executed by the HEART and HEARTDROID inference engines. The engine implements new inference algorithms for the XTT2 network. The textual rule language syntax is described in Sect. 9.6, and the engine design and implementation is given in Sect. 9.7.

System Integration

This stage involves integrating the XTT2 control logic with the external application environment and other design approaches. Several scenarios are considered, including:

- providing a set of attribute callback functions and rule actions that allow for the use of XTT2 logic as a separate control component, see the next section for a discussion of this architecture,
- rule base translation where XTT2 rules are exported into several other rule formats, using the HaThoR translation framework,
- translating the XTT2 structure to the visual UML representation to be used as a business logic design, described in Chap. 11, where the full translation and practical application are given; or generating test cases for software unit testing described in Chap. 12.
- integrating a set of XTT2 tables with a BP and execute it using a hybrid BP/BR environment as described in Chap. 13.
- integrating the XTT2 logic with Description Logic, and providing a hybrid reasoner for such a system. Chapter 14 presents this approach, including the mapping, as well as a practical knowledge engineering solution,
- using XTT2 control logic with additional dedicated callbacks to control mobile robots as discussed in Chap. 16, and finally
- implementing RBS with XTT2 on mobile devices as presented in the final chapter (Chap. 17).

This complete design process is described using the formalized design methods described in the second part of this book, and supported by design tools described in the next sections of this chapter.[7] It is also aimed at providing application built in a specific architecture, which is presented next.

9.3   Architecture of Rule-Centric Applications

In Sect. 2.6 different architectures for integration of rule-based systems where discussed. The SKE approach provides a heterogeneous solution through a clear *separation of Core Business Logic*. One of the main goals of the approach presented here is to offer an efficient platform for the so-called executable design. It could provide a bridge for a *semantic gap* between a design and its implementation as

---

[7]Considering the whole life cycle of a system more phases of the process could be identified, e.g. refinement, tuning, maintenance, etc. However, these are not directly addressed here, because the focus is on building the system.

discussed by Mellor in [3]. Eventually it can shorten the development time by trans-forming the "implementation" into the runtime-integration with a rule-based model. The approach introduces a strong separation of the core application logic from the interfaces and other parts. In fact, it is assumed that the MVC-like software design pattern is used, (Model-View-Controller) [4]. The intelligent application is decomposed into a Model that captures the logic, a View that corresponds to different interfaces, and a Controller that links these two. The SKE architecture provides means for the design and implementation of software logic and the integration of this logic with the presentation layer. The emphasis is on a rich and formally designed and ana-lyzed knowledge-based model. It is important to observe that as opposed to standard software engineering approaches there are no differences in the semantics of design methods. Thanks to the XTT2-based logic core, the knowledge base is represented using a formalized knowledge model. This allows for the use of a formal analysis of the model and avoiding common evaluation problems. Such an analysis can be provided at the design stage, which allows for a gradual refinement of the designed system.[8]

As discussed earlier, the concept of the model considered here is based on certain concepts related to *dynamic system modeling*. The primary assumption is that the rule-based model is a model of a dynamic system having a certain *state*. The state is described using attributes that represent important properties of the system. A statement that an attribute has a given value can be interpreted as a fact in terms of classic expert systems. The concept of the state is similar to the one in dynamic systems and state-machines. The current state of the system is considered as a complete set of values of all the attributes at the instant of time. The *dynamics* of the system (transitions between states) is modeled with the use of *rules* described by a logical representation. The conditional part of a rule is an expression related to the state to be matched. The decision part includes statements that modify a system state in case the rule is fired (the proper decision) and actions that do not change the attribute values, thus the state. This is a *declarative* model (Fig. 9.2).

In general, the values of the XTT2 attributes in the model can be modified by an independent external system (or user). This case concerns attributes representing some process variables, which are taken into account in the inference process, but depend only on the environment and external systems. As such, the variables cannot be directly changed by the XTT2 system. Values of those variables are obtained as a result of some measurement or observation process, and are assumed to be put into the inference system via a *blackboard* communication method discussed by Hayes-Roth in [5]. In fact, they are written directly into the internal memory whenever their values are obtained or changed.

To connect the internal system memory with the external environment, *callbacks* are used. These are dedicated functions related to system attributes. Callbacks are

---

[8]It is worth emphasizing that the principal idea is not to model the whole application in the rule-based manner. Instead, it is asserted that a clear separation of the declarative rule-based logic is possible. Interfaces are identified and possibly designed in another, more common way e.g. using object-oriented frameworks.

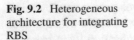

Fig. 9.2 Heterogeneous architecture for integrating RBS

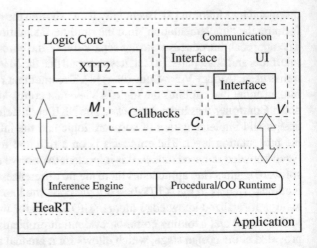

invoked to get and send attribute values from and to the environment. The values of internal attributes can only be modified by the inference process itself. In such a case, values are obtained at certain stages of reasoning as the result of the operations performed in the decision part of the XTT2 rules.

The *integration of a heterogeneous system* is considered mainly at the at the runtime, service and design levels.

- *runtime level*: the application is composed of the rule-based model run by the inference engine integrated with the external interfaces using the callback mechanisms. The model is run by the XTT2 inference engine that uses callbacks to communicate with front-ends implemented with other languages e.g. Java.
- *service level*: the rule-based core is exposed to external applications using a network protocol. This allows for a SOA-like integration [6] where the XTT2 logic is designed and deployed using the dedicated inference engine. The state of the XTT2-based system can be modified with the use of callbacks triggered by attribute changes. The system can additionally invoke actions thus influencing the external system. This solution is provided by the HEART engine and its communication protocol.
- *Design level*: integration considers a scenario, where the application has a clearly identified logic-related part, which is designed using the XTT2 method, and then translated to a domain-specific representation. As an example, in Chap. 11 the translation from XTT2 to the UML notation is discussed.

An important feature of the approach is that the knowledge engineering process emphasizes the semantics of the knowledge representation and processing methods.

We now discuss a set of development tools supporting the SKE methods.

9.4 Overview of the HADEs+ Tool Framework

The design process introduced is Sect. 9.2. is supported by a number of practical tools, as well as custom file formats needed to exchange the knowledge base. The SKE design process has two important aspects:

- *level of abstraction* – the system design is being specified from the *conceptual* level, through the *logical* one, to the *physical* specification, and
- *form of representation* – the system logic can be *visualized* for design purposes, and at the same time it has a well-defined *formal* representation, which can be automatically transformed into an *executable* prototype.

The process is supported by HADEs+ toolkit (Hybrid Design Environment). The complete framework is depicted in Fig. 9.3. It is an incremental enhancement of the original HADEs (HEKATE Design Environment) as described in [7] and considered in [8].

Knowledge Serialization Languages

The ARD+ and XTT2 design tools store and exchange the design model in a dedicated file format based on XML. The format is also used to translate the model into other XML-based languages. HML (HEKATE Markup Language) [9] (see [10] for some early ideas) consisting of three logical parts: attribute specification in ATTML (ATTribute Markup Language), attribute and property relationship specification in ARDML (ARD Markup Language), and rule specification in XTTML (XTT Markup Language). The attribute specification regards attributes present in the system. It includes attribute names and data types used to store attribute values. The rule specification stores actual structured rules. These sub-languages can be used in different scenarios.

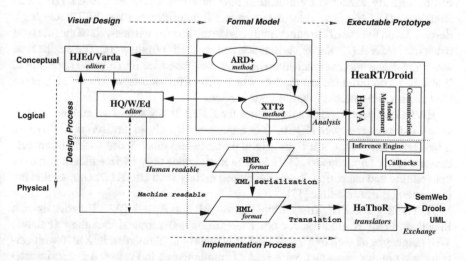

Fig. 9.3 The complete HADEs+ environment

The logical design of the rule base made with HQED can be serialized to HMR. It is textual format used to store XTT2 rules. Moreover, it is directly executable by the HEART engine. In fact, in the new generation of tools HWED generates only HMRwith additional annotations (specifically HMR+) needed by visual editor and ignored by rule engine.

HADES+ *Tools*

The ARD+ conceptual design process is supported by VARDA (Visual ARD+ Rapid Development Alloy) and the HJED (HeKatE Java Editor) visual editor. VARDA [11] is a prototype semi-visual editor for the ARD+ diagrams implemented in Prolog, with an on-line model visualization with Graphviz. The tool also supports prototyping of the XTT2 model [12], where table schemas including a default inference structure are created. The ARD+ design is described in Prolog, and the resulting model can be stored in HML. HJED [9] is a cross-platform tool implemented in Java. Its main features include the ARD+ diagram creation with the design history available. Once created, an ARD+ model can be saved in a HML file. The file can be then imported by the HQED design tools supporting the logical design. ARDML with ATTML is used for the description of ARD+ in VARDA and HJED. HQED uses the complete specification, including XTTML.

HQED (HeKatE Qt Editor) [13, 14] provides support for the logical design with XTT2. It can import a HML file with the ARD+ model and generate an XTT2 prototype. It is also possible to import prototypes generated by VARDA. HQED allows for editing the XTT2 structure with support for syntax checking at the table level. Attribute values are checked with domain specification, and the use of proper relational operators is assured, so that potential anomalies are detected and eliminated.

HQED can automatically generate the textual executable representation of the XTT2 knowledge base in HMR format (HEKATE Meta Representation). HMR files can be directly executed by a custom inference engine called HEART (HEKATE RunTime) [15, 16]. The role of the engine is twofold: run the XTT2 rule logic designed with the use of the editor, and provide on-line formal analysis of the rule base using the HALVA (HeKatE Verification and Analysis) framework. The engine runs as a stand-alone application. Moreover, it can be embedded into a larger application. It also has a communication module which allows for an integration with HQED as well as providing network-based logic service.

HWED is a prototype of a new editor for XTT2. It is web-based, and has a much simplified interface over HQED. It is implemented only in Javascript and runs in a web browser. It makes it lightweight, portable and much more usable on mobile devices. It does not support XML, instead it generates only HMR+ files suitable for both editing and execution. It was developed mainly with HEARTDROID in mind but is a general purpose editor [17].

HATHOR (HEKATE Translation Framework for Rules) [9, 18] provides rule import and export modules for other languages and formats, including Semantic Web languages as well as Drools [19]. It is mainly implemented in XSLT with certain extra plugins integrated with HEART implemented in Prolog. An experimental module allows for the translation of visual XTT2 representation to a dedicated UML representation using an XMI-based (XML Metadata Interchange) serialization, as described in [20].

The above mentioned components of HADES+ will be discussed in the subsequent sections using system cases introduced in the following section.

Design Case Studies for SKE

In this book we use several illustrative use examples to demonstrate different features of SKE. We started with BOOKSTORE on p. 87 to show basic ideas of the ALSV(FD) and XTT2, it will also be used in the research related to the Semantic Web. Then we introduced the PLI on p. 98 during the discussion of XTT2. In order to present HADES+ tools, in this chapter we will use the CASHPOINT example.

The Cashpoint system is composed of a till (cash withdrawal box) which can access a central resource containing the detailed records of bank accounts of a customer. The till is used by inserting a card and typing in a Personal Identification Number (PIN) which is encoded by the till and compared with a code stored on the card. After successful identification, customers may either make a cash withdrawal or ask for a balance of their account is to be printed. Withdrawals are subject to a user resources, which means the total amount that a user has on account. Another restriction is that a withdrawal amount may not be greater than the value of the till local stock. Tills may keep illegal cards after three failed tests for the PIN. The specification considered here is based on the system presented by Denvir et al. in [21] and by Poizat and Royer in [22]. It has served as a valuable benchmark for several formal approaches [23, 24].

The development of the use cases related to the SKE methodology was active for several years. Illustrative system cases were selected, analyzed, designed and implemented. During the HEKATE project important cases were made available online in the *Online* HEKATE *Cases Repository*.[9] Cases were selected in order to investigate and possibly boost the selected language features of XTT2. Moreover, the goal of selection was to exploit and demonstrate the features of the methodology, and to be able to compare it to other solutions. To create the cases repository, a wiki system has been used. It supports storing and document design files, it also allows for a collaborative work of a team of designers. The CASHPOINT system is also available there.[10]

Important details of the system design will be presented in the following sections to demonstrate the features of the methods and the supporting tools.

---

[9]See http://ai.ia.agh.edu.pl/wiki/hekate:cases:start.

[10]See http://ai.ia.agh.edu.pl/wiki/hekate:cases:hekate_case_cashpoint.

9.5   Visual Design of a Modularized Rule Base

Knowledge Base Structure Prototyping

In this phase the ARD+ method, previously presented in Chap. 6, is used. VARDA (Visual ARD+ Rapid Development Alloy) is a prototyping environment for ARD+ designed and presented in [11, 25, 26]. The VARDA tool is designed in a multi-layer architecture, including: (1) a knowledge base to represent the design: attributes, properties, dependencies; (2) the main predicates: adding and removing attributes, properties and dependencies; (3) transformations: finalization and split including defining dependencies and automatic TPH creation; (4) low-level visualization: generating data for the visualization tool-chain; (5) high-level visualization: displaying a dependency graph among the properties and TPH. Prolog facts represent ARD+ attributes, properties, and dependencies. Predicates generate input for Graphviz tool which renders appropriate graphs representing diagrams, which are subsequently displayed by ImageMagick package. The XTT2 schema prototyping algorithm was also implemented as a part of VARDA in [12].

Based on experiences with VARDA, HJED tool was created. HJED, described in [9] is a cross-platform editor for ARD+ implemented in Java. It is an alternative to VARDA with a superior graphical interface. Its main features include the ARD+ diagram creation with the design history available. It can export the model to HML. Furthermore, HQED also supports basic editing of the ARD+ model.

Rule Models Integrated with Business Processes

In Sect. 6.4 we described an important extension of the original ARD+ design. It allows for an integration of a structured rule base prototype with ARD+ with a BP model in BPMN. In this case, the ARD+ model created with the methods described above can be transformed into two parallel yet integrated models. The first one is the regular prototype of XTT2 tables. The second contains prototypes of business rule tasks to be used in a BP. In fact they contain the above mentioned XTT2 tables. The whole process is supported by an enhanced version of the visual editor Oryx and will be described in more detail in Sect. 13.2, see [27].

Logical Design of the XTT2 Rule Base

The HQED editor [13, 14] allows for the visual design of the XTT2 knowledge bases. Moreover, it provides a mechanism for checking the rule model against the syntax and logical anomalies. The tool is able to detect important syntax errors, such as: inaccessible rules, attribute values out of domain, etc. The logical anomalies are detected using HEART.

HQED was implemented as platform independent, using the Qt programming library[11] a cross-platform application development framework. It also provides an intuitive and user friendly graphical interface. Each dialog window has appropriate controls preventing the user from entering incorrect data.

---

[11] See http://www.trolltech.com.

The HQED architecture [14] includes three main components. The *controller* is the most important element, as it enables the data flow between the layers. The *model* consists of two layers that are responsible for the internal models representation. The *ARD Model* stores an ARD+ model and the *XTT Model* stores an XTT2 model. The remaining layers are included in the *view* that maps the model to other formats:

- User Interface – the visual model representation that is appropriate for the user. This layer renders the visual representation of the XTT2 diagram.
- XML mapping – translates the *model* data to HML. This layer allows for storing the state of the model using files.
- Plugins API – provide the communication to other services.

The editor can be integrated with HEART which provides on-line execution and verification. The main aspects of knowledge modeling with HQED are discussed below.

The complete XTT2 diagram consists of tables which contain rules and links. The design starts with a definition of the attribute types. The sequence in defining the XTT2 model with HQED includes definitions of types, attributes, tables, rules, and links. The tool partially enforces this design sequence. For instance: a table can be defined only when there exists at least one attribute.

The logical design starts with the specification of domains of attributes. The *Type Editor* window dialog assists in this process. The next step involves a definition of XTT2 tables and their rows representing rules. The table schema is defined first. For a given table, it groups the attributes in the condition or decision parts of the rule. The *Table editor* dialog allows for defining the table schemas. The last stage of the table definition includes rule specification. Every table row corresponds to a single rule. A rule is defined when all the cells in the row are specified. Cells correspond to the ALSV(FD) triples present in rule conditions and conclusions. *Cell editor* window dialog allows for the creation and editing of the rows and cells (see Fig. 9.4). The last

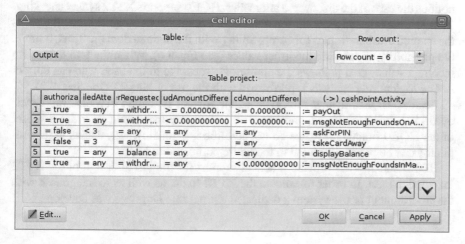

Fig. 9.4 HQED cell editor dialog

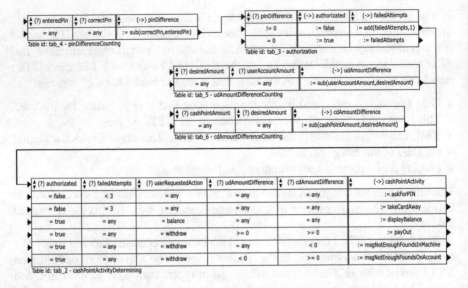

Fig. 9.5 Cashpoint XTT2 model exported from HQED

XTT2 design stage consists in linking the tables. There are two ways to define a link: using the *Connection editor* dialog, or *drag&drop*. The complete XTT2 diagram for the CASHPOINT case (introduced in the next section) is shown in Fig. 9.5.

The description given above concerns the case when the XTT2 model is designed from scratch. However, HQED is able to import the model generated by VARDA. Such a model contains the preliminary definitions of attributes. It also contains rule schemas and inference links. In such a model, HQED allows for filling the XTT2 tables with rules. It also allows for refining the inference network.

The tool checks the model quality during the whole design process. It immediately notifies the user of the discovered anomalies. HQED supports XTT2 syntax checking. That includes detection of all the anomalies except for the logical ones. It checks the entered attribute values for compatibility with the defined domain. Thus, it limits the possibility of entering incorrect values (see Fig. 9.6). When the tool detects an incorrect value, then it shows the appropriate message to the user, who can correct it. The logical rule analysis is performed with the use of HEART (see Sect. 9.8).

Besides HQED, a new prototype editor for XTT2 called HWED was developed [17]. It is web-based, and has a simplified interface. As it provides a subset of the HQED functionality, we do not describe it here. However, thanks to its simplicity, can be better supported in the future compared to HQED.

HQED and HWED provide practical support for the visual design of the XTT2 knowledge base. In order to execute the XTT2 logic, as well as to integrate it with the environment of the system a dedicated textual rule language was developed.

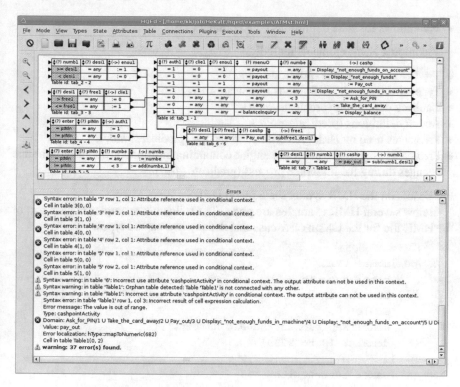

Fig. 9.6 XTT2 model in HQED with anomalies detected

9.6 HMR Rule Language

The visual XTT2 model is represented using a human readable notation, called the XTT2 presentation syntax, or HMR (HEKATE Meta Representation) [15]. The representation can be directly run by the HEKATE RunTime environment – HEART. The HMR file is in fact a legal Prolog code that can be interpreted directly (the number of custom Prolog operators are defined). HMR can be used to manually write rules, or it can be generated by the HQED editor from the visual XTT2 model.

The HMR language was created to provide simple, transparent and human-readable text representation for XTT2 diagrams that would be also easy to parse by a machine. The popular XML notation can be easily parsed, but it is hardly readable.

HMR syntax provides the appropriate predicates that allow for definition of XTT2 concepts:

- *xattr* and *xtype* allow for attribute definitions (see Formulae (4.3) and (4.4)).
- *xschm* defines the table and rule schema (see Formulae (4.20) and (4.21)).
- *xrule* corresponds to the rules (see Formula (4.15)) and allows for defining links between rules (see Formula (4.23)).

- *xstat* allows for state definitions (see Formula (4.5)).

  Additional HMR elements include:

- Type group definitions (predicate *xtpgr*),
- Type attribute group definitions (predicate *xatgr*),
- Callback definitions (predicate *xcall*),
- Action definitions (predicate *xactn*),
- Predicates used by HALVA verification reports (predicate *xhalv*),
- Predicates used by the inference engine containing information about system state changes during the inference process (predicate *xtraj*).

For details on the HMR syntax see a report [15].

Below several HMR examples are given. The following example shows a part of the HMR file for the CASHPOINT case that defines two named types:

```
xtype [name: tPin,
       base: numeric,
       length: 4,
       scale: 0,
       desc:'Represents the PIN numbers',
       domain: [0 to 9999]
      ].
xtype [name: tUserActions,
       base: symbolic,
       desc:'The set of user actions',
       domain: [withdraw,balance]
      ].
```

The first type is named tPin and is based on the *numeric* primitive type. It contains integer values from 0 to 9999. The two parameters length and scale correspond to the maximal length and decimal places of a value. The second type is named tUserActions and is based on the *symbolic* primitive type. Attributes of this type allow for two values: withdraw and balance.

The following example presents a definition of an attribute:

```
xattr [name: correctPin,
       abbrev: corre,
       class: simple,
       type: tPin,
       comm: in,
       callback: [xpce_ask_numeric,[correctPin]]
      ].
```

The attribute has a defined name correctPin and an abbreviation corre. The third line defines the attribute class (see Formula (4.2)). In this case the attribute can take only one value at a time. In the next line, a type of the attribute is defined. According to the previous example, the set of allowed values of the attribute contains the integer values from 0 to 9999. The fifth line defines the attribute relation to an external system (see Formulae (4.6)). In this case, the value of the attribute is provided by a system. The sixth line defines a callback that provides a value of the attribute from an external system. The next example shows a definition of a table schema:

```
xschm 'isPinCorrect':
       [enteredPin,correctPin] ==> [pinDifference].
```

The decision table defined by this schema is named isPinCorrect. The table can contain rules that have three attributes: enteredPin, correctPin and pin Difference. The first two must be placed in a conditional part of the rules and the last one in a decision part.

In the following example, a single rule is defined. The rule belongs to the table defined by the previously defined schema:

```
xrule 'isPinCorrect'/1:
       [enteredPin eq any,
        correctPin eq any]
    ==>
       [pinDifference set (correctPin-enteredPin)]
    :'authorization'.
```

The first line contains information about rule schema. This information can be considered redundant, but from an implementation point of view it allows for convenient processing of rules. After the / character the rule identifier is defined. The second and third lines constitute a conditional part of the rule whereas the fifth line a decision part. The last line defines a link to the authorization table.

The last example presents the HMR syntax that allows for state definition:

```
xstat init0: [enteredPin, 1111].
xstat init0: [correctPin, 1234].
```

Both lines define a value of an attribute for the state init0. HMR allows for the definitions of any number of states. This can be especially useful during system testing. The test cases can be generated in a form of states.

The HMR representation provides a callback mechanism. This mechanism is related with the attributes and allows for exchanging values with the environment. The HEART tool supports this mechanism in two ways: all the callbacks are executed before the inference starts, or a single callback is executed during the inference when the value of an attribute is unknown.

HMR is a dedicated textual serialization of the XTT2 representation. Therefore, it needs a dedicated rule engine (interpreter). In the following section the design and implementation aspects of the HEART engine are discussed.

A more recent development of SKE led to some important extensions of HMR. The so-called HMR+ language extends HMRby introducing: uncertainty modeling in rules, statistical and time-based operators in rule formulae. Moreover, it supports semantic annotations used by the mediation mechanism in context-aware systems. It is supported by the newer HEARTDROID engine and will be described in Sect. 17.3.

9.7   HEART and HEARTDROID Rule Engines

HEART (HEKATE RunTime, later simply Hybrid Rule RunTime) [15, 16, 28], is a dedicated inference engine for the XTT2 rule bases. The engine is implemented in Prolog, using the SWI-Prolog stack. The main HMR parser is based on the Prolog operator redefinition [29]. To implement custom rule inference modes, a dedicated forward and backward chaining meta interpreter is provided. The engine has a modularized architecture composed of several modules. The following modules are identified:

- *The inference module* implements the four inference modes offered by the system: Data-Driven, Goal-Driven, Token-Driven and Fixed Order (see Sect. 4.5).
- *The* ALSV(FD) *logic module* is responsible for evaluating rule conditions according to the ALSV(FD) logic (see Sect. 4.1).
- *The communication and integration module* is responsible for providing a network communication protocol.
- *The rules processing module* is responsible for selecting rules which have true firing conditions, and then firing them.
- *The states management module* provides a mechanism for printing, adding and removing system states, and implementing a mechanism for tracking states of the system during the inference process.
- *The verification and analysis module* with the verification predicates is exposed through the HALVA framework (see Sect. 9.8).

Below the most important details of the implementation are given.

Inference Control Modes

HEART implements inference strategies as specified in Sect. 4.5. The main task of this module is to build a stack of tables to be run by the rules processing module. Depending on the inference mode, different predicates are used to build the stack. The inference is controlled by HEART differently for each inference mode. The *Fixed-Order mode* does not require building a stack, because it is the user who determines the order and number of tables to be fired. The list passed by the user is directly run by the rules processing module.

The *Data-Driven mode* requires one or more start tables from which the inference should begin. If no tables are specified, all tables without predecessors are treated as start tables. Based on them, a stack of successor tables is created and then processed. The order of the tables in the list is not relevant. The successor table is designated as the one that requires an attribute value that the predecessor table produces.

The *Goal-Driven mode* is similar to the Data-Driven manner. It needs at least one goal table, which is supposed to be the final table to fire in the inference process. If no tables are specified, all tables without successors are treated as finals. Based on the goal table (or tables) a stack of predecessor tables is created and then processed. The predecessor table is designated as the one that produces an attribute value required by the successor table. When the stack is created, all tables within the stack are run.

The *Token-Driven mode* also needs one or more final tables to be fired in the inference process. Based on these, a stack of predecessor tables is created. Then the required number of tokens for each table is calculated, and the stack of tables is processed. The order of the tables in the list is not relevant. Predecessor table designation is based on the links between tables. When the stack is created, the tokens for all tables are set to zero, and they are fired.

The inference module uses the rule processing module to run single rules. The rule processing module is responsible for selecting rules having true firing conditions, and then firing them. In order to evaluate rule conditions support for all the ALSV(FD) formulae was implemented.

ALSV(FD) *Support Implementation*

The ALSV(FD) logic module implements all of the ALSV(FD) operators and inference rules (see Sect. 4.1). Moreover, it provides a large set of predicates that allows for the validation of attribute values within states, rule conditions, and the decision parts. This module is also responsible for evaluating and executing transitions within the decision part of the XTT2 rules. To allow the proper work of all mentioned predicates and for the computation of the generalized attributes, a set of special predicates was implemented. These predicates allow for the computation of intersection, difference, union and complement of a sets defined in the HMR notation.

For evaluating rule conditions written in ALSV(FD) logic the predicate called `alsv_valid(+Expr,+StateVal)` is used. An example clause for the atomic formula $A \in \langle L, U \rangle$ is presented below:

```
alsv_valid(Att in [L to U], State) :-
    alsv_attr_class(Att,simple),
    alsv_values_get(State,Att,StateValue),
    !,
    alsv_values_check(Att,[StateValue]),
    alsv_values_check(Att,[L]),
    alsv_values_check(Att,[U]),
    normalize(Att,[StateValue],[NormStateValue]),
    normalize(Att,[L],[NormL]),
    normalize(Att,[U],[NormU]),
```

```
NormStateValue =< NormU,
NormStateValue >= NormL,
heart_debug(2,['Valid:',Att,in,[L to U]]).
```

The `alsv_values_check(+AttName, +AttValue)` predicate is used for checking if the attribute values used in the condition and the decision part of the rule is within a domain of the attribute defined in the HMR file.

```
alsv_values_check(Att,[LS to US|Rest]) :-
    alsv_domain(Att,Domain,numeric),
    heart_debug(2,['Check:',Att,numericreg,LS to US]),
    math_member([LS to US],Domain),
    alsv_values_check(Att,Rest),
    !.
```

For evaluating a decision part of the rule the `alsv_transition(+AttName, +AttVal)` predicate is used.

```
alsv_transition(Att set Expr) :-
    \+ member(_,Expr),
    alsv_domain(Att,_,numeric),
    alsv_attr_class(Att,simple),
    alsv_eval_simple(Expr, Val),
    heart_debug(2,['Adding new fact:',Att,' set to',Val]),
    alsv_new_val(Att,Val).
```

There are two ways of tracking HEART operations: (1) using the built-in information system that depends on a value of the `debug_flag/1` predicate that produces information about system activity, and (2) using trajectory a projection mechanism that depends on a value of the `trajectory_mode/1` predicate and saves system states over time when the inference process takes place. The HEART debug mechanism is a simple information system that reports system current activity *on-the-fly*. The details of information printed to the Prolog console depends on a value of the `debug_flag/1`.

The state management module provides a mechanism for printing, adding and removing system states. It also implements a mechanism for tracking system states during the inference process (the trajectory of the system). Every time the inference process is run its trajectory is saved.

HEART *Communication Protocol*

To make HEART more flexible, and to introduce another integration level, the TCP-based network communication protocol was developed to allow remote access to the inference engine. Moreover, several libraries were written in popular programming languages to support it. The protocol part of the communication module was designed

Fig. 9.7 Example of GUI
for HEART callbacks

 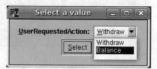

to allow many clients to work with one HEART instance acting as an inference server. This means that several different models owned by several different clients can be stored in the HEART memory.

Using this protocol, HEART can operate in two network modes, stand-alone and as a server, offering network-based integration. In particular, it allows for the integration with a complete rule design and verification environment as discussed in [9]. In fact, the communication of HQED with HEART during the design process is implemented using the protocol.

To make integration of HEART easier, three libraries JHeroic, PHeroic or YHeroic were implemented. JHeroic is written in Java. Based on it one can build applets, desktop application or even JSP services. YHeroic provides an API for Python. PHeroic was created in PHP5. For more details and examples, and full implementation of these libraries see [15].

Callbacks Framework

HEART supports Prolog, Java, and C runtime environment integration. Callbacks can be used to create GUI with JPL [12] and SWING in Java. Another option is to use the SWI Prolog XPCE GUI.[13] The callback mechanism is a part of the HMR language discussed in Sect. 9.6. To help in integrating Java with HEART a simple callback library was implemented that supports basic data exchange between HEART and the user. A library of callback dedicated to different XTT2 data types was provided. An example of execution of callback functions for the XPCE GUI is shown in Fig. 9.7.

Rule-Based System Execution

In HEART an XTT2 rule base can be executed in several scenarios by: invoking HEART from HQED, executing HEART directly from the operating system (OS) shell, and accessing a running HEART instance in the server mode using the communication protocol.

From a design point of view the first solution is the most useful one. An example of such a design session is shown in Fig. 9.8. HQED is able to invoke HEART (in fact it communicates using a plugin accessing the engine through the network protocol), and sending the XTT2 model to the engine, executing it, and then visualizing the execution trajectory (which rules/tables have been fired). Moreover, HQED invokes the HALVA verification framework, described in Sect. 9.8, to provide an on-line logical verification of the system.

---

[12] See http://www.swi-prolog.org/packages/jpl.

[13] See http://www.swi-prolog.org/packages/xpce.

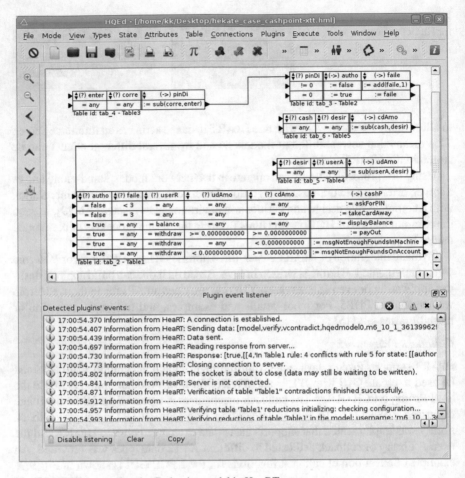

Fig. 9.8 HQED executing the Cashpoint model in HEART

A direct execution of the engine is especially useful for the deployment of the system. Then HEART operates as a stand-alone application. In this mode HEART can be run from the OS shell with the rule model supplied as an external file, together with inference specification. The engine can be used in an interactive session giving the user full control of its functions. Below a simple session is presented.

```
$ ./heart cases/hekate_case_cashpoint-clb.pl
Starting HeaRT...

Copyright (C) by G. J. Nalepa and A. Ligeza and the HeKatE Project
HeaRT is free software: you can redistribute it and/or modify it under
the terms of the GNU General Public License as published by the
Free Software Foundation, either version 3 of the License,
or (at your option) any later version.
```

```
This is HeaRT, the HeKaTE Run Time engine.
Use 'hlp.' to get some help

?- gox(init1,['Table1'],gdi).
[enteredPin,1111],[correctPin,1112],[failedAttempts,0],
[desiredAmount,100.0],[cashPointAmount,100000.0],
[userAccountAmount,10000.0],[userRequestedAction,withdraw].


[Table3,Table2,Table4,Table5,Table1]

State "current":
[enteredPin,1111],[correctPin,1112],[failedAttempts,0],
[desiredAmount,100.0],[cashPointAmount,100000.0],
[userAccountAmount,10000.0],[userRequestedAction,withdraw],
[pinDifference,1],[udAmountDifference,9900.0],
[cdAmountDifference,99900.0]
true.

?- xtraj T.
T = t1:[[start, s1], ['Table3'/1, s2], ['Table4'/1, s3],
        ['Table5'/1, s4]].
```

In the above example the HMR file containing the *Cashpoint* example is loaded
from the command line. The inference process is started from state init1 in the
GDI mode. The algorithm generates the execution order of the tables (3, 2, 4, 5, 1)
and starts the inference. The state current is reached. Trajectory t1 is recorded,
containing the list of tables and subsequent states. In the case of large systems the
trajectory saving mechanisms slow down the inference and can be disabled.

In the stand-alone mode, it is possible to statically compile the engine with the
SWI-Prolog environment, so a single executable file with no external library dependencies is produced. The static compilation is available for multiple system platforms.

The HEARTDROID *Mobile Rule Engine*

The HEARTDROID inference engine was designed as a standalone, self sustainable
reasoning unit that can be deployed on any platform that is equipped with Java Virtual Machine [30]. It is a re-implementation of the original HEART in Java, but with
portability and optimization in mind. Moreover, it implements the extended HMR+
rule language. The core components of its architecture are the inference engine core
module, responsible for performing reasoning, model manager module responsible
for switching between different models, and working memory component responsible for historical states management.

Additionally, HEARTDROID offers two interfaces that allow for different types
of interaction with the reasoning mechanism: interactive and programming. The
interaction is provided by the HAQUNA shell. It offers a command line interface that
allows to dynamically create and modify the HMR+ models and to simulate inference
cycles. The programming interface was designed to allow for the integration of the
HEARTDROID inference engine with mobile applications. This integration can be
performed twofold, with an usage of two mechanisms: (1) API-level binding – a
communication with the application's logic is done via programming interface of

HEARTDROID, and (2) Model-level binding – HMR+ offers two possibilities of connecting the XTT2 model with the application's logic: callbacks and actions.

HEARTDROID offers three basic inference strategies of HEART, i.e. FOI, DDI and GDI, with the TDI mode skipped. Furthermore, HEARTDROID allows for a hybrid reasoning mode that combines two different inference mechanisms [31]: deterministic reasoning supported by certainty factors algebra [32], and probabilistic reasoning supported by the Bayesian interpretation of XTT2 models [33]. Finally, it implements statistical and time-based operators of HMR+. More details will be provided in Chap. 17.

9.8   Rule Analysis with HALVA

HEART provides an *analysis framework*, called HALVA (HEKATE Verification and Analysis). The framework implements detection methods as plugins. They can be run from the interpreter or from the design environment using the communication module. In fact, two generations of HALVA algorithms have been implemented. The first version called HALVA 1 was described in [34]. It mainly used the basic approach based on the Cartesian product of domains. The improved version, HALVA 2, implements the more optimal algorithms. It was described in [35]. The rule verification process can be performed during the design, from HQED integrated with HEART using the communication protocol. In the listing below, part of the modified Cashpoint model has been verified by checking table completeness:

```
Starting HeaRT...
Copyright (C) by G.J. Nalepa and A. Ligeza and the HeKatE Project
HeaRT is free software: you can redistribute it and/or modify
it under the terms of the GNU General Public License
as published by the Free Software Foundation, either version 3
of the License, or (at your option) any later version.
?- ['hekate_case_cashpoint-clb.pl'].
?- vcomplete('authorization').
State: pinDifference State: authorizated State: failedAttempts
?- vcomplete('cashPointActivityDetermining').
State: authorizated      State: failedAttempts State: userRequestedAction
State: udAmountDifference State: cdAmountDifference
'In Table1 uncover state: '(authorizated) in ([0.0])(cdAmountDifference)
in ([-100000.0 to -1.0])(failedAttempts) in ([4.0 to 5.0])(udAmountDifference)
in ([-100000.0 to -1.0])(userRequestedAction) in ([withdraw])''
'In Table1 uncover state: '(authorizated) in ([0.0])(cdAmountDifference)
in ([-100000.0 to -1.0])(failedAttempts) in ([4.0 to 5.0])(udAmountDifference)
in ([-100000.0 to -1.0])(userRequestedAction) in ([balance])''
'In Table1 uncover state: '(authorizated) in ([0.0])(cdAmountDifference)
in ([-100000.0 to -1.0])(failedAttempts) in ([4.0 to 5.0])(udAmountDifference)
in ([0.0 to 1e+06])(userRequestedAction) in ([withdraw])''
'In Table1 uncover state: '(authorizated) in ([0.0])(cdAmountDifference)
in ([-100000.0 to -1.0])(failedAttempts) in ([4.0 to 5.0])(udAmountDifference)
in ([0.0 to 1e+06])(userRequestedAction) in ([balance])''
'In Table1 uncover state: '(authorizated) in ([0.0])(cdAmountDifference)
in ([0.0 to 1e+06])(failedAttempts) in ([4.0 to 5.0])(udAmountDifference)
in ([-100000.0 to -1.0])(userRequestedAction) in ([withdraw])''
```

Fig. 9.9 Inconsistency between conditions

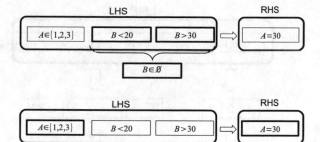

Fig. 9.10 Inconsistency between condition and conclusion

```
'In Table1 uncover state: '(authorizated) in ([0.0])(cdAmountDifference)
in ([0.0 to 1e+06])(failedAttempts) in ([4.0 to 5.0])(udAmountDifference)
in ([-100000.0 to -1.0])(userRequestedAction) in ([balance])''
'In Table1 uncover state: '(authorizated) in ([0.0])(cdAmountDifference)
in ([0.0 to 1e+06])(failedAttempts) in ([4.0 to 5.0])(udAmountDifference)
in ([0.0 to 1e+06])(userRequestedAction) in ([withdraw])''
'In Table1 uncover state: '(authorizated) in ([0.0])(cdAmountDifference)
in ([0.0 to 1e+06])(failedAttempts) in ([4.0 to 5.0])(udAmountDifference)
in ([0.0 to 1e+06])(userRequestedAction) in ([balance])''
```

The verification shows uncovered states –incomplete rule specification.[14]

In the following paragraphs practical anomaly detection methods implemented in HALVA are described. They implement the corresponding concepts previously formally described in Sect. 4.6.

Inconsistency of a Single Rule

This anomaly can be understood in two ways: the presence of inconsistent conditions (Fig. 9.9), and inconsistency between one of the conditions and conclusion (Fig. 9.10). To discover inconsistency between conditions within a single rule the following steps should be performed:

1. Create a list of all attributes from the conditional part of the rule.
2. Take the first attribute from the list and find all other conditions containing the attribute. If there are no attributes left on the list, stop.
3. Calculate the intersection of the sets that are covered by the selected conditions. If the intersection is an empty set, report inconsistency.
4. Delete previously selected attribute and go to step 2.

A warning about potential inconsistency between rule conditions and conclusion is generated if the same attribute exists in both the conditional and decision part of the rule.

---

[14]In fact the original case study does not have this deficiency. To show the discovery of uncovered states, the model domain has been changed and the number of failed attempts have been increased to 5.

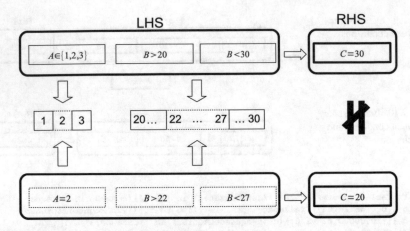

Fig. 9.11 Inconsistency between a pair of rules

Inconsistency of a Pair of Rules

It is defined as a case when decision parts of two rules are different, but there exists a state when preconditions of both rules are true (Fig. 9.11). To discover inconsistency between a pair of rules, the following steps should be performed:

1. Designate the states covered by the two rules from the same context.
2. Calculate the intersection of these states.
3. In the case that intersection is not an empty set and decision parts of both rules differ, report inconsistency.

Incompleteness of a Group of Rules

The most difficult part of logical verification is the completeness test. To check if the system is complete, all possible input data must be passed to the system, and depending on the response a conclusion about completeness is produced. The completeness tests described here are limited to a group of rules – the single XTT2 table, not the entire system. Checking the entire system for completeness is a far more complicated issue, and it is practically hardly possible. Three approaches for verifying the completeness on the single table level to this problem were developed and tested: Cartesian product of domains, Cartesian product of partition of domains, and decision trees (Fig. 9.12).

The first approach is generally inefficient. The one with the Cartesian product of partitions of domains is also inefficient for bigger problems, where domains were partitioned into many parts. An example of generating a combination of partition of domains is presented in Fig. 9.13. To enhance performance of the completeness test, the approach using decision tree representing system states was proposed. An idea is to create a tree, where each branch represents a state to be covered. Every level of the tree corresponds to attributes that covers states denoted by it. The completeness test is based on the Depth-First Search algorithm. The enhancement lies in the fact

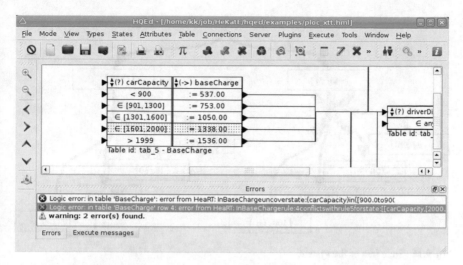

Fig. 9.12 The verification process of the model in the HQEd editor

$(-inf ; 1)$	$(-inf ; 20]$	20	$(-inf ; 20]$
$(-inf ; 1)$	$(20 ; 30)$	20	$(20 ; 30)$
$(-inf ; 1)$	$[30 ; 100)$	20	$[30 ; 100)$
$(-inf \cdot 1)$	$[100 ; inf)$	20	$[100, inf)$
$[1 ; 10]$	$(-inf ; 20]$	$(20 ; inf)$	$(-inf ; 20]$
$[1 ; 10]$	$(20 ; 30)$	$(20 ; inf)$	$(20 ; 30)$
$[1 ; 10]$	$[30 ; 100)$	$(20 ; inf)$	$[30 ; 100)$
$[1 ; 10]$	$[100 ; inf)$	$(20 ; inf)$	$[100 ; inf)$

Fig. 9.13 Combination of states

that when the branch is found that is not covered by any condition of rules from the analyzed XTT2 table, this branch, with all its successors are pruned and are not analyzed. An example of this situation is presented in Fig. 9.14.

Subsumption of a Pair of Rules

One rule subsumes the other if its conditional part is more general and the decision part is the same. Such a situation is presented in Fig. 9.15. A test for the presence of this kind of subsumption is done as follows:

1. Designate states that are covered by rules from a given context (XTT2 table).
2. Calculate intersections of this states.

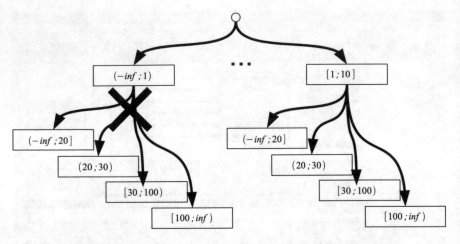

Fig. 9.14 State tree of the system

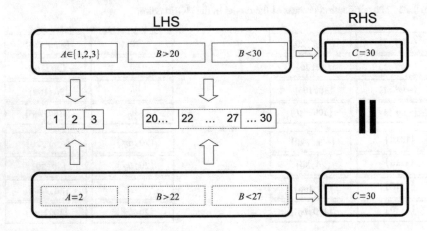

Fig. 9.15 Subsumption of a pair of rules

3. In the case when the intersections are not an empty set and decision parts of the
   rules are the same, report the subsumption error.

   The algorithms described above were practically implemented in HALVA.

9.9 Summary

Overview of SKE

In this chapter the SEMANTIC KNOWLEDGE ENGINEERING approach was discussed.
The first version of this approach was introduced in [8]. However, here we introduced

number of important extensions that have been developed since then. In fact, this chapter bridges the second, and the third part of the book. In the first part we gave a short and selective introduction to the issues related to the development of RBS. They served as a motivation for introducing a number of important formal models in the second part of the book. Based on them, SKE was introduced in this chapter that opens the third part. Different applications of SKE will be discussed in the next chapters in the form of short case studies.

The main objectives of SKE can be summarized as follows: First of all formalized models for knowledge representation, processing and interchange are proposed. This mostly includes the XTT2 and ALSV(FD), but also models discussed in the second chapter of the book. A second feature is the consideration of a structured knowledge bases. This includes not only decomposition in the form of decision tables, but also hierarchization in the form of business process-based inference control. This also results in the consideration of reasoning modes alternative to the classic ones (e.g. Rete in RBS). A third aspect is related to the design of RBS. We advocate for the use of visual modeling methods that can be supported by CASE tools. Finally, the fourth aspect is the integration of rule-based logic with other classes of systems.

The SKE design approach roughly follows the three phase design of data base systems and includes conceptual, logical, and physical phase. We demonstrated how the models introduced in the second part of the book fit in these phases. The core of the design process is the development of a structured rule base. It is represented in a formal way using the XTT2 notation. The XTT2 model can be automatically processed and executed by an inference engine, or translated and integrated with other systems as a logical core.

The design process is supported by a set of tools we call HADES+. These tools, and corresponding cases studies were developed in three phases. The first included prototype tools for ARD+ design: VARDA, HJED, first versions of HQED supporting XTT2, and also HATHOR for syntactic knowledge translation. The second phase included mature versions of HQED that also included basic ARD+ support, as well as the development of HEART and HALVA. Furthermore, selected case studies discussed in next chapters allowed to develop UML translation of XTT2, DAAL for Semantic Web, and first versions of LOKI. This is roughly the stage described in [8]. The third important phase included the work business processes and mature version of LOKI, semantic level knowledge interchange, introduction of uncertainty and finally the development of tools oriented on mobile systems, mostly HEARTDROID and HWED.

Knowledge Engineering at the Semantic Level

The primary motivation for SKE was to provide a coherent notation for a knowledge representation language that can be used is several domains. Regarding the challenges mentioned at the beginning of the chapter, this approach considers:

Semantics of the knowledge representation The rule language used to build the
    knowledge-based model is formalized not only on the syntactic, but also on the

semantic level. Moreover, the semantics of language expressions is considered, including the relations between these expressions.

The semantics of the logical inference process A number of inference methods for the same knowledge base build with the language is provided. Their application depends on the needs of the user and is not determined by a single algorithm.

The semantics of the properties of the knowledge-based model The logical aspects of the model quality are considered. It is possible to detect certain semantic anomalies of the model e.g. its redundancy.

The semantics of the design process The design process of the model is precisely described in several consecutive phases. These correspond to different levels of abstraction of the knowledge of the designer. Moreover, translations of the model to other representations are provided. They preserve the original semantics of the model.

In the remaining chapters we will explore these issues in the case studies.

Outlook

The rest of this final part of the book is composed of eight selected case studies. We begin with the application of SKE to improve semantic rule interoperability with rule based system shells in Chap. 10. Then in Chap. 11 we consider design issues, and discuss how the XTT2 can support UML-based modeling of software. In Chap. 12 we briefly show the application of XTT2 to the automation of software testing. As of the the area of business applications, in Chap. 13 we present the integration of business processes with XTT2 rules. The two next chapters regard Semantic Web technologies: In Chap. 14 we integrate of ALSV(FD) with Description Logics. Chapter 15 discusses the practical knowledge engineering tool, LOKI, a semantic wiki platform. The last two chapters are closer to specific hardware platforms. In Chap. 16 we discuss the design of control logic for mobile robots using SKE methods. In the last chapter (Chap. 17) we discuss context-aware systems on mobile platforms, and the newest generation of SKE tools and methods.

References

1. Atzmueller, M., Nalepa, G.J.: A textual subgroup mining approach for rapid ARD+ model capture. In: Lane, H.C., Guesgen, H.W. (eds.) FLAIRS-22: Proceedings of the Twenty-Second International Florida Artificial Intelligence Research Society conference: 19–21 May 2009, Sanibel Island, Florida, USA, Menlo Park, California, FLAIRS, pp. 414–415. AAAI Press (2009)
2. OMG: Semantics of Business Vocabulary and Business Rules (SBVR). Technical Report dtc/06-03-02, Object Management Group (2006)
3. Mellor, S.J., Balcer, M.J.: Executable UML: A Foundation for Model Driven Architecture, 1st edn. Addison-Wesley Professional (2002)
4. Burbeck, S.: Applications programming in Smalltalk-80(TM): How to use Model-View-Controller (MVC). Technical report, Department of Computer Science, University of Illinois, Urbana-Champaign (1992)

5. Hayes-Roth, B.: A blackboard architecture for control. Artif. Intell. **26**(3), 251–321 (1985)
6. Bieberstein, N., Bose, S., Fiammante, M., Jones, K., Shah, R.: Service-Oriented Architecture (SOA) Compass: Business Value, Planning, and Enterprise Roadmap. IBM Press (2006)
7. Kaczor, K., Nalepa, G.J.: HaDEs - presentation of the HeKatE design environment. In: Baumeister, J., Nalepa, G.J. (eds.) 5th Workshop on Knowledge Engineering and Software Engineering (KESE2009) at the 32nd German Conference on Artificial Intelligence: September 15, 2009, Paderborn, Germany, pp. 57–62 (2009)
8. Nalepa, G.J.: Semantic Knowledge Engineering. A Rule-Based Approach. Wydawnictwa AGH, Kraków (2011)
9. Nalepa, G.J., Ligęza, A., Kaczor, K., Furmańska, W.T.: HeKatE rule runtime and design framework. In: Giurca, A., Nalepa, G.J., Wagner, G. (eds.) Proceedings of the 3rd East European Workshop on Rule-Based Applications (RuleApps 2009) September 21, 2009, Cottbus, Germany, pp. 21–30 (2009)
10. Nalepa, G.J., Wojnicki, I.: XML-based knowledge translation methods for XTT-based expert systems. In: Tadeusiewicz, R., Ligęza, A., Szymkat, M. (eds.) CMS'07: Computer Methods and Systems 21–23 November 2007, Kraków, Poland, AGH University of Science and Technology, Cracow, Oprogramowanie Naukowo-Techniczne, pp. 77–82 (2007)
11. Nalepa, G.J., Wojnicki, I.: VARDA rule design and visualization tool-chain. In: Dengel, A.R., et al. (eds.) KI 2008: Advances in Artificial Intelligence: 31st Annual German Conference on AI, KI 2008: Kaiserslautern, Germany, September 23–26, 2008. Lecture Notes in Artificial Intelligence, vol. 5243, pp. 395–396. Springer, Berlin (2008)
12. Nalepa, G.J., Wojnicki, I.: Hierarchical rule design with HaDEs the HeKatE toolchain. In: Ganzha, M., Paprzycki, M., Pelech-Pilichowski, T. (eds.) Proceedings of the International Multiconference on Computer Science and Information Technology, vol. 3, pp. 207–214. Polish Information Processing Society (2008)
13. Kaczor, K., Nalepa, G.J.: Design and implementation of HQEd, the visual editor for the XTT+ rule design method. Technical Report CSLTR 02/2008, AGH University of Science and Technology (2008)
14. Kaczor, K., Nalepa, G.J.: Extensible design and verification enviroment for XTT rule bases. In: Tadeusiewicz, R., Ligęza, A., Mitkowski, W., Szymkat, M. (eds.) CMS'09: Computer Methods and Systems: 7th conference, 26–27 November 2009, Kraków, Poland, Cracow, AGH University of Science and Technology, Cracow, Oprogramowanie Naukowo-Techniczne, pp. 99–104 (2009)
15. Nalepa, G.J., Bobek, S., Gawędzki, M., Ligęza, A.: HeaRT Hybrid XTT2 rule engine design and implementation. Technical Report CSLTR 4/2009, AGH University of Science and Technology (2009)
16. Nalepa, G.J.: Architecture of the HeaRT hybrid rule engine. In: Rutkowski, L. et al. (eds.) Artificial Intelligence and Soft Computing: 10th International Conference, ICAISC 2010: Zakopane, Poland, June 13–17, 2010, Pt. II. Lecture Notes in Artificial Intelligence, vol. 6114, pp. 598–605. Springer (2010)
17. Bobek, S., Nalepa, G.J., Babiarz, P.: Web-based edtitor for structured rule bases. In: Rutkowski, L., et al. (eds.) ICAISC 2017 Proceedings. LNCS, Springer (2017)
18. Szostek-Janik, J.: Translations of knowledge representations for rule-based systems. AGH University of Science and Technology. MSc Thesis (2008)
19. Kluza, K., Nalepa, G.J., Łysik, Ł.: Visual inference specification methods for modularized rulebases. Overview and integration proposal. In: Nalepa, G.J., Baumeister, J. (eds.) Proceedings of the 6th Workshop on Knowledge Engineering and Software Engineering (KESE6) at the 33rd German Conference on Artificial Intelligence September 21, 2010, Karlsruhe, Germany, pp. 6–17 (2010)
20. Nalepa, G.J., Kluza, K.: UML representation for rule-based application models with XTT2-based business rules. Int. J. Softw. Eng. Knowl. Eng. (IJSEKE) **22**(4), 485–524 (2012)
21. Denvir, T., Oliveira, J., Plat, N.: The cash-point (ATM) 'Problem'. Form. Asp. Comput. **12**(4), 211–215 (2000)

22. Poizat, P., Royer, J.C.: Kadl specification of the cash point case study. Technical report, IBISC, FRE 2873 CNRS - Universite d'Evry Val d'Essonne, France, Genopole Tour Evry 2, 523 place des terrasses de l'Agora 91000 Evry Cedex (2007)
23. Poizat, P., Royer, J.C.: A formal architectural description language based on symbolic transition systems and temporal logic. J. Univers. Comput. Sci. **12**(12), 1741–1782 (2006). http://www.jucs.org/jucs_12_12/a_formal_architectural_description
24. Olderog, E.R., Wehrheim, H.: Specification and (property) inheritance in csp-oz. Sci. Comput. Program. **55**(1–3), 227–257 (2005). Formal Methods for Components and Objects: Pragmatic aspects and applications
25. Nalepa, G.J., Wojnicki, I.: An ARD+ design and visualization toolchain prototype in Prolog. In: Wilson, D.C., Lane, H.C. (eds.) FLAIRS-21: Proceedings of the Twenty-First International Florida Artificial Intelligence Research Society conference: 15–17 May 2008, Coconut Grove, Florida, USA, pp. 373–374. AAAI Press (2008)
26. Nalepa, G.J., Wojnicki, I.: ARD+ a prototyping method for decision rules. method overview, tools, and the thermostat case study. Technical Report CSLTR 01/2009, AGH University of Science and Technology (2009)
27. Kluza, K., Kaczor, K., Nalepa, G.J.: Enriching business processes with rules using the Oryx BPMN editor. In: Rutkowski, L. et al. (eds.) Artificial Intelligence and Soft Computing: 11th International Conference, ICAISC 2012: Zakopane, Poland, April 29–May 3, 2012. Lecture Notes in Artificial Intelligence, vol. 7268, pp. 573–581. Springer (2012)
28. Bobek, S., Kaczor, K., Nalepa, G.J.: Overview of rule inference algorithms for structured rule bases. Gdansk Univ. Technol. Fac. ETI Ann. **18**(8), 57–62 (2010)
29. Bratko, I.: Prolog Programming for Artificial Intelligence, 3rd edn. Addison Wesley, Harlow (2000)
30. Bobek, S., Nalepa, G.J., Ślażyński, M.: Heartdroid – rule engine for mobile devices and context-aware systems. Pervasive and Mobile Computing (2016)
31. Bobek, S., Nalepa, G.J.: Uncertain context data management in dynamic mobile environments. Future Gener. Comput. Syst. **66**, 110–124 (2017)
32. Bobek, S., Nalepa, G.J.: Incomplete and uncertain data handling in context-aware rule-based systems with modified certainty factors algebra. In: Bikakis, A., Fodor, P., Roman, D. (eds.) Rules on the Web. From Theory to Applications. Lecture Notes in Computer Science, vol. 8620, pp. 157–167. Springer International Publishing (2014)
33. Bobek, S., Nalepa, G.: Compact representation of conditional probability for rule-based mobile context-aware systems. In: Bikakis, A., Fodor, P., Roman, D. (eds.) Rules on the Web. From Theory to Applications. Lecture Notes in Computer Science. Springer International Publishing (2015)
34. Ligęza, A., Nalepa, G.J.: Proposal of a formal verification framework for the XTT2 rule bases. In: Tadeusiewicz, R., Ligęza, A., Mitkowski, W., Szymkat, M. (eds.) CMS'09: Computer Methods and Systems: 7th conference, 26–27 November 2009, Kraków, Poland, Kraków, AGH University of Science and Technology, Cracow, Oprogramowanie Naukowo-Techniczne, pp. 105–110 (2009)
35. Nalepa, G., Bobek, S., Ligęza, A., Kaczor, K.: HalVA – rule analysis framework for XTT2 rules. In: Bassiliades, N., Governatori, G., Paschke, A. (eds.) Rule-Based Reasoning, Programming, and Applications. Lecture Notes in Computer Science, vol. 6826, pp. 337–344. Springer, Berlin (2011)

Chapter 10
Rule Interoperability with Expert System Shells

The SKE approach was proposed to address some of the limitations of knowledge engineering with classic expert systems shells. We proposed the introduction of formalized knowledge representation, namely the XTT2 to address them. One of the limitations was related to a lack of interoperability between different implementations of such systems. In Chap. 8 we introduced a formalized model, extending the ideas of XTT2 in order to address the issue of rule interoperability. In this chapter we demonstrate how this model can be used to provide translation of rule base for DROOLS and CLIPS. We will use the model to formalize the main aspects of both rule languages.

In Sect. 10.1 the semantically equivalent features of rule languages in production systems are discussed. Next, in Sect. 10.2 the main features of CLIPS are analyzed, and in Sect. 10.3 the features of Drools. Then, in Sect. 10.4 the formalized model introduced in Chap. 8 is used to formalize the main aspects of both rule languages. Then this formalization is applied in Sect. 10.5 to the PLI case study. Modularization of the rule base is discussed in Sect. 10.6. The chapter is summarized in Sect. 10.7.

10.1 Semantically Equivalent Features of Rule Languages

Rule languages are based on a specific programming paradigm in which system is described by means of facts and rules. Facts define the current knowledge about system, while rules allow for inferring new knowledge according to the existing facts. Besides these two elements, rule languages also provide additional functionalities, described below.

Data Types and Facts

Features related to data types and facts are supported by both of the considered rule languages. They correspond to data types and objects that are known from object oriented (OO) programming paradigm. The object oriented approach together with

© Springer International Publishing AG 2018

G.J. Nalepa, *Modeling with Rules Using Semantic Knowledge Engineering*,
Intelligent Systems Reference Library 130,
https://doi.org/10.1007/978-3-319-66655-6_10

type system is used for the modeling of facts structure. Within the discussed rule languages, the representation of facts usually has a complex internal structure that corresponds to the structure of objects. Depending on language, the expressiveness of syntax allowing for defining facts is different.

Rules

Rules are the most important element of a rule language. In general, the structure of a rule is very similar in each language, and takes the IF-THEN form. The IF part (Conditional or Left-Hand (LHS)) contains the number of logical expressions which are evaluated during the inference process. When the logical value of the LHS part is evaluated to TRUE, then the actions placed within the THEN part (Action/Decision or Right-Hand (RHS)) are invoked. It is worth to notice that the IF-THEN structure fits better the procedural programming languages where it is evaluated only when the program control flow reaches it. From the RBS point of view, the more appropriate structure is WHENEVER-THEN because the inference engine always watches the rules which have their LHS satisfied. This is done by pattern-matching algorithm (e.g. RETE) which provides a pattern-matching network allowing for the efficient evaluation of rules conditional part. Satisfied rules are placed in a conflict set and later are scheduled for execution by moving them to the agenda. It contains rules and defines an order in which they must be executed. This order is in turn determined by a selected conflict set resolution strategy. Sometimes, a *Declare* part is used, that contains the additional rule properties.

Variables

These are supported by the majority of the rule languages. A variable can be used for storing a certain type of values and is considered as a non knowledge element i.e. it does not have an impact on the inference process. Similarly to facts and rules, the rule languages provide disparate support for variables: from the only global variables having a dynamically assigned type to the possibility of defining local variables with precisely defined type and domain.

Rule Base Modularization

It is a mechanism that allows for creating rule bases that have an internal structure. An ordinary rule base consists of rules and facts. The large rule base can contain thousands or even more rules. This is why, most of the rule languages provide a mechanism that facilitates the maintenance of such a number of rules. Thanks to this mechanism rules can be divided into so-called *modules*. The detailed discussion concerning this issue can be found in [1].

Features of the rule languages discussed in this chapter are crucial from the point of view of the knowledge semantics. Therefore, an efficient interoperability method must take them into account in order to assure the correct translation of the knowledge-based aspects of the system. In the following parts of this section, an analysis of the CLIPS and DROOLS rule languages is provided by means of these features. The short examples presented in the remaining part of the chapter, come from the PLI use case.

10.2   Analysis of the CLIPS Rule Language

CLIPS is considered as one of most successful implementation of expert systems shells in AI [2] as it has received widespread acceptance throughout the, industry, and academia. It provides an expressive rule language that supports all the important aspects of building of the rule-based expert systems.

Data Types

The type system provided by CLIPS supports eight primitive types for representing information. These types are `float`, `integer`, `symbol`, `string`, `external-address`, `fact-address`, `instance-name` and `instance-address`.

Facts

CLIPS provides three ways for representing facts: *ordered facts*, *unordered facts* and *objects*.

Ordered facts consist of a number of fields. The first field specifies a *relation* that involves the remaining fields in the ordered fact. Fields in an ordered fact may be referred by its index and may be of any of the primitive types (with the exception of the first field which must be a symbol). For example, the `driver` type of facts may be defined using an ordered fact in the following way: (`driver 5, 12, 5`), where the first field corresponds to the driver class, the second provides information concerning driver age and the last one indicates how long this driver has had a driving license.

Non-ordered facts provide the user with the ability to abstract the structure of a fact by assigning names to each field in the fact. Such a type of fact can be created using a `deftemplate` construct, analogous to a record. It allows the name of a fact template to be defined along with zero or more definitions of named fields (*slots*). Unlike ordered facts, the slots of a `deftemplate` fact may be constrained by type, value, and numeric range. What is more, a slot can be multivalued i.e. it can take on more than one value at one time. The example below shows a definition of non-ordered version of `driver` fact that, in comparison to an ordered version, provides additional information about allowed values of the fields describing each driver:

```
(deftemplate driver
  (slot class        ; driver class
    (type INTEGER)
    (range -1 9)
    (default 1)
  )
  (slot age          ; driver age
    (type INTEGER)
    (range 18 120)
  )
  (slot licage       ; time of holding driving licence
```

```
    (type INTEGER)
    (range 0 102)
  )
)
```

Objects are defined as an instance of a certain class. Objects in CLIPS are divided into two important categories: primitive types and instances of a user-defined classes. Primitive type objects have no names or slots and are handled by CLIPS. In turn, an instance of a user-defined class is referenced by name or address and is created and deleted explicitly via messages and special functions. The properties of an instance of a user-defined class are expressed by a set of slots, which the object obtains from its class. The primary difference between objects and non-ordered facts lies in the notion of inheritance that allows the properties and behavior of a class to be described in terms of other classes. CLIPS supports multiple inheritance: a class may directly inherit slots and message-handlers from more than one class.

Rules

Each rule consists of three parts: the two most important are the conditional part and conclusion part. Additionally CLIPS provides a third part where the properties of a rule can be defined. The template of rule definition can be written as follows:

```
(defrule <name> [<comment>]
  [<declaration>]              ; Rule Properties
  <conditional-element>*       ; Left-Hand Side (LHS)
=>
  <action>*                    ; Right-Hand Side (RHS)
)
```

The conditional part consists of a set (possibly empty) of *Conditional elements* (CEs) which typically consist of patterns that are matched against fact entities. There are eight types of CEs: *pattern*, test, and, or, not, exists, forall, and logical. The *pattern* is the most basic and commonly used conditional element. It defines constraints which are used by an inference engine to determine the set of fact entities that satisfy the pattern. There are several types of constraints that can be provided by pattern CEs. Depending on this type, the following subtypes can be distinguished: *literal* pattern, *wild-card*-based pattern, *connective* pattern, *predicate* pattern, *return value* pattern, *object*-based pattern, *addresses*-based pattern. They differ in terms of types of constructs used for constraint definition.

The conclusion part of the CLIPS rule contains a set of actions that are performed when the LHS is satisfied. CLIPS allows for using any function that is supported by the language. The assert function allows for creating a new fact entity. retract is a complementary function, as it allows for removing facts from a knowledge base. The fact that it is intended to be removed can be specified by its identifier or variable containing reference bound in the LHS. The retraction of a fact has a twofold impact on the knowledge base: (1) it removes all activation of rules that depend on this fact and (2) retracts of other facts which are logically depended from this fact. The modify statement allows for modification of the existing facts.

The declare part is the last part of CLIPS rule that allows the properties of a rule to be defined. The most important is the salience property that allows for assigning a priority to a rule. The priority has impact on the order of rule firing: when multiple rules are intended to fire, the rule with the highest priority will fire first.

An example of CLIPS rule that generates 10% of discount if the driver age is between 40 and 55 can be written as follows:

```
(defrule base-charge-modifiers::decrease-driver-age
    (driver (age ?age&:(>= ?age 40)&:(<= ?age 55)))
=>
    (assert (base-modifier (value -10)))
)
```

Variables

Variables in CLIPS are similar to those that are known from such languages like C and can be used for storing values. They can be considered as non knowledge elements because the change made in its value does not invoke a pattern-matching algorithm. The tool supports both global and local variables. The global ones are visible and can be accessed anywhere in the environment while local variables are accessible only in a module in which they are defined. What is more, the local variables are removed when a reset function is executed, while the global are set to their initial value. The CLIPS variables are not restricted to holding a value of a well-defined data type i.e. CLIPS supports only weakly typed variables. This means that a value of a variable is automatically converted according to the context of usage.

10.3 Analysis of the DROOLS Rule Language

DROOLS is a one of the most commonly used implementation of BRMS. It introduces the Business Logic integration Platform divided into five subprojects. From the perspective of this chapter the most important is DROOLS Expert because it is a rule engine providing forward-chaining inference mode and execution control features. It uses a dedicated rule language (DRL) which is analyzed next.

Data Types

In comparison to CLIPS, where a type of a global variable is assigned dynamically and where already defined types can be used only for facts, types in DROOLS can be used for defining both global variables and facts. Thanks to the tight connection between the JAVA and DRL languages, any type, that is defined within JAVA, can also be used within DRL. A new type in DRL can be defined using declare keyword followed by the list of typed fields, and the keyword end. The following example shows the definition of the Driver type described by three elements:

```
declare Driver
   DriverClass : int
   Age         : int
   LicAge      : int
end
```

Similarly to CLIPS, the DRL syntax makes the definition of a complex facts possible in a similar manner.

The DRL language supports inheritance by using an `extends` construct. A new type can be defined as an extension of an already existing type that is defined in JAVA or DRL. A class hierarchy, that is built thanks to the inheritance, is supported by the pattern-matching algorithm and allows rules to act polymorphically. Rules containing a fact of a base class in their LHS are considered during pattern-matching also when a facts of a derived class are asserted.

Facts

Facts in DROOLS can be defined using already defined types. Since DROOLS is written in JAVA, the DRL types and fact instances are mapped to JAVA classes and objects, respectively. The fact instances are then inserted to working memory and thus are handled in a different way than ordinary objects.

Ordered facts DROOLS provides ordered-like syntax for facts that do not require specification of the names of the fact fields, e.g.:
`Driver(DriverClass == 9, Age == 18, LicAge == 5)` can also be written as `Driver(9, 18, 5)`. When such syntax is used, DROOLS matches the arguments to the order of the fact fields defined within `declare...end` statement.

Non-ordered facts in DROOLS are supported in a native way. Actually, every fact instance can be referred to as a non-ordered one including ordered facts. The last issue concerns fields of facts that can store multiple values. A DROOLS type definition construct does not support any syntax that semantically corresponds to multislots. Nevertheless, DROOLS allows for using any existing class as a type of a fact field. This allows for defining fields as being a collection (e.g. lists) that can act similarly to multislots.

Rules

Rules in DROOLS have a similar format to that of CLIPS.

```
rule "name"
   <attributes>*              // Rule Properties
   when
      <conditional element>* // Left-Hand Side (LHS)
   then
      <action>*               // Right-Hand Side (RHS)
   end
```

The conditional part consists of zero or more CEs. If it is empty, it will be considered as a condition element that is always TRUE and it will be activated once, when a new working memory session is created. Each CE works on at least one pattern that can correspond to one fact instance of a given type. Each pattern can additionally impose some other restrictions on a fact, mainly related to values of fact fields. For this purpose, DROOLS allows for using the following operators: <, <=, >, >=, ==, ! =, contains, not contains, memberof, not memberof, in, not in, matches, not matches, str, soundslike and inline version of eval.

The conclusion part provides a set of actions that are performed when the LHS of a rule is satisfied. The strong connection of DROOLS and JAVA allows for using any JAVA function and expression within RHS. The most important actions having impact on knowledge processing are described next. The insert action allows for the creation of a new object in the working memory. The retract action allows for removing some fact(s), given as argument, from the working memory. The result of modify is logically equivalent to the sequence of retract and insert actions. The inference engine can be also notified about changes in the working memory by using update action.

The declare part in DROOLS allows for providing many different rule attributes. The majority of the rule attributes are related to rule execution control and rule base modularization. The most important are the salience and auto-focus properties that play the same role as in CLIPS. Their value can be retrieved from different sources e.g. from a global variable or function call. The remaining properties like no-loop, lock-on-active, activation-group, agenda-group and ruleflow-group are related to rule base modularization.

An example of complete rule written in DROOLS is presented below. This is a corresponding rule to those defined with the help of CLIPS:

```
rule "driver-age"
   agenda-group "base-charge-modifiers"
   lock-on-active true
when
   Driver($Age : Age, Age >= 40, Age <= 55)
then
   insert(new Modifier(-10));
end
```

Variables

They play in DROOLS a similar role as in CLIPS. Nevertheless, they may only be defined within global scope thanks to the global keyword. The main goal of global variables in DROOLS is to make objects from the JAVA application available in rules. Typically, they are used to provide data or services that the rules use.

10.4   Formalization of Selected Features of Rule Languages

The identified language features are divided into six categories related to data types and facts, rule LHS, rule RHS, rule attributes, variables, and rules impact on inference. Each category is presented in a separate table that provides a set of features and briefly describes how they are supported in each rule language and the formalized model proposed in Chap. 8.

Data types and facts The model provides a complete description of types. It defines two primitive ones called *Real* and *Smbl* that correspond to numeric and symbolic values, respectively. Each type T defined in the model consists of two elements determining the structure of the type and set of allowed values of its instances:

$$T \overset{def}{=} (\mathbf{S}_T, \mathbf{A}_T) \tag{10.1}$$

where:

- $\mathbf{S}_T$ defines a set of *composite* types of type T and determines structure of facts in terms of other types. It is represented as a tuple of types: $\mathbf{S}_T \overset{def}{=} (T_1, T_2, \ldots, T_n)$.
- $\mathbf{A}_T$ defines a set of attribute interpretation functions of facts (facts attributes for short). It is represented as a tuple of attribute functions: $\mathbf{A}_T \overset{def}{=} (A_1, A_2, \ldots, A_n)$.

The model provides several operators working with types that allows for extending a certain type with the help of other types (inheritance – operator "::") or defining instances of the type (facts – operator ":").

The type constraints the set of allowed values of facts in two ways. Firstly it defines the structure that each fact must have. Secondly, the codomain of each attribute defines a set of allowed values of each element of a fact structure.

An attribute is a fact interpretation function that maps a single fact to the set of allowed values D_{A_i} (domain). An attribute function is defined as:

$$A_i : \mathbb{F}_T \to D_{A_i} \tag{10.2}$$

where:

- A_i is an element of the $\mathbf{A}_T$ tuple.
- $\mathbb{F}_T$ is the set of all facts being of type T.
- The D_{A_i} set is called an *attribute domain* (in fact this is a codomain of the attribute interpretation function).

In context of identification of corresponding elements, ten aspects of rule languages are considered in this category:

t1 *Primitive Data types* – the list of primitive data types provided by the language.
t2 *Data types definitions* – means of defining own data types based on the primitive ones.

t3 *Complex data types* – means of defining complex data types i.e. data types being a composition of fields of other types.

t4 *Unordered data types* – indicates if the fields of facts can be referred by their name.

t5 *Ordered data types* – indicates if the language allows for using facts as tuples where the order of the fields is important.

t6 *Derivation* – corresponds to inheritance of types.

t7 *Multivalued fields* – denotes fields taking more than one value at once i.e. collections/sets of values.

t8 *Constraints* – denotes if the rule language allows for defining a set of admissible values for the fields within certain type.

t9 *Meta data* – additional annotations made during type definition that allows for changing type interpretation.

t10 *Fact comparison* – denotes how the objects are compared.

The summary of comparison of the two selected rule languages in terms of data types and facts is provided in Table 10.4.

LHS *of a rule* Considering rule-based systems as a dynamic systems, rules are the elements that define the dynamics of this system by specifying possible transitions between system states. The proposed model considers a rule as a transition function that is defined with the help of knowledge formulae that describe how the transitions between system states must be performed. Each single rule defines a single transition that can be made always when a certain condition is satisfied. The model defines the condition (LHS) of a rule as its domain. It can be defined with the help of two main types of formulae: *pattern* formulae build with the help of $\mathcal{P}$ operator and *constraint* formulae expressed using $\ominus$, $\textcircled{\#}$ and $\odot$ operators.

Pattern formulae allows for defining a set of facts having a common feature or value i.e. facts that satisfy the requirements expressed by logical formula defined by the operator. For example the following pattern formulae $\mathcal{P}(S, T, \phi_l)$ defines a set of facts belonging to the current system state S, being of type T and satisfying logical formulae ϕ_l (Table 10.1).

Pattern formulae works on the fact level whereas constraint formulae on the set of facts level. This type of formulae allows for providing a simple condition for the set of facts as a whole and can be evaluated to *true* or *false* values. The definition of constraint operators allows only pattern formula to be an argument of this operator. The provided interpretations of constraint operators aim at checking the relation between two sets of facts – a set defined by a pattern formulae without constraints: $\mathcal{P}(S, T, true)$ and its subset containing those facts that satisfy provided formula ϕ_l. In this context, the $\ominus$ operator means that there is at least one fact belongs to $\mathcal{P}(S, T, true)$ that satisfy ϕ_l. In turn, $\odot$ allows for checking if each fact belonging to $\mathcal{P}(S, T, true)$ satisfy ϕ_l. In order to make the definition of the constraint operators clearer, one can write their interpretations in a more formalized way. For the $\ominus$ operator: $\ominus \mathcal{P}(S, T, \phi_l) = true \Leftrightarrow \exists F \in \mathcal{P}(S, T, true): \phi_l(F) = true$.

Table 10.1 Summary of the support for the type system provided by the considered rule languages and proposed model

Feature	Tool		
	CLIPS	DROOLS	Model
t1. Primitive data types	`float`, `integer`, `symbol`, `string`	All JAVA primitive types	*Smbl*, *Real*
t2. Data types definitions	`deftemplate` construct	`declare ... end`	Definition of T type
t3. Complex data types	`deftemplate` construct	`declare ... end`	Usage of existing types in $\mathbf{S}_T$
t4. Unordered data types	`slot`	Names of fact fields	Names of attributes
t5. Ordered data types	In the form of tuple (`a,b,c,...`)	In the form of object construct `a(b,c,...)`	In the form of tuple (`a,b,c,...`)
t6. Derivation	No[a]	`extends` keyword	`::` operator
t7. Constraints	`allowed-values`, `range`	JAVA enums	D_{A_i} defines allowed values
t8. Multivalued fields	`multislot`	JAVA collections	D_{A_i} may contain sets as allowed values
t9. Meta data	No	Yes	No
t10. Fact comparison	Equal fact must have the same type and values	Equal fact must have the same value of its hash[b]	Equal fact must have the same type and values

[a]Inheritance in CLIPS is supported only in COOL
[b]Using meta data for a certain type, one can select these fields of the type that are relevant for hashing function

In an analogous way an interpretation of an ⓐ operator can be provided. In turn, the ⓥ operator can be defined in the following way:

$$\bar{\vee}\mathcal{P}(S, T, \phi_l) = true \Leftrightarrow \forall F \in \mathcal{P}(S, T, true): \phi_l(F) = true.$$

The domain of a rule may contain several pattern and constraint expressions. All these expressions are connected with the help of ⓐ operator that denotes a logical conjunction.

The following nine aspects regarding the LHS formulation, since they determine the expressiveness of rule statement.

l1 *Negation as failure* – due to the fact that all of the selected languages support only one type of negation, this feature also denotes if the provided language supports open or closed word assumption.

l2 *Fact references* – indicates if language allows for referring to facts instantiating rules.

l3 *Complex expressions* – denotes if rule language allows for expressing constraints using complex expressions i.e. expressions containing at least one operator or function call.

14 *Pattern expression* – specifies if the LHS of a certain rule allows for selecting only facts that satisfy provided constraint.

15 *Conjunction expression* – indicates if the conditional expressions can be connected by using conjunction logical operator.

16 *Disjunction expression* – denotes if the conditional expressions can be connected by using disjunction logical operator.

17 *Quantified expressions* – denotes if rule language allows for expressing constraints involving set of facts instead of only single fact.

18 *Aggregation expressions* – specifies if the rule language supports constructs allowing for defining constraints involving value aggregation.

19 *Any Boolean expressions* – is related to the possibility of creating a rule conditional parts from any expressions that can be evaluated *true* or *false*.

A comparison of the support for LHS elements within the selected rule languages can be found in the Table 10.4.

RHS *of a rule* Features related to conclusion part (RHS) are complementary to the features involving a conditional part of the rules. The rule RHS provided by the model is defined with the help of knowledge formulae. In turn, these formulae are composed of three different operators that allow for manipulating knowledge. Two of them, $\mathcal{A}$ (assert) and $\mathcal{R}$ (retract) operators, can be used for adding and removing facts from the fact base. Additionally, the assert operator allows for specifying logical support for a newly added fact with the help of a definition of its third argument. The third, assignment operator, allows for changing the interpretation functions of the facts and variables, and thereby, modification of the current knowledge. Apart from knowledge manipulation capabilities, the RHS of a rule allows for changing the scope of the evaluated rules during the next iteration of an inference algorithm. In this way it allows for changing the default inference control (Table 10.2).

The following seven elements have the greatest impact on capabilities of rule language connected to rule base modification (Table 10.3):

r1 *Complex expressions* – denotes if the rule language allows for expressing constraints using complex expressions i.e. expressions containing at least one operator or function call.

r2 *Facts assertion* – indicates if the rule language allows for modifying knowledge based on creating and inserting new facts.

r3 *Truth maintenance* – represents if rule language supports truth maintenance mechanisms i.e. allows for logical assertions that are automatically removed if the conditional part of a rule becomes unsatisfied.

r4 *Facts retraction* – indicates if rule language allows for modifying knowledge base by removing facts.

r5 *Facts modification* – means that the conditional part of a rule may contains instructions that modify existing facts.

r6 *Facts modification* (no pattern matching) – similar to the previous one, besides the case that such modification does not invoke a new iteration of pattern-matching.

r7 *Inference control* – points if conclusion a part of a rule may contain constructs that affect inference control e.g. changing current focus.

Table 10.2 Summary of rule LHS expressiveness in considered rule languages and proposed model

Feature	Tool		
	CLIPS	DROOLS	Model
11. Negation as failure	Yes	Yes	Yes
12. Fact references	`<-` operator	`:` operator	Reference by rule arguments
13. Complex expressions	Yes	Yes	Yes
14. Pattern expressions	Yes	Yes	$\mathcal{P}$ operator
15. Conjunction expressions	`(and ...)` construct	`(and ...)`, `and` operators	$\otimes$ operator
16. Disjunction expressions	`(or...)` construct	`(or...)`, `or` operators	No[a]
17. Quantified expressions `forall`	`(forall ...)` construct	`forall(...)` operator	$\otimes$ operator
17. Quantified expressions `exists`	`(exists ...)` construct	`exists(...)` operator	$\exists$ operator
17. Quantified expressions `not`	`(not ...)` construct	`not(...)` operator	$\nexists$ operator
18. Aggregation expressions	No[b]	`accumulate`, `collect` operators	No[b]
19. Any Boolean expressions	`(test ...)` construct	`eval(...)` operator	With the of $\mathcal{P}$ operator

[a]Disjunctions can be easily replaced by additional rule
[b]Aggregations can be easily replaced by additional rule(s)

Table 10.4 provides comparison of these features in terms of the selected rule languages:

Rule attributes The fourth group of features involves properties of rules that are not related to inference control. In the considered rule languages only two rule properties are considered: a1: priority and a2: module assignment.

Priority is mainly used by the inference engine in order to determine the order of rules within a conflict set. The model provides a corresponding feature in the form of the function that maps a given rule into a real number. This function is denoted as π and is defined as follows:

$$\pi: \Gamma \to \mathbb{R} \tag{10.3}$$

where Γ is the set of all rules. The Γ set can be divided into several non-empty pairwise disjoint subsets that are called modules:

$$\Gamma = M_1 \cup M_2 \cup \cdots \cup M_n \tag{10.4}$$

where M_i is a non-empty set called *module* and $\forall_{i \neq j} M_i \cap M_j = \varnothing$. According to the provided definition a single module is a set of rules. In turn, M_i belongs to the ordered

Table 10.3 Summary of rule RHS expressiveness in the considered rule languages and the proposed model

Feature	Tool		
	CLIPS	DROOLS	Model
r1. Complex expressions	Yes	Yes	Yes
r2. Facts assertion	`(assert ...)` construct	`insert(...)` function	$\mathcal{A}$ operator
r3. Truth maintenance	`(logical ...)` construct[a]	`LogicalInsert` function	Specified in third argument of the $\mathcal{A}$ operator
r4. Facts retraction	`(retract ...)` construct	`retract(...)` function	$\mathcal{R}$ operator
r5. Facts modification	`(modify ...)` construct	`modify(...)` function	$\ominus$ operator
r6. Facts modification (no pattern matching)	No	Modification of fact without `modify` function	No
r7. Inference control[b]	Yes	Yes	Yes

[a] This construct must be used within LHS of a rule
[b] Details concerning inference control are provided within Table 10.6

set of all modules ($\mathbb{M}$, $<_M$), where the $<_M$ relation defines the order of the modules. The maximal, in terms of $<_M$, element is always referred as M_a and is called the *autofocus* module. This is the only module that can be empty and contains rules that simultaneously must belong to any of the remaining modules. It has a special purpose that is related to the *autofocus* rule property.

These two rule properties play a significant role in knowledge interoperability:

- *Priority* – indicates if a language provides mechanisms allowing for rule prioritizing.
- *Module assignment* – denotes how rules are assigned to modules. *Manual* assignment means that each rule must define the appropriate property that points to the target module. In turn, *automatic* means that rules modularization is done in an automatic way.

Table 10.4 provides information concerning how the rule attributes are supported by the selected languages.

Variables In general, variables that can refer to instances of already defined types (e.g. to facts). However, the most significant differences between support for variables in rule languages are related to their scope and a way of typing. Therefore, in this context two important aspects are considered:

v1 *Global/local variables* – indicates that the rule language supports global/local variables.

Table 10.4 Summary of support for rule attributes provided by the considered rule languages and proposed model

Feature	Tool		
	CLIPS	DROOLS	Model
a1. Priority	`salience` construct	`salience` keyword	π function
a2. Module assignment	Manual – rule name preceded by `::` operator and module name	Manual – `agenda-group` or `ruleflow-group` rule properties	Manual – Γ set is divided into modules

Table 10.5 Summary of support for variables provided by the considered rule languages and proposed model

Feature	Tool		
	CLIPS	DROOLS	Model
v1. Global/local variables	Both	Both	Global, set of all variables $\mathbb{V}$
v2. Typing	Weak	Strong	Strong

v2 *Typing* – denotes how the concept of variable works with a type system. There are considered only two cases: *weak* typing and *strong* typing. Weak typing means that the type of variable is defined dynamically after value assignment. In turn, in the case of strong typing the assigned value must be compatible with type of variable that is defined earlier.

The comparison of rule languages with respect to these features can be found in Table 10.5.

Inference control In order to provide support for features of rules related to inference control like *lock-on-active* or *activation-group*, each single module in the model has its own structure. Each module is divided into several submodules that correspond to these features of rules. This is why, a structure of a single module M_i is defined as follows:

$$M_i = M_i^o \cup M_i^l \cup M_i^{\oplus_1} \cup M_i^{\oplus_2} \cup \cdots \cup M_i^{\oplus_n} \qquad (10.5)$$

where:

- M_i^o is called an *ordinary* submodule and contains only rules that can be processed in an ordinary way.
- M_i^l is called a *lock-on-active* submodule and contains rules that can be instantiated only by facts that are not modified by other rules from the module M_i.
- $M_i^{\oplus_j}$ is called a *xor* submodule and contains rules that can be activated only when there is no pending activation of rule from the submodule $M_i^{\oplus_j}$. In contrast to the *lock-on-active* submodule, there can be more than one *xor* submodule within a single module M_i.

Table 10.6 Summary of rules impact on the inference process provided by the rule languages and model

Feature	Tool		Model
	CLIPS	DROOLS	
i1. Self-activation control	No	`no-loop` feature	Modification of M rule argument in the rule RHS: $M^1 := M \backslash r$
i2. Re-activation control	No	`lock-on-active` feature	M^l submodule
i3. Exclusive execution	No	`activation-group` feature	$M^{\oplus i}$ submodule
i4. Module autoswitching	`auto-focus` rule property	`auto-focus` rule property	M_a module

- n is the number of *xor* submodules.
- It is assumed that an *ordinary* submodule is disjoint with all the remaining submodules, while all the submodules, besides an *ordinary* one, can contain the same rules i.e. they are allowed to have a non-empty intersection.

The following group of language features involves aspects related to the impact of rules on inference control.

i1 *Self-activation control* – corresponds to such properties of rule that allows for locking rule reactivation caused by changes made by this rule.
i2 *Re-activation control* – is more general than the *self-activation* feature and is related to rule property that prevents rules from activating caused by rules from the same module.
i3 *Exclusive execution* – corresponds to mechanism that allows for combining rules into groups in which only one rule at a given point of time can be scheduled for firing.
i4 *Module autoswitching* – indicates if a rule language provides an additional execution control triggered by rule activation.

Table 10.6 provides information concerning how rule attributes are supported by the selected languages.

This section provided identification of the corresponding features of the considered rule languages in terms of their semantics expressed by the formalized model. These features are considered as the most important ones among those that have a direct impact on the knowledge semantics. This is why, these features play a crucial role in this work in the context of semantic knowledge interoperability. Tables from 10.4 to 10.6 show how the features from one representation are translated into another in order to preserve the original semantics of rules in the original language.

10.5 Rule Translation Using a Case Study

For evaluation purposes, the PLI use case is used. This use case is considered as non trivial since it provides a structured rule base consisting of four modules having a different nature.

Definitions of Types

The PLI use case presents a system for determining the price of the liability insurance, which protects against third party insurance claims. The final price of the insurance is determined by several factors that can be classified into groups related to the client, insured car and the type of insurance. Therefore, the model of the system introduces three fact types that correspond to these groups: T_{Driver}, T_{Car}, $T_{Insurance}$.

The model-based definition of the T_{Driver} can be written as follows:

$$T_{Driver} = \left(\mathbf{S}_{T_{Driver}}, \mathbf{A}_{T_{Driver}} \right) = \left((Real, Real, Real), (A_{DriverClass}, A_{Age}, A_{LicAge}) \right)$$
(10.6)

In order to refine this definition a specification of the attributes and their domains must be provided:

$$A_{DriverClass} : \mathbb{O}_{T_{Driver}} \rightarrow \left\{ x \in \mathbb{Z} : x \in [-1, 9] \right\}$$
$$A_{Age} : \mathbb{O}_{T_{Driver}} \rightarrow \left\{ x \in \mathbb{N} : x \in [18, 120] \right\} \quad (10.7)$$
$$A_{LicAge} : \mathbb{O}_{T_{Driver}} \rightarrow \left\{ x \in \mathbb{N} : x \in [0, 102] \right\}$$

The corresponding definition of the T_{Driver} type expressed in CLIPS rule language is as follows:

```
1  (deftemplate driver
2    (slot class  (type INTEGER) (range -1 9) (default 1) )
3    (slot age    (type INTEGER) (range 18 120) )
4    (slot licage (type INTEGER) (range 0 102) )
5  )
```

Type definitions in DRL are translated by the DROOLS engine into JAVA classes. Thus, the DRL definition markup does not provide any constructs for domain specification and allows only for type assignment:

```
1  declare Driver
2    DriverClass : int
3    Age         : int
4    LicAge      : int
5  end
```

All the previous rule languages provide support for complex types. XTT2 supports only primitive types and thus the definition of the T_{Driver} type must be expressed in terms of three separate XTT2 types:

```
 1  xtype [name: typeDriverAge,
 2     base: numeric, length: 3,
 3     scale: 0, domain: [18 to 120]
 4  ].
 5  xtype [name: typeDriverClass,
 6     base: numeric, length: 1,
 7     scale: 0, domain: [-1 to 9]
 8  ].
 9  xtype [name: typeDriverLicage,
10     base: numeric, length: 3,
11     scale: 0, domain: [0 to 102]
12  ].
```

All the provided XTT2 types are based on the `numeric` primitive type. This type corresponds to `double` type known from DROOLS or the NUMBER type provided by CLIPS or JESS. An integer type can be expressed in XTT2 with the help of the `length` and `scale` parameters provided by the type definition markup. The `length` parameter denotes how many digits in total (including decimal ones) the value can have. In turn, the `scale` parameter defines how many decimal places are allowed. Setting this parameter to 0 constraints the type to only integer values (decimal places are forbidden). Moreover, all the XTT2 types constraint the set of allowed values also with the help of the `domain` attribute.

Initial State

The decision process must start from a well-defined initial state. The initial state(s) can be defined just after the type definitions by asserting facts into a knowledge base that may be retrieved from e.g. the external environment. In the context of this case study, the appropriate initial state must provide information concerning driver, car and insurance. Therefore, at the beginning the system needs three facts: the first fact must be of T_{Driver} type, the second one of T_{Car} type and the last one of $T_{Insurance}$ type.

The following definition provides three objects having exemplary identifiers that correspond to specific instances of the T_{Driver}, T_{Car} and $T_{Insurance}$ types, respectively. Thus, the initial state of the $\mathbb{O}$ set can be defined as follows:

$$\mathbb{O} = \{o_{tom}, o_{vw}, o_{ins}\} \tag{10.8}$$

It is assumed that the o_{tom} object corresponds to a person whose name is *tom* and it is an instance of the T_{Driver} type. The o_{vw} object corresponds to a car which brand name is *VW* and it is an instance of the T_{Car} type. Finally, the o_{ins} object corresponds to an insurance agreement and it is an instance of the $T_{Insurance}$ type. For each of the objects, the initial values of their components must be provided according to the external system. All this can be expressed in terms of the proposed model in the following way:

$$o_{tom} : T_{Driver} \quad \text{and } o_{tom} = (7, 50, 10)$$
$$o_{vw} : T_{Car} \quad \text{and } o_{vw} = (1900, 14, \textit{true}, 5, \textit{false}, 0) \qquad (10.9)$$
$$o_{ins} : T_{Insurance} \text{ and } o_{ins} = (\textit{true}, 2, \text{single}, \textit{false}, \textit{false})$$

The initial state of each object is defined by a tuple containing values in the order corresponding to the order of attributes in the type definition. Having these objects, the required initial facts, related to them, may be introduced:

$$\mathbb{F} = F_{o_{tom}} \cup F_{o_{vw}} \cup F_{o_{ins}} \qquad (10.10)$$

Both in CLIPS and JESS, the initial state of the system can be defined by a dedicated command (deffacts).

```
1   (deffacts initial-facts
2      (driver (class 1) (age 29) (licage 2))
3      (car (capacity 997) (age 12) (historic false) (seats 5)
4         (technical true) (accidents 0))
5      (insurance (continue true) (cars 1) (payment single)
6         (otherins false) (certificate true))
7   )
```

In contrast DROOLS does not provide dedicated syntax for initial state definition. The initial facts can be defined two ways: (1) with the help of rule that asserts required facts or (2) in JAVA code that is used for the execution of the system. From this thesis point of view, the first approach is better as the initial state can be defined in the DRL rule language (not in JAVA). The following listing shows a fragment of a rule asserting three initial facts:

```
1   rule "initial"
2     agenda-group "initial"
3   when
4   then
5     ...
6     insert(new Driver(1, 29, 2));
7     insert(new Car(977, 12, false, 5, true, 0));
8     insert(new Insurance(true, 1,
              PaymentType.single, false, true));
9     ...
10  end
```

XTT2 provides a concept of *states* which allows for specifying the initial states of the system. In fact, a single state is a set of assignments of values to attributes, where each assignment is labeled with the same name e.g. `input/1`. Thus, a state definition consists of three elements: state name (label), attribute(s) and value(s).

```
1  xattr [name: driverClass,
2     abbrev: driv4,
3     class: simple,
4     type: typeDriverClass,
5     comm: inter,
6     desc: 'The class of a client'
7  ].
```

This definition combines attributes, that describe the system, with already defined types (see 4th line). The following code defines an initial state named `input/1` that corresponds to the previously presented states:

```
1  xstat input/1: [driverClass,1].
   xstat input/1: [driverAge,29].
   xstat input/1: [driverLicage,2].
2  xstat input/1: [carCapacity,997].
   xstat input/1: [carAge,12].
   xstat input/1: [carHistoric,0].
3  xstat input/1: [carSeats,5].
   xstat input/1: [carTechnical,1].
   xstat input/1: [carAccidents,0].
4  xstat input/1: [insuranceContinue,1].
   xstat input/1: [insuranceCars,1].
   xstat input/1: [insurancePayment,single].
5  xstat input/1: [insuranceOtherins,0].
   xstat input/1: [insuranceCertificate,1].
```

Definitions of Rules

The selected rule involves many nontrivial aspects for the rule interoperability method and thus it gives an intuition as to how the remaining rules can be interchanged. The following rule is responsible for modification of the value of the $A_{DriverClass}$ attribute according to the current value of this attribute and the number of accidents during the last year:

$$r_{bonus-malus1-2}(M, H_1, H_2) == \begin{array}{l} \text{if} \begin{cases} H_1 \in \mathcal{P}\left(\mathbb{F}, T_{Driver}, \left(A_{DriverClass}(H_1)\oplus\{2,3,4,5,8,9\}\right)\right) \\ H_2 \in \mathcal{P}\left(\mathbb{F}, T_{Car}, A_{Accidents}(H_2)\ominus1\right) \end{cases} \\ \text{then} \begin{cases} A_{DriverClass}(H_1)\ominus A_{DriverClass}(H_1)\ominus3 \\ M \end{cases} \end{array}$$

$$(10.11)$$

This rule can be read as follows:

If the current[1] value of the $A_{DriverClass}$ attribute of a certain fact of the T_{Driver} type, belongs[2] to the set $\{2, 3, 4, 5, 8, 9\}$ and[3] the current value of the $A_{Accidents}$ attribute of a certain fact of the T_{Car} type, is equal[4] 1, then decrease the value of the $A_{DriverClass}$ attribute of the fact of the T_{Driver} type by 3.

The translation of this rule requires the translation of two conditions and one decision constructs. Translation to CLIPS/JESS are equivalent and can be done by expressing the first condition with the help of a predicate constraint in a pattern conditional element by using member$ construct. In turn, the second condition can be written by a simple literal constraint in a pattern conditional element. The translation of a decision statement is straightforward and can be done by the (modify) function. The complete CLIPS/JESS rule is as follows:

```
1   (defrule bonus-malus::bonus-malus1-2
2     ?driver <- (driver
      (class ?class&:(member$ ?class (create$ 2 3 4 5 8 9))))
3     (car (accidents 1))
4   =>
5     (modify ?driver (class (- ?class 2)))
6     (pop-focus)
7   )
```

Translation to DROOLS rule language can be considered as more intuitive than to CLIPS/JESS because of syntax of the DRL. The membership of the value of the DriverClass in a given set can be checked by using the in operator. In turn, in order to test if the value of Accidents is equal to 1, the == operator can be used. The decision part is analogous as in CLIPS and is expressed with the help of modify function:

```
1   rule "bonus-malus1-2"
2     agenda-group "bonus-malus"
3     lock-on-active true
4     activation-group "bonus-malus-ag"
5   when
6      $driver : Driver($DriverClass : DriverClass,
7                          DriverClass in (2, 3, 4, 5, 8, 9))
8      Car(Accidents == 1)
9   then
10    modify($driver){setDriverClass($DriverClass - 2)}
11  end
```

---

[1] *current* – because the H_1 and H_2 belong to $\mathbb{F}$ i.e. current state of the system (see Definition (8.25) on p. xxx).

[2] *belongs* – due to the operator $\ominus$.

[3] *and* – as the single rule conditions are connected by $\oslash$ operator (see Definition (8.35) on p. xxx).

[4] *equal* – because of the operator $\ominus$.

Translation to XTT2 can be done in a straightforward way by the direct use of operators having corresponding semantics. The first condition may be written by using `in` operator, while the second one by using an `eq` operator. In turn, the value of the `driverClass` attribute can be changed by the `set` operator:

```
1   xrule bonus-malus/5:
2     [driverClass in [2,3,4,5,8,9]],
3      carAccidents eq 1
4   ==>
5     [driverClass set (driverClass-2)]
6     :base-charge.
```

In the case of the CLIPS, JESS and DROOLS, a rule engine searches for facts that satisfy the provided conditions. In general, there can be several facts matching each condition. In such a case, every combination of them is used by the engine for rule instantiation. Due to the different nature of XTT2, such behavior cannot be observed during inference in this representation. This is because, the inference engine tests conditions which involve specific attributes and does not search for other ones. Therefore, rules in this language do not refer to any attribute of a given type but to a specific one. This difference may make the automatic translation difficult as in XTT2 the number of attributes must be known in advance, see [3] for more details.

In the case study this problem can be observed within all rules belonging to the $M_{base-charge-modifiers}$ module. All the rules from this module assert a new fact of the $T_{Modifier}$ type that provides a value of the insurance base charge modification. All these facts (base charge modifications) are later aggregated and removed by the $r_{calculation}$ rule (see Definition (10.12)) until at least one modification exist in the fact base.

$$r_{calculation}(M, H_1, H_2, H_3) == \begin{cases} \text{if} \begin{cases} H_1 \in \mathcal{P}\left(\mathbb{F}, T_{Base}, true\right) \\ H_2 \in \mathcal{P}\left(\mathbb{F}, T_{Modifier}, true\right) \\ H_3 \in \mathcal{P}\left(\mathbb{F}, T_{Result}, true\right) \end{cases} \\ \text{then} \begin{cases} A_{Value}(H_3) \ominus \\ \quad \ominus\left(A_{Value}(H_3) \oplus A_{Value}(H_1) \odot (A_{Value}(H_2) \ominus 100)\right) \\ \mathcal{R}\left(H_2\right) \\ \{M \end{cases} \end{cases}$$

$$(10.12)$$

Such an aggregation of values from the existing facts cannot be expressed in XTT2. However, there are two ways to provide semantically equivalent constructs in XTT2. In the first way, there must be a separate attribute for each reason of modification which takes 0 as the initial value. If the rule considering a particular reason of modification is executed, it changes the value of such an attribute to the desired value. After an evaluation of all of the rules within the $M_{base-charge-modifiers}$ module an aggregation rule may be executed in order to sum all the attributes storing modification values. This rule may be defined in the following way:

```
1   xrule 'MAIN'/1:
2    [base-modifierValue_1 eq any,
3     base-modifierValue_2 eq any,
4     ...
5     base-modifierValue_n eq any,
6     baseValue eq any]
7   ==>
8    [resultValue set (baseValue+(baseValue*(
9     (base-modifierValue_1+base-
        modifierValue_2+...+base-modifierValue_n)
10    /100.000)))].
```

The second way to express the aggregation is to make it earlier. For example, it can be done by the rules belonging to the $M_{base-charge-modifiers}$ module. Using this way only one attribute that stores total a modification value is required. Initially it takes the value of 0 and later it is modified by every executed rule that provides base charge modification. At the end, it is taken by the $r_{calculation}$ rule into account in the following way:

```
1   xrule 'MAIN'/1:
2      [base-modifierValue eq any,
3       baseValue eq any]
4   ==>
5      [resultValue set (baseValue+
        (baseValue*(base-modifierValue/100.000)))].
```

In this case, the second approach was applied. Nevertheless, an implementation of the translation method that uses any of these ways is a non trivial issue as it requires a semantic analysis of the aggregation.

10.6   Translation of Module Structure

Defining the modules determines the structure of a rule base. A rule base in the provided case study consists of four modules were each module has a different nature i.e. rules in every module must be processed in different way. Therefore, the following example discusses how the defined modules must be interpreted and presents one exemplary rule from each module in all the considered rule languages (the remaining rules may be translated in an analogous way).

The first of the distinguished modules contains rules that are intended to determine the new client class. This module is called $bonus - malus$. The second module provides rules responsible for the calculation of the base price of an insurance and is named $base - charge$. Rules from the third module, that is called $base - charge - modifiers$, provide information as to how the base charge can be increased or

Fig. 10.1 Order of module evaluation in the PLI use case

decreased. The last module is called *main* and it provides a rule that calculates the final price of the insurance. Taking the provided names into account, the set of all modules identifiers can be defined as follows:

$$\mathbb{X}_M = \{main, bonus - malus, base - charge, base - charge - modifiers\}$$
$$(10.13)$$

A complete definition of the set of all modules requires a definition of the ordering relation that reflects the default order of the modules evaluation performed by the inference engine:

$$(\mathbb{M}, <_M) = \{M_{bonus-malus}, M_{base-charge}, M_{base-charge-modifiers}, M_{main}\} \quad (10.14)$$

where the ordering relation is defined in a way that the following holds:

$$M_{bonus-malus} <_M M_{base-charge} <_M M_{base-charge-modifiers} <_M M_{main}$$

This relation can also be depicted in a graphical form (see Fig. 10.1).

Modules evaluation order can also be defined within considered rule languages. Definitions in CLIPS and JESS are the same and can be written with the help of dedicated function (focus) in the following way:

```
1  (focus base-charge bonus-malus base-charge-modifiers)
```

This definition specifies the content of the focus stack where the modules are pushed in the reverse order they are listed. The current module is set to the last pushed module i.e. the current module is on the top of the stack. Such an approach can also be observed in the DROOLS function drools.setFocus("module name") which pushes the "module name" module to the top of the stack. This function can be used only within the decision part of a rule:

```
1  rule "initial"
2     agenda-group "initial"
3  when
4  then
5     drools.setFocus("base-charge-modifiers");
6     drools.setFocus("base-charge");
7     drools.setFocus("bonus-malus");
8  end
```

During the inference, the modules may be pushed on the stack (e.g. by an executed rule) or removed from it when there are no rule activations in the module, or if a certain rule forces such removal. In XTT2, the order may be determined manually, by connections between tables or by functional dependencies between attributes. It is assumed that the interoperability method uses mainly connections because they allow for the most flexible modeling of modules evaluation order. Figure 10.3 shows a visual representation of the XTT2-based model of PLI use case. This model consists of four tables corresponding to modules. The connections between the tables define the order of their evaluation.

Each of the modules has a homogeneous structure i.e. all the rules within a certain module also belong to the same submodule. Rules in first module called *bonus − malus* are intended to update the value of the driver class according to the current value of this attribute and the number of accidents during the last year. The rules are defined in a way that the intersection of LHS of each of the two of them is empty. Thus, in order to update the current value of this attribute, only one rule can be executed during the evaluation of the module. On the other hand, each rule modifies a fact that simultaneously instantiates it. Such modification causes the next iteration of the inference against the modified value of the driver class attribute. During this iteration not desirable rule activations may appear. Therefore, in order to ensure that rules are not evaluated against modifications made by themselves, each of them must be placed in the *lock-on-active* submodule. Furthermore, in order to assure only one execution, all the rules must also be placed in the *xor* submodule. Finally, the definition of the $M_{bonus-malus}$ module can be written as follows:

$$M_{bonus-malus} = M^l_{bonus-malus} = M^{\oplus bonus-malus-ag}_{bonus-malus} = \{r_{bonus-malus0}, r_{bonus-malus1-1},$$
$$r_{bonus-malus1-2}, r_{bonus-malus1-3}, r_{bonus-malus2-1}, r_{bonus-malus2-2},$$
$$r_{bonus-malus2-3}, r_{bonus-malus2-4}, r_{bonus-malus3-1}, r_{bonus-malus3-2}\}$$
$$(10.15)$$

The translation of the presented structure to the considered languages is presented with the help of one exemplary rule that is defined in terms of the model in the following way:

$$r_{bonus-malus0}(M, H_1, H_2) == \quad \text{if} \begin{cases} H_1 \in \mathcal{P}\left(\mathbb{F}, T_{Driver}, \left(A_{DriverClass}(H_1)\oslash 9\right)\right) \\ H_2 \in \mathcal{P}\left(\mathbb{F}, T_{Car}, A_{Accidents}(H_2)\ominus 0\right) \end{cases}$$
$$\text{then} \begin{cases} A_{DriverClass}(H_1)\ominus A_{DriverClass}(H_1)\oplus 1 \\ M \end{cases}$$
$$(10.16)$$

Translation of such module structure to DROOLS is straightforward. In order to assign a certain rule to a given module the `agenda-group` rule property can be used. Moreover, DROOLS provides `lock-on-active` and `activation-group` rule features corresponding to *lock-on-active* and *xor* submodules, respectively. Therefore, the $r_{bonus-malus0}$ rule can be translated to DROOLS as:

```
1  rule "bonus-malus0"
2    agenda-group "bonus-malus"
3    lock-on-active true
4    activation-group "bonus-malus-ag"
5  when
6    $driver : Driver($DriverClass : DriverClass,
                DriverClass < 9)
7    Car(Accidents == 0)
8  then
9    modify($driver){setDriverClass($DriverClass + 1)}
10 end
```

Module assignment in CLIPS or JESS can be done by the addition of module name as prefix followed by : : to name of a rule. Nevertheless, translation of the internal module structure is not so obvious as in case of DROOLS. The default rule execution within this module is constrained by two conditions. The first one forbids the execution of rules caused by the modifications made by themselves. In turn, the second condition allows for only one rule execution. It can be easily noticed that the second condition is stronger as the first execution of a rule in a certain module cannot be caused by modification made by other rules from this module. Therefore, in order to satisfy both of the conditions, each rule belonging to this module must move the focus to the next (in terms of the default order) module. Considering the translation of the *lock-on-active* submodule it is important to notice that all the rules belong to the L-O-A submodule while the other rules submodule is empty. The translation is depicted in Fig. 10.2.

In the case of CLIPS and JESS, the module can be changed by a (pop-focus) function that removes the module from the top of the focus stack, and thus moves the focus to the next one. Therefore, in order to interchange the internal structure of this module, each *bonus-malus* rule must provide this function:

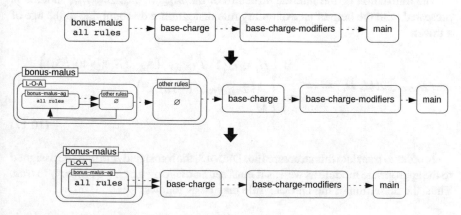

Fig. 10.2 Translation of the *bonus-malus* module structure

```
1   (defrule bonus-malus::bonus-malus0
2     ?driver <- (driver (class ?class&:(< ?class 9)))
3     (car (accidents 0))
4   =>
5     (modify ?driver (class (+ ?class 1)))
6     (pop-focus)
7   )
```

In order to move the inference to another table in XTT2, the executed rule must provide a connection to the desired destination. This is why, all the rules belonging to the *bonus-malus* table are connected to the *base-charge* table (see Fig. 10.3). An XTT2 rule that corresponds to the $r_{bonus-malus0}$ rule (see Definition (10.16)) can be defined in HMRformat in the following way:

```
1   xrule bonus-malus/1:
2     [carAccidents eq 0,
3      driverClass in [-1 to 8]]
4   ==>
5     [driverClass set (driverClass+1)]
6     :base-charge.
```

The $M_{base-charge-modifiers}$ module contains rules that specify increases and decreases of the insurance price. A conditional part of each rule may depend on a different set of attributes describing different reasons of the price change e.g. the number of seats in the insured car, how long has the driver had the driving license, etc. In turn, the conclusion part of each rule adds a new fact to the knowledge base that determines the percentage change of the insurance base charge. It is assumed that a driver can receive multiple increases and decreases of the insurance rate, however each change must be given according to a different reason. Therefore, each rule that belongs to this module, may be executed at most once.

The translation of the internal structure of the $M_{base-charge-modifiers}$ module is presented with the help of an exemplary rule that grants a discount due to the age of a driver.

$$r_{driver-age}(M, H_1) == \begin{array}{l} \text{if} \left\{ H_1 \in \mathcal{P}\Big(\mathbb{F}, T_{Driver}, \big(A_{Age}(H_1)\ominus[40, 55]\big)\Big) \right. \\ \\ \text{then} \left\{ \begin{array}{l} \mathcal{A}\Big(T_{Modifier}, (-10), \varnothing\Big) \\ M \end{array} \right. \end{array}$$

$$(10.17)$$

In order to translate this structure into DROOLS, the translated rule must be assigned to the appropriate module as well as it must set lock-on-active property to *true*. Thus, the definition of the DROOLS rule may be as follows:

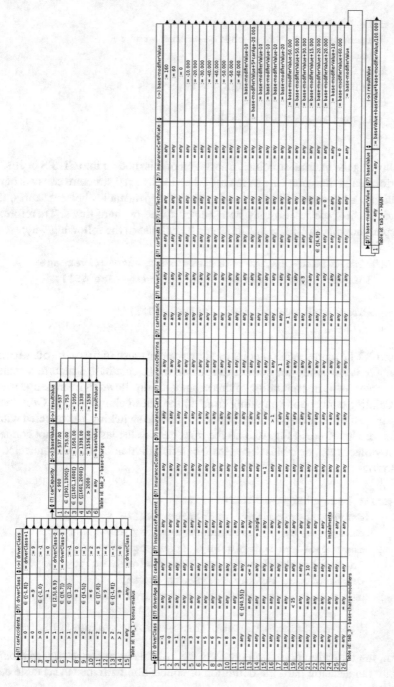

Fig. 10.3 Model of PLI case study in XTT2

```
1   rule "driver-age"
2     agenda-group "base-charge-modifiers"
3     lock-on-active true
4   when
5     Driver($Age : Age, Age >= 40, Age <= 55)
6   then
7     insert(new Modifier(-10));
8   end
```

Following the internal structure, translation of this module into CLIPS or JESS can be performed in the same way as in case when the internal structure consists of only an *ordinary* submodule. Because of the lack of dependencies between rules, there is no risk that some rule(s) may be executed twice or more times. Therefore, the corresponding rule in CLIPS and JESS can be defined in the following way:

```
1   (defrule base-charge-modifiers::decrease-driver-age
2     (driver (age ?age&:(>= ?age 40)&:(<= ?age 55)))
3   =>
4     (assert (base-modifier (value -10)))
5   )
```

From XTT2 representation point of view, the existence of a lock-on-active submodule is necessary for correct translation of the rule base structure semantics. This is because the interpretation of this submodule disallows an interchange method from defining connections outgoing from all the rules belonging to it. Only the last rule in the table that is responsible for default module switching is connected with the next table. Therefore, in Fig. 10.3 all the rules besides the last one are not connected with any other rules or tables. The HMR-based definition of the not connected rule is as follows:

```
xrule base-charge-modifiers/12:
    [driverClass eq any, driverAge in [40 to 55],
     carAge eq any, insurancePayment eq any,
     insuranceContinue eq any, insuranceCars eq any,
     insuranceOtherins eq any, carHistoric eq any,
     driverLicage eq any, carSeats eq any,
     carTechnical eq any, insuranceCertificate eq any]
==>
    [base-modifierValue set (base-modifierValue-10)].
```

In turn, the following definition presents the last rule in the *base-charge-modifiers* table. In the decision part, the definition of connection with the MAIN table can be observed:

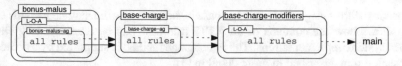

Fig. 10.4 Complete translation of the rule base structure for the PLI case

```
xrule base-charge-modifiers/26:
  [driverClass eq any, driverAge eq any,
   carAge eq any, insurancePayment eq any,
   insuranceContinue eq any, insuranceCars eq any,
   insuranceOtherins eq any, carHistoric eq any,
   driverLicage eq any, carSeats eq any,
   carTechnical eq any, insuranceCertificate eq any]
==>
  [base-modifierValue set base-modifierValue]
  :'MAIN'/1.
```

The presented model of rule base of the PLI use case consists of four modules. The last one is called `main` and it contains only of one rule that calculates the final price of the insurance taking the base charge and all the increases and decreases into account (see Definition (10.12)). The internal structure of this module consists of only an ordinary submodule and thus it can be considered as a module without internal structure. Therefore, the definition of the M_{main} module can be written as follows:

$$M_{main} = M^o_{main} = \{r_{calculation}\} \tag{10.18}$$

The *calculation* rule belonging to this module can be easily translated into CLIPS, JESS and DROOLS and thus this issue is not discussed here. However, its translation to XTT2 is not a trivial issue as it aggregates values of price modifications. Depending on the approach to interchange of the aggregations, it may be required to define the internal structure of this module with the help of a *lock-on-active* submodule instead of an *ordinary* one. In such a case, translation of internal structure of the *main* module can be done in the same way as the translation of the *base-charge-modifiers* module. Figure 10.4 depicts a complete rule base structure after the translation of all modules.

10.7 Summary

In this chapter we demonstrated how the introduction of a formalized model introduced in Chap. 8 of the rule base can support rule interoperability. Thanks to it is possible to translate between different rule-based languages, in our case the ones of DROOLS, CLIPS, and XTT2. Formalization allows for preserving the operational semantics of the designed knowledge base in different specific languages. This work also shows how the contents of an XTT2 rule base can be interpreted by classic expert system implementations. The interchange method is supported by a dedicated translation tool described in more detail in [3].

The next chapter is devoted to the use of the SKE approach as a supporting design solution in software modeling.

References

1. Kaczor, K., Nalepa, G.J.: Encapsulation-driven approach to interchange of knowledge base structure. Lect. Notes Softw. Eng. **4**(1), 66–72 (2016)
2. Giarratano, J., Riley, G.: Expert Systems. Principles and Programming. 4th edn. Thomson Course Technology, Boston (2005). ISBN 0-534-38447-1
3. Kaczor, K.: Knowledge formalization methods for semantic interoperability in rule bases. Ph.D. thesis, AGH University of Science and Technology (2015) (Supervisor: Grzegorz J. Nalepa)

Chapter 11
Visual Software Modeling with Rules

In the last decades visual design methods have been gaining popularity and importance in SE (Software Engineering) [1] as a mean of coping with the increasing complexity of software. Nowadays, visual modeling has become an essential part of the design in the SE processes. When it comes to practical software design, Unified Modeling Language (UML) [2] is *de facto* the standard for modeling software applications [3]. UML attempts to be a universal visual notation for software design. Although the semantics of UML 2.0 is more precise than the one of UML 1.5, there are still many ambiguities. There is ongoing research, on precising, extending or redefining the UML semantics to overcome its limitations. Today, UML diagrams are typically not detailed enough to describe every aspect of the modeled system. Moreover, there are concepts e.g. constraints, which cannot be expressed in pure UML. A complete system model requires the use of the Object Constraint Language (OCL) [4] to provide consistency of the design. OCL is a formal language, which has been developed in order to avoid ambiguous constraint expressions. OCL syntax is a simple, textual notation whereas UML is a visual one.

This chapter concerns practical design issues of rule-based models integrated with business applications built using the Model-View-Controller (MVC) [5] architectural pattern and designed in UML. In Sect. 11.1, the detailed motivation for the research aimed at applying Semantic Knowledge Engineering in the field of SE is given. The main idea consists in the introduction of a visual UML representation for business rules modeling the application logic. The representation is based on XTT2. A translation from the ARD+ model is discussed in Sect. 11.2. A complete bidirectional translation between XTT2 and the UML representation is also presented. The translation preserves the semantics of XTT2 in a UML-friendly fashion, allowing UML designers to approach the XTT2-based rule logic model in a unified way. The implementation of practical translators is described in Sect. 11.3. The evaluation of this approach, using a practical example is presented in Sect. 11.4. The chapter ends with a summary and directions for future works in Sect. 11.5.

© Springer International Publishing AG 2018   275
G.J. Nalepa, *Modeling with Rules Using Semantic Knowledge Engineering*,
Intelligent Systems Reference Library 130,
https://doi.org/10.1007/978-3-319-66655-6_11

11.1   Integration of Rules with UML Design

The presented rule-based approach has several important features namely that
the core logic of the application (the model) is clearly identified and separated.
Moreover, it is built in a declarative way, which makes the design transparent and
easier to follow, develop and update in the case of changing user requirements (which
is often the case in real-life projects). Thanks to these features the approach increases
the agility of the design process.

In the approach considered in this research the application is designed in the MVC
design pattern and the model is designed with the use of Business Rules. The remaining parts of the application (Views and Controllers) including its interfaces are usually
designed using UML. In this approach, the core logic of the application (Model) is
clearly identified and separated. Moreover, it is built in a declarative way, which
makes the design transparent, easier to follow and develop.

This approach has number of qualities. However, its practical application faces
three important challenges: (A) the rule visualization problem, (B) the design representation mismatch, and (C) the model quality assurance problem.

The first problem concerns the fact that in the BR approach rules are captured using
a textual notation, with no visualization. This limitation becomes a major obstacle
once the number of rules grows, and dependencies between them become complex.
Such a design specification is also incompatible with the visual nature of UML.

The second issue is related to the fact that concepts on which the rule representation
is based cannot be directly modeled using UML constructs. In fact, the extended
semantics of UML 2.0 diagram types does not directly corresponds to rule semantics.
There is no UML diagram to model rules and there is no direct way to relate the rule-
based model to the object-oriented one. In classic software engineering practice,
both analytics and programmers use UML to specify and document the project.
When some parts of the application are designed in a different way communication
problems may occur.

The last challenge consists in providing effective verification techniques for the
rule-based application logic. While in the software industry there are numerous methods and techniques aiming at providing certain quality assurance of software (mainly
based on software verification and validation through testing, code review, etc. [6, 7]),
they cannot be directly applied to a rule-based model. On the other hand, in the area
of classic RBS a number of well-defined formalized verification methods exist. The
use of these methods should be made available during the design process of RBS by
some practical tools.

The research presented in this chapter aims at addressing all three problems
(A, B, C). The solution to these problems consists in:

- using an expressive visual rule design formalism (XTT2) (addresses problem A),
- defining a direct translation between XTT2 and selected UML diagrams (addresses
  problem B),
- implementing practical translators between the logical XTT2 model and the MOF-
  based (Meta-Object Facility) [8] UML model,

- using verification features provided with the XTT2 framework to assure the quality of the rule base (addresses problem C).

It is important to emphasize that the proposed approach is not about modeling the whole application using rules. The semantics of a rule-based paradigm does not match the methods and tools of SE. Nevertheless, RBS logic specification can be complementary to the higher level specification of the system in SE [9]. Moreover, this approach must not be confused with ideas to use the OCL to provide the consistency of the UML design. OCL is a formal language, which has been developed in order to avoid ambiguous constraint expressions. It aims at precising existing UML models whereas the approach presented here aims at modeling certain parts of the application for which existing UML diagrams are not suitable. This research considers a method of designing rule bases, which will be proper and consistent with UML. In order to provide the background for the work presented in this chapter, some important research areas were presented in Sect. 3.5. There, we discussed selected UML-based visual methods of rule representation, and on selected UML formalization attempts as well as methods of quality maintenance.

11.2   Representation of ARD+ and XTT2 with UML Diagrams

UML representation of ARD+ has been proposed in [10, 11]. The representation uses class diagrams to convey ARD+. Other preliminary approaches and more details can be found in [12].

Expressing the ARD+ *Model with UML Class Diagrams*

In UML, classes describe sets of objects sharing the same specifications of features, constraints, and semantics. Attributes are one of class features. This also applies to properties in ARD+ which are described by attributes. According to the UML Superstructure [13], the purpose of classes is to specify: a classification of objects, and features that characterize the structure and behavior of these objects. Because ARD+ does not describe the behavior, there is no need to use operations in the UML representation of ARD+.

Elements of the proposed UML model correspond to one-to-one to the properties in ARD+ (it is a bijective relation). In the proposed model:

- *class* corresponds to the simple property in ARD+,
- *abstract class* with attributes corresponds to ARD+ complex property,
- *«derive» dependency* corresponds to ARD+ dependency between attributes,
- *«refine» dependency* corresponds to ARD+ finalization transformation,
- *aggregation* corresponds to ARD+ split transformation.

The difference between a conceptual and a physical attribute is presented in the same form as in ARD+. The name of a conceptual attribute begins with a capital, while the name of a physical attribute begins with a lower case.

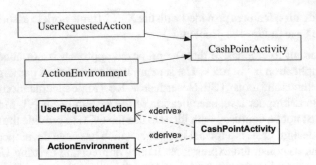

Fig. 11.1 Example of UML representation of ARD+ diagram

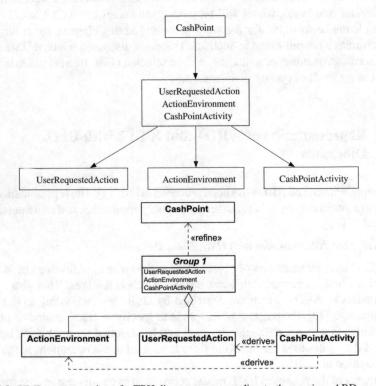

Fig. 11.2 UML representation of a TPH diagram corresponding to the previous ARD model

Figure 11.1 shows an example of a simple ARD+ diagram and its UML representation. The corresponding representation of a TPH diagram is shown in Fig. 11.2. The interpretation of ARD+ is as follows: the property CashPointActivity can be derived from UserRequestedAction and ActionEnvironment. The TPH model shows the history of transformations: CashPoint is finalized into three properties UserRequestedAction, ActionEnvironment, and CashPointActivity. Then a split is made – a complex property is separated into two properties and a dependency between them is identified.

The proposed UML model is similar to the original ARD+. It is transparent and intuitive. This approach does not introduce custom elements, only standard UML artifacts are used.

Regarding the relations between the artifacts, UML is a much richer language than ARD+. There are some relationships which can be observed in ARD, but they are not distinguished. The first not distinguished relation is the relation of realization. If a conceptual attribute is finalized to a physical attribute, this is a quite different case than if it is finalized to some other conceptual attribute or attributes. The first case is an example of realization. It is so, because the physical attribute realizes the concept of the conceptual attribute. The second one is an example of some refinement.

The second relation which is not distinguished is a «trace» dependency. A trace relationship among model elements can be used in a TPH diagram to show elements representing the same concept at different levels of the design.

Metamodel of the ARD+ Model

The abstract syntax of UML is defined by the metamodel. As the proposed model uses only a subset of UML artifacts and relationships, its metamodel can be created as a subset of the UML metamodel with constraints imposed [13].

Figure 11.3 shows the metamodel of the UML representation for ARD+ diagrams. The representation uses only UML classes (with or without attributes) and the «derive» dependencies. Moreover, there is a need to define some OCL constraints in the metamodel, because the metamodel itself is not stated precisely enough (see [12] for details).

There is a need to define a separate metamodel for TPH diagrams, based on the previous metamodel. Figure 11.4 shows the metamodel of the UML representation for

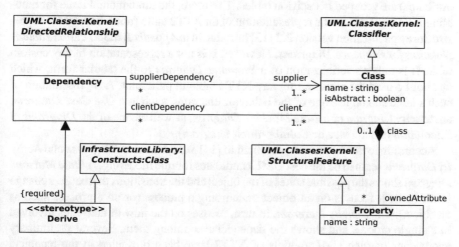

Fig. 11.3 Metamodel for the UML representation of ARD+ diagrams

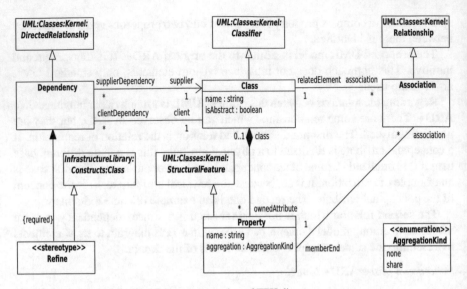

Fig. 11.4 Metamodel for the UML representation of TPH diagrams

TPH diagrams. The representation uses UML classes (with or without attributes), aggregations and the «refine» dependencies. This metamodel also requires some OCL constraints to specify the exact syntax of the TPH representation. All OCL expressions from the ARD+ metamodel are valid for the TPH metamodel (see [12] for details).

XTT2 *Model with UML Activity Diagrams*

In the XTT2 representation rules working in the same context (having the same attributes) are grouped in decision tables. Therefore, the fundamental issue for modeling XTT2 in UML is the representation of an XTT2 table (unit). The UML Superstructure Specification version 2.2 [13] introduced in *Appedix E* the so-called *Tabular Notation for Sequence Diagrams*. However, it is not a representation of any custom table. It is rather a serialization of a *Sequence Diagram* to the tabular form, which can not be used for representing any XTT2 table in particular. A representation of such a table should use one of the behavior diagrams, such as a *Use Case Diagram*, an *Activity Diagram* or a *State Machine Diagram*, as well as one of the *Diagrams of Interaction* (a *Sequence* or *Collaboration Diagram*).

According to the discussion presented in [14] *State Machine Diagrams* and *Activity Diagrams* seem to be the best UML candidates for rule modeling. A *State Machine Diagram* shows the possible states of the object and the transitions that cause a change in the state. It focuses on an object undergoing a process (or on a process such as an object). The *Activity Diagram*, in turn, focuses on the flow of activities involved in a single process and shows the dependencies among them. Several preliminary approaches of such UML models of XTT2 have been presented in the technical report [12]. Such a representation is not as expressive as XTT2. Especially, in the

(?) pinDifference	(->) authorized	(->) failedAttempts
!= 0	:= false	:= add(failedAttempts,1)
= 0	:= true	:= failedAttempts

Table id: tab_3 - authorization

Fig. 11.5 Example of an XTT2 table

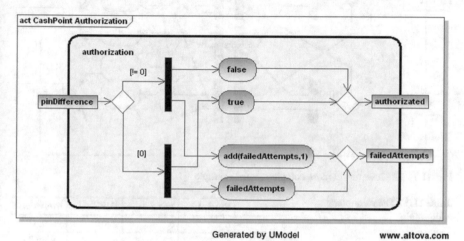

Generated by UModel   www.altova.com

Fig. 11.6 UML representation of the XTT2 table from Fig. 11.5

case of larger systems with a huge number of rules, using UML poses practical problems. However, it is possible to use it to express rules in the case of smaller systems.

The XTT2 representation in UML uses *Activity Diagrams* to express the XTT2 model (see [10] for more details on preliminary modeling attempts). These diagrams are related to flow diagrams and illustrate the activities in the system. At the lower level of the knowledge base a single XTT2 table is represented. An example of a UML representation (corresponding to the table in Fig. 11.5) is shown in Fig. 11.6. In the proposed UML model of the XTT2 table the following node types are used:

- *Activity Parameter* – the parameter representing an XTT2 attribute,
- *Action* – sets the value of the output parameter,
- *Decision Node* – enables checking the distinct values of the input parameter,
- *Merge Node* – corresponds to logical *or* operation for flows,
- *Fork Node* – enables to branch out and manifold the flow, and
- *Join Node* – corresponds to logical *and* operation for flows.

Every XTT2 attribute is represented by an (input or output) *Activity Parameter* node. The whole *Activity* resembles a logical gate system, where *Merge/Join Nodes* behave like logical *or/and* gates. Values of attributes are checked in guard conditions by the *Decision Nodes*, and values of output parameters are set by *Actions*. In more

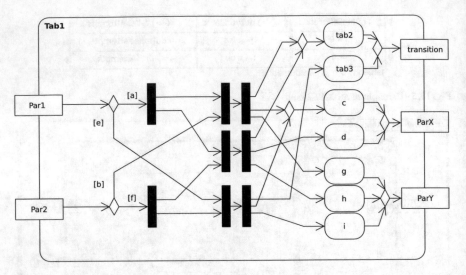

Fig. 11.7 The lower level more complex model example

Table 11.1 The exemplary XTT2 table

Par1	Par2	ParX	ParY	*transition*
a	b	c	g	→ tab2
a	f	d	h	→ tab2
e	f	c	i	→ tab3

complex cases, there is a possibility of different transitions from one table. In such a situation, a new *transition* output *Activity Parameter* is introduced, see Fig. 11.7. The diagram corresponds to the exemplary Table 11.1, where *Par* are attribute names.

At the higher level the whole XTT2 model consisting of linked tables is represented. In the model, every table has a corresponding *Action*, which calls the behavior modeled at the lower level. These *Actions* are connected as previously at the lower level model. If an action has an *Output Pin* on its output, it means that the corresponding table has transitions to more than one table (in accordance with each distinct value of *transition* attribute from the lower level). An example of such representation is shown in Fig. 11.8.

Metamodel of the XTT2 *Model*

The metamodel of the proposed model constitutes a subset of the UML *Activity Diagram* metamodel [13] with additional constraints imposed. Figure 11.9 shows the metamodel of the UML representation for the XTT2 diagrams. The presented metamodel is not yet precisely stated, because the UML diagrams themselves are typically not detailed enough to describe every aspect of a specification. To ensure the accuracy of models, some additional OCL constraints for the metamodel have to be provided.

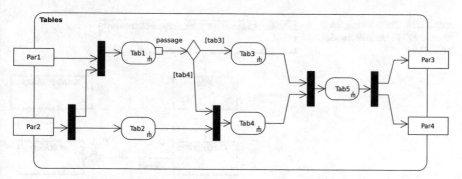

Fig. 11.8 The higher level model example

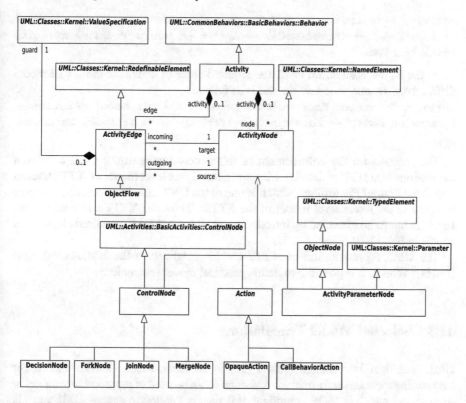

Fig. 11.9 Metamodel for XTT2 diagrams

The metamodel does not enforce the order of nodes. It can be observed in the metamodel that it allows for generating models which do not match the XTT2 ones. The required order of nodes for the XTT2 model at the lower abstraction level is as follows (nodes which can occur optional are in brackets):

Fig. 11.10 Introducing the
Output Pin to the metamodel
of XTT2

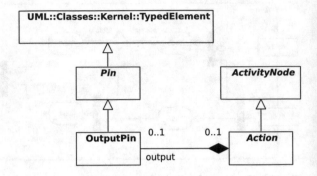

Activity Parameter Node → Decision Node → (Fork Node) → Join Node[1]
→ Fork Node[2] → (Merge Node) → Action → (Merge Node) → Activity
Parameter Node.

At the higher abstraction level, the required order of nodes for the XTT2 model differs from the one stated above and is as follows:

Activity Parameter Node → (Fork Node) → X (Join Node) → Action →
((Decision Node)[3] → back to X) → (Fork Node) → Activity Parameter
Node.

The solution for the enforcement of this node order is to use OCL constraint expressions (see [12] for details). Moreover, the higher level model of XTT2 uses an additional *Output Pin* artifact (which is one of the UML standard artifacts; however not used in the lower level model of the XTT2). Thus, the XTT2 metamodel from Fig. 11.9 has to be extended by introducing an additional UML standard element, as shown in Fig. 11.10.

The UML representation of XTT2 can be serialized to the XML-based XMI format.[4] With it, it is possible to define practical model translations.

11.3   Selected Model Translations

XML Metadata Interchange (XMI) [15] is an XML-based OMG standard for exchanging metadata information. Although it can be used to represent any model or meta-model which is MOF compliant, it is mostly applied to encode UML models for model exchange between various tools. There is a variety of XMI formats and many UML tools store models in XMI using their own tool-specific XMI format.

---

[1]Element occurs when there is more than one input *Activity Parameter* node.

[2]Element occurs when there is more than one output *Activity Parameter* node or there is the transition parameter node (which enables the transition to different tables).

[3]Element occurs when previous *Action* has *Output Pin*.

[4]See http://www.omg.org/spec/XMI.

Moreover, the XMI Specification [15] does not define XML tags for every UML artifact, but specifies how to create them for the metamodel concepts.

For the purpose of the implementation the newest XMI 2.1 version has been adapted. This version is supported by several UML tools (e.g. *Altova UModel,*[5] *Visual Paradigm for UML*[6]). A detailed description of the XMI tags that represent UML artifacts can be found in [12]. Although XMI is predominantly used to exchange model data between different UML tools [16], in this case it is considered as a format for serializing models for further translation and code generation.

The UML representation of XTT2 is serialized to the XMI file and the XTT2 model is stored in the HML file. Because both are XML compliant, the implementation of the proposed translation algorithms is done with the use of XSLT. Although there are some disadvantages of using XSLT (such as poor readability or poor error reporting), among various methods, XSLT is recommended for a couple reasons [17, 18]. It should be used when both source and target formats are XML-based and there is a need for matching elements from the source document with using its structure and associating them with the target document structure. But the key advantage of using XSLT is that there are many XSLT engines available. Algorithms used for the implementation of these translators are presented next.

Translation from XMI to ARD+

In the case of XMI to ARD+ translation the algorithm is simple, and elements can be easily mapped as it is described in Sect. 11.2. This is done according to the following rules:

1. Every class without attributes becomes an ARD+ simple property.
2. Every abstract class becomes an ARD+ complex property.
3. Every «derive» dependency becomes an ARD+ dependency.

In the case of the TPH, in addition, one has to use two following rules:

4. Every «refine» dependency becomes an ARD+ finalization.
5. Every aggregation becomes a part of an ARD+ split.

Translation from ARD+ to XMI

In the case of ARD+ to XMI translation the algorithm is very similar to the XMI to ARD+ algorithm. The mapping between elements is the same, and the translation can be done according to the following rules:

1. Every ARD+ property becomes a class with attributes (for complex property) or without (for simple property).
2. Every ARD+ dependency becomes a «derive» XMI dependency.

---

[5] See http://www.altova.com/umodel.html.
[6] See http://www.visual-paradigm.com/product/vpuml.

In the case of the TPH, in addition, one has to use the two following rules:

1. Every ARD+ finalization becomes the «refine» XMI dependency.
2. Every ARD+ split becomes an XMI aggregation.

Translation from XMI to XTT2

In the XTT2 UML representation (serialized to XMI) UML artifacts have their own semantics, and the translation to XTT2 (serialized to HML) is not simple. It is not sufficient to perform the mapping of elements (between XMI and HML) – additional transformations have to be done. Figure 11.11 shows the flowchart of the translation.

The algorithm for the lower level representation is as follows. For every table represented by the UML diagram:

1. Name of the table is saved.
2. *Activity Parameter Nodes* become attributes of the table.
3. In order to transform the rules, there is a need to simplify the representation. This can be done in the following way (the process is illustrated in Fig. 11.12):

   3.1. For every input *Activity Parameter Node*, its id is stored in the ref attribute of the following *Decision Node* and the *Parameter Node* is removed.
   3.2. For every *Object Flow* with the source in a *Fork Node*, the source of the incoming edge of the node becomes the source of the *Object Flow* (with nested guard element preserved) and the *Fork Node* is removed.
   3.3. For every *Object Flow* of which a target element is a *Merge Node*, the target of the *Merge Node* becomes a target of the *Object Flow*, and the *Merge Node* is removed.
   3.4. For every output *Activity Parameter Node*, its *id* is stored in the *ref* attribute of the precedent *Action Node*, and the *Parameter Node* is removed.

4. After the simplification of the model, it is much easier to transform the UML representation of the rules to HML.

   4.1 If there is more than one input *Activity Parameter Join Nodes* become rule elements in the following way. For every Join Node:
      4.1.1. Every guard value in the incoming edges to the *Join Node* becomes the expression in the condition part for the proper rule attribute.
      4.1.2. Every action name in the outgoing edge of the *Join Node* becomes the expression in the decision part for the proper rule attribute.
   4.2. If there is only one input *Activity Parameter*, *Object Flows* which have the source in the *Decision Node* become rule elements in the following way:
      4.2.1. Every guard value in the flow becomes the expression in the condition part for the proper rule attribute.
      4.2.2. Every action name in the outgoing edge of the Join Node becomes the expression in the decision part for the proper rule attribute.

At this stage of the algorithm the table representation is finished. However, there are no links between tables. The algorithm for the higher level representation is:

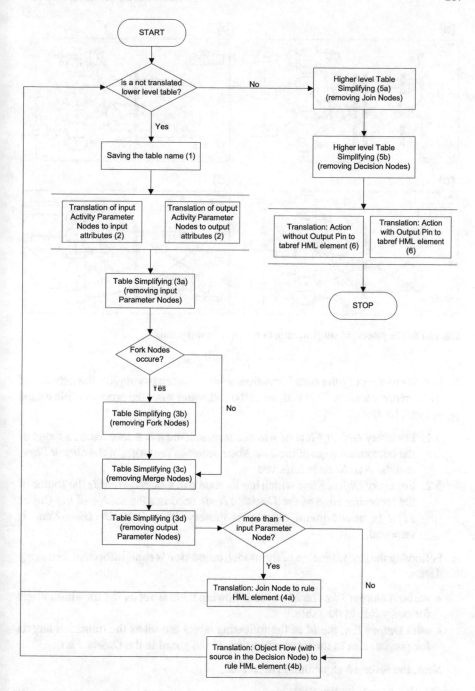

Fig. 11.11 The flowchart presenting the translation from XMI to XTT2

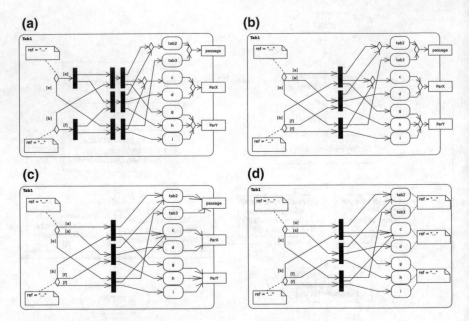

Fig. 11.12 The process of simplification of the table representation

5. In order to simplify the transformation, there is a need to simplify the network of tables representation. This is done in the following way (the process is illustrated in Fig. 11.13):

   5.1. For every *Object Flow* of which a target element is a *Join Node*, a target of the outcoming edge of the *Join Node* becomes the target of the *Object Flow*, and the *Join Node* is removed.

   5.2. For every *Object Flow* which has its source in a *Decision Node* the source of the incoming edge of the *Decision Node* becomes the source of the *Object Flow* (a nested guard element is preserved), and the *Decision Node* is removed.

6. Following the simplification of the model, translation is straightforward. For every *Action*:

   • without *Output Pin*, the *id* of the following table is set as the transition target for every rule in the table.
   • with *Output Pin*, the *id* of the following tables are set as the transition targets for proper rules in the table, according to the guard in the *Object Flow*.

   Next, the reversed algorithm is described.

Translation from XTT2 to XMI

The translation algorithm from XTT2 to XMI is more complicated than the previous ones. This section contains the refined version of the algorithm presented in [10]

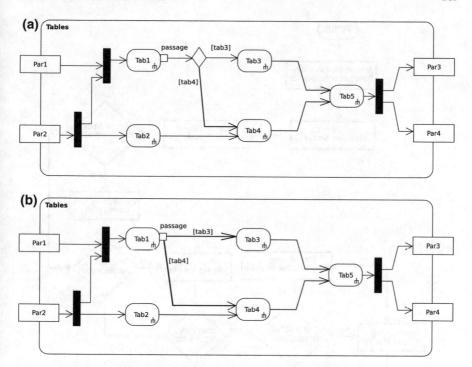

Fig. 11.13 The process of simplification of the network of tables representation

for XTT2 tables. Because of the semantics of the UML artifacts, some complex transformations have to be done.

The algorithm for the lower level representation is as follows (it is illustrated in Fig. 11.14). For every XTT2 table, *Activity* is created, and:

1. All condition attributes become input *Activity Parameter Nodes* and decision attributes become output *Activity Parameter Nodes*.
2. For each condition attribute (input *Activity Parameter Node*) the *Object Flow* goes to a decision node and for each unique value of the attribute:

   2.1. The *Object Flow* with this unique value as a *Guard* condition is introduced,
   2.2. If the value occurs frequently, the flow is finished with a *Fork Node*.

3. If there is more than one condition attribute, for each XTT2 rule a *Join Node* is created, and:

   3.1. For each condition attribute the proper *Object Flow* is connected to the *Join Node* (if the value of the considered condition attribute occurs frequently, the flow starts from the proper *Fork Node*),
   3.2. Depending on the number of decision attributes:
      3.2.1. If there is only one decision attribute: the proper *Object Flow* starts from the *Join Node* and is connected to the *Action* having a value

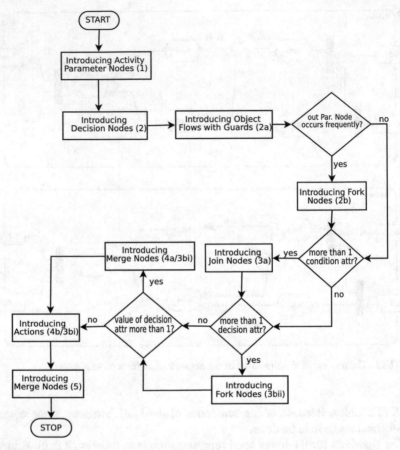

Fig. 11.14 The translation from XTT2 to XMI for the lower level representation

    corresponding to the decision attribute (if the value of the decision attribute occurs in the table more than once, a *Merge Node* is introduced and the connection is through this *Merge Node*),

  3.2.2. Otherwise: a *Fork Node* is introduced and the *Object Flow* from the *Join Node* to this *Fork Node* is created.

4. If there is more than one decision attribute, from each *Fork Node* (corresponding to XTT2 rule) for each decision attribute the proper *Object Flow* from the *Fork Node* is introduced and connected to:

  4.1. The proper *Merge Node*, if the value of the decision attribute occurs in the table more than once (if the proper *Merge Node* does not exist yet, it should be introduced). Then, the *Object Flow* from the *Merge Node* is introduced and connected to:

    4.2. The *Action* (having a value corresponding to the value of the decision attribute in the rule).

5. For every decision attribute, outputs of *Actions* (having the values of particular attribute) are connected to a *Merge Node* and an *Object Flow* from the *Merge Node* leads to the corresponding output *Activity Parameter Node*.

The algorithm for the higher level representation is as follows. The activity diagram for the whole system is created, and:

1. For every XTT2 input and output system attributes, corresponding input and output *Activity Parameter Nodes* are created.
2. For every XTT2 table, the *CallBehaviorAction* is created, and:

   - if more than one table has the link to this table, a *Join Node* is created and the proper *Object Flows* from those tables to the *Join Node* are created,
   - if only one table has the link to this table, an *Object Flow* from that table to this table is created,
   - if the condition attribute of this table is the input system attribute, the *Object Flow* from the corresponding *Activity Parameter Node* to this table is created (if the attribute is the condition attribute in more than one table, a *Fork Node* is introduced and the connection is through this *Fork Node*).

3. For every output *Activity Parameter Node* an *Object Flow* from the proper *CallBehaviorAction* is created, according to the decision attributes of the XTT2 tables, and:

   - if the attribute is the decision attribute in more than one table, a *Join Node* is introduced and the connection is through this *Join Node*,
   - if it is not the only one decision attribute in the table, a *Fork Node* is introduced and the connection is through this *Fork Node*.

The application of translators is demonstrated next.

11.4 Evaluation Using a Case Study

To evaluate the discussed approach the CASHPOINT system is used. The BL is modeled with rules represented in XTT2. Then, the UML representation is given for this model. Finally, the UML design for the user interface (View) and the Controller connecting it with the Model is given.

Let us consider the XTT2 rule model as shown in Fig. 9.5. The network of decision tables represents the complete model for the system. The XTT2 model can be serialized to the XML-based HML notation suitable for the XSLT translation. Using XTT2UML XSLT translator the model can be transformed from HML to XMI and then opened in a UML tool. A set of diagrams for the UML model of the presented XTT2 *Cashpoint* model is shown in Fig. 11.15. These diagrams correspond to the

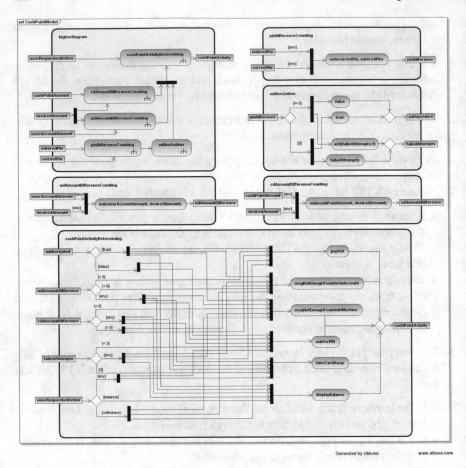

Fig. 11.15 UML representation of the Cashpoint XTT2 Model

previously defined decision tables. The domain model for the rule attributes can be observed in Fig. 11.16. The design conforms to the MVC pattern. In this case the model is responsible for storing domain data (see Fig. 11.16) as well as the system behavior (see Fig. 11.15). The detailed behavior description allows for responding to requests concerning information about the system state.

The View part of the system is responsible for a Graphical User Interface (GUI) and user interaction. This separation of the View from the Model allows for independent development and easy substitution of each layer. The exemplary View implementation specification is shown in Fig. 11.17. The GUI for this implementation is shown in Fig. 11.18 (PIN, Amount and Action Windows). It is an example implemented using the Qt library that heavily uses the MVC architectural pattern for the application design. For brevity, here we omit the detailed specification of the Controller.

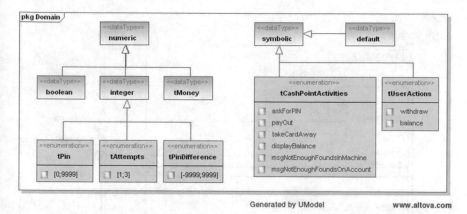

Generated by UModel   www.altova.com

Fig. 11.16 UML representation of the domain for the *Cashpoint* XTT2 Model

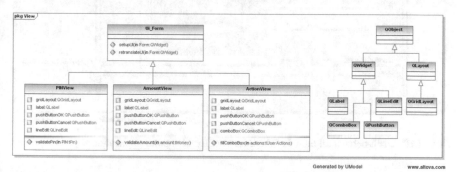

Generated by UModel   www.altova.com

Fig. 11.17 UML model of the View for *Cashpoint*

Fig. 11.18 Qt GUI for the *Cashpoint* application

However, the logic of its behavior is presented in Fig. 11.19. Using features provided by Qt (or other application frameworks) it can be easily implemented.

There is also a possibility of translating backwards, from the UML model to the XTT2 model. In this case, the UML model has to be defined and designed in a UML tool, which enables the generation of the XMI code. The model serialized to XMI can be translated using the UML2XTT2 XSLT translator to the HML representation, which allows for refining the design in HQED and verification with HEART.

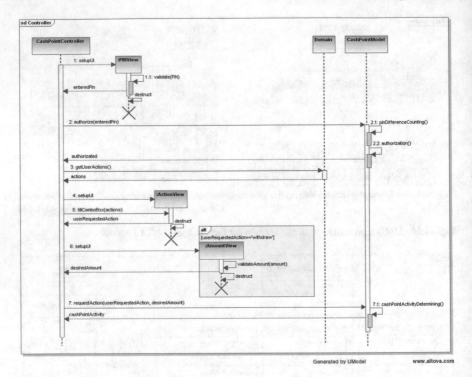

Fig. 11.19 UML behavioral model of the Controller for *Cashpoint* application

It can be observed that this benchmark case study has 11 rules, from which 6 constitute the final decision table. To show the quantitative difference between compared notations, one can observe that the final XTT2 decision table (*cashPointActivityDetermining*) has 6 rules and 6 attributes. This makes 36 cells with values in the decision table to fill in. In the presented UML model there is no redundancy of attribute values, therefore there is only 21 various values to fill in. It is worth noting that in textual notations, such as OCL, this would require the use of every 36 values and writing separate rules.

Evaluation of the Translation

The proposed UML model for rules is MOF compliant. It does not introduce any custom element, but uses only standard UML artifacts. This allows a user to use any of the existing UML tools, which supports model exporting to XMI. Although the model is not similar to the original XTT2, it shall be intuitive when one knows either the UML Activity Diagram semantics or the electronic logical gates (which perform the logical operations on logic inputs and produce a logic output).

This representation, as all solutions mentioned earlier, does not scale well. When the number of rules in the XTT2 table grows, the diagram becomes poorly readable. Due to the fact that visual rule representation faces the problem of scalability, in this field there is still space for new research. Although there exist methods for designing single rules, designing rule groups constitutes a problem. Groups of similar rules often can be modeled as a decision table and if it comes to readability, tables seem to be irreplaceable. Hence, they are so widely used in Knowledge Engineering.

To summarize, the presented UML representation has several advantages:

- It uses only standard UML elements, so any UML tool (that supports XMI generation) can be used for modeling the system.
- In the model there is no redundancy of attribute values as it is in the XTT2 table.
- The model transparently shows inputs and outputs of the system.
- Decomposition of the rule base is provided through a bi-level hierarchy of diagrams in the model. The lower level model for the visual design of the rule set and the higher level model which stores the knowledge about control. This allows for the creation of more complex XTT2 trees, without the risk of losing readability and to ensure the efficiency of the system.

It should also be pointed out that the UML representation for XTT2 seems to be more useful when the rule-based logic is designed with XTT2 and then translated to UML. The reverse translation (from UML to XTT2) is also fully possible. However, it seems to be less useful in a practical design, because the semantics of XTT2 speeds up the rule design process. On the other hand, it could prove valuable in cases where there are some existing Business Logic models in UML the Activity Diagrams.

The entire design process is based on UML, including the specification of application logic. Because of the direct translation between XTT2 and UML representation, and the possibility of automatic logic implementation, the problem of *design representation mismatch* is also addressed. The proposed UML representation uses only standard UML artifacts and allows for representing the whole decision tables, not just single rules like in the case of URML. Its weakness is the visual scalability issue – the size of the activity diagrams grows with the size of the decision tables. On the other hand, this follows the limitations of the diagrams provided with UML 2.0.

Finally, because the rule model is expressed in a formal way, the SKE methodology provides an online verification of the model, thus dealing with *the quality assurance problem*. The provided verification checks are optimized for the rule representation. Since the XSLT model translation preserves the semantics, it does not alter any formal properties of the model. Thus, the obtained UML model preserves the previously verified formal properties.

11.5 Summary

The focus of the chapter is on the visual methods for designing software with knowledge-based components. The assumption is made that the UML-based representation should capture the whole system design. On the other hand, the core parts of the Business Logic could be designed in an optimal way using the Business Rules approach. In fact, it has been a common approach in recent years. However, the approach has several important limitations concerning an appropriate visual design specification for Business Rules. Such a representation should be semantically compatible with UML, while keeping all the qualities of the rule-based Model. At the same time the quality issues of the resulting hybrid Model should be considered as well as the appropriate knowledge integration problems addressed.

The proposed solution to these problems uses the XTT2 rule representation to design the application core logic. This logic is integrated with the remaining parts of the application using the MVC architectural pattern. Moreover, the XTT2 rule design can be translated to a corresponding visual UML Model. Simultaneously, formal properties of the designed Business Rules logic can be verified. In the presented rule-based approach the core logic of the application (the Model) is clearly identified and separated. The Model is built in a declarative way, which makes the design transparent, easier to follow, and update in the case of changing user requirements. Thanks to these features the approach increases the agility of the design process.

Based on the evaluation presented in this chapter, it can be observed that this solution is superior both to existing visual rule notations, as well as visual design tools for rules. It does not introduce any custom UML artifacts and can be used with standard UML tools. The resulting UML design can be analyzed by software designers familiar with UML. On the other hand, it can be easily refined by application architects using Business Rules semantics.

Currently, the Business Logic Model can be integrated with application interfaces that use the SKE framework. In this case the Business Logic is executed by HEART and linked with the interfaces on the runtime. Future works include possible genera-tion of object-oriented code from the resulting UML Model. In this case, the complete application would be implemented in the same programming language (e.g. Java).

References

1. Sommerville, I.: Software Engineering. International computer science, 7th edn. Pearson Edu-cation Limited, Boston (2004)
2. OMG: Unified Modeling Language version 2.1.2. infrastructure. specification. Technical Report formal/2007-11-04, Object Management Group, November 2007. http://www.omg.org/cgi-bin/doc?formal/2007-11-04.pdf
3. Hunt, J.: Guide to the Unified Process Featuring UML. Java and design patterns. Springer, Berlin (2003)
4. Object Management Group: Object Constraint Language Version 2.0. Technical Report, OMG, May 2006
5. Burbeck, S.: Applications programming in Smalltalk-80(TM): How to use Model-View-Controller (MVC). Technical Report, Department of Computer Science, University of Illinois, Urbana-Champaign (1992)
6. Ammann, P., Offutt, J.: Introduction to Software Testing. Cambridge University Press (2008)
7. Xu, D., Xu, W., Wong, W.E.: Automated test code generation from class state models. Int. J. Softw. Eng. Knowl. Eng. **19**(4), 599–623 (2009)
8. OMG: Meta object facility (MOF) version 2.0, core specification. Technical Report formal/2006-01-01, Object Management Group, January 2006. http://www.omg.org/cgi-bin/doc?formal/2006-01-01.pdf

9. Wan-Kadir, W.M.N., Loucopoulos, P.: Relating evolving business rules to software design. J. Syst. Arch. **50**, 367–382 (2004)
10. Nalepa, G.J., Kluza, K.: UML representation proposal for XTT rule design method. In: Nalepa, G.J., Baumeister, J. (eds.) 4th Workshop on Knowledge Engineering and Software Engineering (KESE2008) at the 32nd German conference on Artificial Intelligence: September 23, 2008, pp. 31–42. Kaiserslautern, Germany, Kaiserslautern, Germany (2008)
11. Kluza, K., Nalepa, G.J.: Metody i narzędzia wizualnego projektowania reguł decyzyjnych. In: Grzech, A., et al. (eds.): Inżynieria Wiedzy i Systemy Ekspertowe. Problemy Współczesnej Nauki, Teoria i Zastosowania. Informatyka, Warszawa, Akademicka Oficyna Wydawnicza EXIT , pp. 197–208 (2009)
12. Kluza, K., Nalepa, G.J.: Analysis of UML representation for XTT and ARD rule design methods. Technical Report CSLTR 5/2009, AGH University of Science and Technology (2009)
13. OMG: Unified Modeling Language (OMG UML) version 2.2. superstructure. Technical Report formal/2009-02-02, Object Management Group, February 2009
14. Nalepa, G.J., Wojnicki, I.: Using UML for knowledge engineering – a critical overview. In Baumeister, J., Seipel, D. (eds.): 3rd Workshop on Knowledge Engineering and Software Engineering (KESE 2007) at the 30th Annual German conference on Artificial intelligence: September 10, 2007, pp. 37–46. Osnabrück, Germany, September 2007
15. OMG: MOF 2.0/XMI mapping version 2.1. specification. Technical Report formal/2005-09-01, Object Management Group, September 2005. http://www.omg.org/cgi-bin/doc?formal/2005-09-01.pdf
16. Daum, B., Merten, U.: System Architecture with XML. Morgan Kaufmann (2002)
17. Gerber, A., Lawley, M., Raymond, K., Steel, J., Wood, A.: Transformation: The missing link of MDA. In: Graph transformation: first international conference, ICGT 2002, pp. 252–265, Barcelona, Spain (2002)
18. Fong, J., Shiu, H., Wong, J.: Methodology for data conversion from xml documents to relations using extensible stylesheet language transformation. Int. J. Softw. Eng. Knowl. Eng. **19**(2), 249–281 (2009)

Chapter 12
Using Rules to Support Software Testing

Software engineering seeks novel methods and approaches for dealing with growing challenges, such as the quality control of software. The growing scale of software systems stimulates the use of automated tools for performing a number of repetitive tasks. Preserving and monitoring the quality of the execution of these tasks is of great practical importance. Hence formalized, and declarative methods are preferred. Moreover, there is a growing interest in using a range of intelligent tool to support them.

Testing is an important area in the software lifecycle. First of all, it is one of the most common activities related to the quality assurance of software. While, definitely it is not the only one, or even should not be the main one, it is important to note, that it extensively uses automation. In the classic V model of the software lifecycle [1] several types of tests that correspond to the subsequent phases of the lifecycle are considered. Most of them, especially on the lower level can be fully automated in terms of execution.

In this chapter we present a practical rule-based method for supporting the unit testing process. In Sect. 12.1 our approach to the use of rules in software unit testing is presented. Then we focus on decision table based testing in Sect. 12.2. Moreover, a practical tool implementing the method was developed and discussed in Sect. 12.3. The evaluation of our approach using examples is discussed in Sect. 12.4. The chapter is summarized in Sect. 12.5.

12.1 Software Unit Testing with Rules

Unit testing is one of the most basic and broadly used type of testing [2]. In mature programming approaches writing tests is closely related to writing the code itself. In some approaches [3] tests are written before the actual code is created. There are

© Springer International Publishing AG 2018
G.J. Nalepa, *Modeling with Rules Using Semantic Knowledge Engineering*,
Intelligent Systems Reference Library 130,
https://doi.org/10.1007/978-3-319-66655-6_12

several standardized tools supporting the creation of unit test. A common example is JUnit[1] for the Java language [4]. It also has a number of counterparts for other programming languages.

Unit testing frameworks such as JUnit have well developed facilities for the implementation and execution of tests. In this area especially execution can mostly be automated. However, the main challenge lies in the actual preparation of test cases. In most of the cases this is the activity that requires knowledge of a human programmer or tester. While data for test cases can be semi-automatically inferred from the code or system specification (where it is available), the proper preparation and validation of test cases requires multiple manual activities.

As an example of such an approach, automated rule-based tests generation was discussed in [5]. In that work, system specification is captured as an Excel spreadsheet. This document is parsed using Java Library (e.g. J-Excel). Test cases are generated and saved as an XML document. This file is parsed and JUnit test cases are generated. The authors emphasize the redundancy elimination achieved by the equivalence classes analysis. Their framework generates minimal yet a complete set of test cases.

In this setting, the main motivation for this research presented in this chapter is to provide an automated method and tool supporting the generation of test cases from a formalized system specification. The assumption is that this specification is described with the use of decision rules. The method itself combines a black-box testing technique based on decision tables (DT) with the formalized design of business rules [6, 7]. In this approach the specification of a unit (in terms of unit testing, e.g. a class) is described by a sets of business rules combined into decision tables. Then, the tables are used to automatically generate test cases for the JUnit. The formalized description of the tables allows for the generation of complete test specifications.

12.2   Decision Table-Based Testing

In order to automate the creation of test cases, the specification of requirements for a software system has to be at least partially formalized. In this research we do not require the use of formal methods (e.g. Petri Nets). Instead, we assume the use of explicit description of certain requirements on the level of a unit. On a general level this description can be given by sets of input values to the system and the corresponding output values. This is the so-called black-box testing technique. It is not concerned with the internal structure of the system. Instead, it is solely based on the system specification and takes into account only the system response for a given input. In *decision tables (DT) based testing* the specification is described by a DT that groups admissible input values for a software unit being tested and specifies the expected output values.

---

[1] See http://junit.org.

Table 12.1 Rules of an exemplary system

No.	Condition	Action
1	*distance* > 5	*means* := *drive*
2	*distance* > 1 ∧ *time* < 15	*means* := *drive*
3	*distance* > 1 ∧ *time* ≥ 15	*means* := *walk*
4	*means* = *drive* ∧ *location* = *city centre*	*decision* := *take a taxi*
5	*means* = *drive* ∧ *location* ≠ *city centre*	*decision* := *drive your car*
6	*means* = *walk* ∧ *weather* = *bad*	*decision* := *take an umbrella and walk*
7	*means* = *walk* ∧ *weather* = *good*	*decision* := *walk*

For simplicity, we would assume here that a unit corresponds to a single class in an OO language (e.g. Java).[2] The contents of DT correspond to the possible combinations of the values of attributes in a class. There are two main concepts involved in the DT testing:

1. *Equivalence classes* group the sets of all possible values for a given attribute. An equivalence class defines a subset of values that should be processed by a system in the same manner. Thanks to that, the number of value combinations can be significantly decreased. It is assumed that DT contains input values from all possible combinations of equivalence class.
2. *Boundary values* are considered an extension of the equivalence classes. Having the set of equivalence classes, the boundary values can be specified according to ends of the equivalence class ranges. This technique is efficient, because many errors are caused by such values. A properly built DT should contain conditions related to each equivalence class and boundary values.

As an illustration, let us consider a simple system in [8]. Based on distance, location, available time and weather, a system determines how to reach a particular location. Possible actions are: walk, take an umbrella and walk, take a taxi and drive your car. The system logic is presented in the form of seven rules in a decision table (Table 12.1). An example of a boundary value is 5 for *distance*.

Our work is related to SKE, the conceptual and visual design of the rule-based intelligent systems, including XTT2. It uses DTs for knowledge representation where each row of a table corresponds to a single rule. Additionally, it provides features facilitating the modeling process. These include:

- The underlying formalism – the ALSV(FD) logic which provides a rigorous and precise definition of the rule language [9],
- Visual representation where DTs are designed in a visual manner by HQEd,
- Logical Quality Analysis (LQA) mechanism allows for discovering the logical anomalies in DTs like in completeness, redundancy, contradictions, etc.,

---

[2]However, in a general situation this is not always the case. While our tool currently supports test case generation for classes, our approach could easily be extended.

- Strong typing with the specification of attribute domains which in turn are used by LQA.

Apart from the DT modeling, we consider the application of the XTT2 method to automate generation of tests based on DTs. In this context, it can be used as a conceptualization method which allows for design of the given application specification using DTs.

Based on the above concepts, we prepared a specification of a framework for automatic unit test generation using DTs. It should support several steps that are performed during the testing process. It is assumed, these steps involve operations that can be summarized as follows:

1. **Creation** of a specification template that identifies input and output parameters for the methods belonging to the tested unit. In this step, empty decision tables (decision table schemas), corresponding to the methods, are created. Fields of the tested unit are mapped to the parameters that are used within conditions and actions of the decision tables. In this work rules are saved in XTT2 notation. Template creation can be done in HQED editor.
2. **Fulfillment** of the specification template by defining rules providing information about initial values of the identified parameters and corresponding return values of the methods. This task can be done in HQED editor.
3. **Verification** of the complete specification with regard to logical errors such in completeness or redundancy of rules, is done in the HEART rule engine with HALVA plugin.
4. **Generation** of tests according to the complete specification. The set of test cases is generated and then saved as a testing code (e.g. JUnit for Java code). This functionality is provided by a new plugin for the HEART rule engine.
5. **Execution** of tests using generated test cases. For example, JUnit tests can be run by the Eclipse with the JUnit plugin.

Here we focus on the generation step which produces a unit testing code in a given programming language. It contains the set of test cases generated from the DT-based specification of the unit. During this step, the provided specification is parsed and, for every single rule, the following operations are performed:

1. A list of the parameters used by the tested methods is extracted from the decision table schemas that were created during the first step of the testing process.
2. A set of test values is generated for each parameter by using a boundary value analysis. For this purpose, a simple algorithm is provided (see Table 12.2).
3. Test cases are generated as tuples (Cartesian product) where each tuple contains one test value for each parameter.
4. For each test case (a single element of the Cartesian product) an expected result is calculated.
5. The set of generated test cases is saved as a tests code.

Furthermore, as an extension of this approach, the generation of additional positive text cases can be provided according to the Table 12.3.

Table 12.2 Test value selection

Case	Tested value(s)
Att = Value	*Value*
Att > Value	*Value* + 1
Att ≥ Value	*Value*
Att < Value	*Value* − 1
Att ≤ Value	*Value*
Att ∈ [L, U]	*L, U*
Att ∈ {Att1, ..., AttN}	*Att1*

Table 12.3 Positive test values

Case	Tested value(s)
Att ≠ Value	Random value from domain ≠ *Value*
Att > Value	Random value > *Value*
Att ≥ Value	Random value > *Value*
Att < Value	Random value < *Value*
Att ≤ Value	Random value < *Value*
Att ∈ [L, U]	Random value from [*LtoU*]
Att ∈ Att1, Att2, ..., AttN	Random value from *Att1, Att2, ..., AttN*
Random value of *Att*	Random value from domain of *Att*

According to the provided specification of the DT-based framework for automatic unit tests generation, an architecture and implementation of a proof of the concept framework were developed.

12.3 Framework for Test Unit Generation

The proposed framework supports all the identified steps of the testing process. The first proposal of such a framework was presented in [10]. In that proposal we made several assumptions concerning the framework:

1. The specification of a tested unit is stored as an XTT2 model. This allows for using tools being part of the HADES+ framework, such as HQED for rule modeling, and HEART for rule processing [11], and test generation.
2. All the test cases are saved in JUnit format. This format is currently very common and thus it is well supported by many tools.

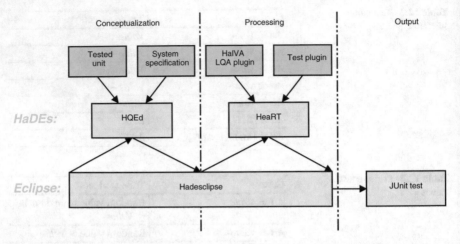

Fig. 12.1 Architecture of testing framework based on HaDEs and Eclipse

3. The framework is based on Eclipse IDE with the HADESCLIPSE plugin [12]. This
   plugin supports communication between the Eclipse and the HADES+ tools as
   well as providing graphical user interface.

In this work, the architecture of the framework is refined (see Fig. 12.1) and
includes the following elements:

- A tested unit — a source code of a tested unit. Based on this code, information
  about a unit is extracted (available fields, existing methods, etc.).
- System specification — knowledge about the behavior of a tested unit. According
  to this specification an XTT2 model is created.
- HADES — a set of tools for modeling and managing rule-based systems.

  – HQED — a visual editor for XTT2 models.
  – HEART — a rule engine for XTT2 models that uses the HALVA plugin [13] for
    providing the LQA mechanism and test plugin for generation of the test cases.

- Eclipse — an integrated development environment (IDE) for many programming
  languages that can be easily expanded via a plugins system:

  – HADESCLIPSE — a plugin for the integration of the Eclipse and HADESframe-
    work, extending Eclipse with new creators and editors.
  – JUnit — a plugin for executing JUnit tests using user-friendly interface.

The set of steps identified previously is supported by the proposed framework.
Figure 12.2 depicts the communication diagram for the testing process by means of
the architecture of the proposed framework. The communication can be summarized
as follows:

1. Creating initial specification in an Eclipse workspace as an empty hml file by
   using the creator provided by the HADESCLIPSE.

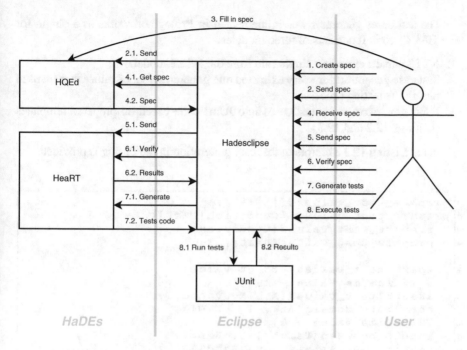

Fig. 12.2 Communication diagram for testing process

2. Sending the specification file to HQED (in server mode).
3. Completing the model in HQED by:

    3.1 Creating types that represent classes of the parameters from the tested unit.
    3.2 Defining attributes that represent parameters from a tested unit.
    3.3 Specifying tables that correspond to methods from a tested unit.
    3.4 Filling tables with rules that coincide input and result from the tested methods.

4. Receiving complete specification from HQED and storing it in a Eclipse workspace
    as hml file (for future editing) and hmr file (for sending to HEART).
5. Sending specification in the hmr format to HEART (in server mode).
6. Verifying a specification in HEART against in completeness, contradictions, sub-
    sumptions and redundancy of rules by using the HALVA plugin.
7. Generating test cases using the test plugin from HEART.
8. Receiving generated test cases from HEART and storing it in Eclipse.
9. Run prepared tests file in JUnit plugin.

The implementation of the DT-based framework for the automatic generation of
unit tests is based on the provided architecture. The implementation process can
be divided in two main implementation tasks: the first is related to the development of the test plugin for HEART while the second is related to the development
of the HADESCLIPSE plugin for Eclipse.

The test cases generator was implemented in Prolog and works as a plugin for the HEART tool. It consists of three modules:

1. XTT2 parser: extracts a single rule from the XTT2 model.
2. Test case generator: for every extracted rule generates a set of values that are used during the testing process.
3. JUnit generator: saves test cases into JUnit code based on prepared templates (Listings 12.2 and 12.3).

In the Listing 12.1 excerpts of test case generation in the Prolog is provided.

```
1  prepare_test_values([]) :- !.
2  prepare_test_values([Condition|Rest]) :-
3    prepare_test_values(Condition),
4    prepare_test_values(Rest), !.
5
6  prepare_test_values(Att gt Value) :-
7    NewValue is Value + 1,
8    assert(test_values(Att,NewValue)),
9    attribute_domain(Att, [_ to U]),
10   Value2 is Value + 2,
11   random_between(Value2, U, Rand),
12   assert(test_values(Att, Rand)),
13   !.
14 prepare_test_values(Att lte Value) :-
15   assert(test_values(Att,Value)),
16   attribute_domain(Att, [L to _]),
17   Value1 is Value - 1,
18   random_between(L, Value1, Rand),
19   assert(test_values(Att, Rand)),
20   !.
21 prepare_test_values(Att in [L to U]) :-
22   assert(test_values(Att,L)),
23   assert(test_values(Att,U)),
24   L1 is L + 1,
25   U1 is U - 1,
26   random_between(L1, U1, Rand),
27   assert(test_values(Att, Rand)),
28   !.
```

Listing 12.1 Example of a test case generation in Prolog.

HADESCLIPSE is the second plugin that was developed. It implements HADES communication protocol in order to provide integration with HEART and HQED tools. It also extends the Eclipse platform by the dedicated creator that facilitates the generation of the test cases as well as the perspective for rule-based testing (see Fig. 12.3) that provides three dedicated views:

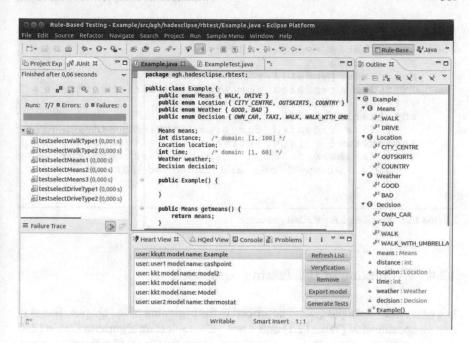

Fig. 12.3 Rule-based testing perspective in Eclipse

```
1  /* insert package here */
2  import static org.junit.Assert.*;
3  import org.junit.Test;
4
5  class [model]Test{
6      [model] obj[model];
7      /*** here JUnit code for rules goes ***/
8  }
```

Listing 12.2 Header template for JUnit generator.

- HQED view for communication with HQED (export/import),
- HEART view for communication with HEART (export/verification/test generation), and
- JUnit view for executing tests.

The software framework was practically evaluated using several examples as discussed next.

```
1  @Test
2  void test[tableName][ruleNo]() {
3      //test case
4      obj = new [Model]();
5      /*** line repeated for every condition: ***/
6      obj.set[attName]([value]);
7      obj.[tableName]();
8      /*** line repeated for every action: ***/
9      assertTrue( obj.get[attName]() == [value] );
10     /*** if there is more than one test case for ↵
           rule, above lines are repeated for each **↵
           */
11  }
```

Listing 12.3   Rule template for JUnit generator.

12.4   Evaluation of the Testing Approach

Let us consider a simple system described in the beginning of the chapter. Based on system specification (Table 12.1) and a test values selection algorithm (Table 12.2), there is the expectation that generated test cases will look like those presented in Table 12.4. There is an assumption that attributes have the following domains: *distance* $\in$ [1, 100], *time* $\in$ [1, 60],
means $\in$ {*drive, walk*}, *location* $\in$ {*city centre, outskirts, country*},
weather $\in$ {*good, bad*},
decision $\in$ {*take a taxi, drive your car, take an umbrella and walk, walk*}.

Table 12.4   Expected test cases for the system presented in Table 12.1

Rule	Test cases	Expected results
1	*distance* $= 6 \wedge$ *time* $= 1$	*means* $=$ *drive*
	distance $= 6 \wedge$ *time* $= 60$	*means* $=$ *drive*
2	*distance* $= 1 \wedge$ *time* $= 1$	*means* $=$ *drive*
	distance $= 1 \wedge$ *time* $= 15$	*means* $=$ *drive*
	distance $= 5 \wedge$ *time* $= 1$	*means* $=$ *drive*
	distance $= 5 \wedge$ *time* $= 15$	*means* $=$ *drive*
3	*distance* $= 1 \wedge$ *time* $= 16$	*means* $=$ *walk*
	distance $= 5 \wedge$ *time* $= 16$	*means* $=$ *walk*
4	*means* $=$ *drive* $\wedge$ *location* $=$ *city centre*	*decision* $=$ *take a taxi*
5	*means* $=$ *drive* $\wedge$ *location* $=$ *outskirts*	*decision* $=$ *drive your car*
6	*means* $=$ *walk* $\wedge$ *weather* $=$ *bad*	*decision* $=$ *take an umbrella and walk*
7	*means* $=$ *walk* $\wedge$ *weather* $=$ *good*	*decision* $=$ *walk*

Now we will go through the test case generation procedure (see Fig. 12.2) to demonstrate how the framework operates. In the beginning, three tools are launched: Eclipse with HADESCLIPSE plugin, HQED (in server mode) and HEART (in server mode). There is also an Eclipse project with class `Example.java` which implements assumed system logic. The step by step description is:

1. Create an empty (`Example.hml`) file in workspace using a creator provided by HADESCLIPSE.
2. Send the specification file to HQED via HQEd View in HADESCLIPSE.
3. Edit the model in HQED editor:

    3.1 Create six types that represent the domains for attributes from the tested unit as presented above.
    3.2 Create six attributes that represent the fields from the tested unit (Fig. 12.4).
    3.3 Create three tables that represents the methods from the tested unit.
    3.4 Fill table headers with attributes that represent conditions and actions for the methods. Then fill rules in the prepared model template (Fig. 12.5).

4. Receive specification from HQED using and save it in workspace as an `Example.hml` file (for future editing) and an `Example.hmr` file (for sending to HEART).
5. Send the specification file (`Example.hmr`) to HEART using HeaRT.
6. Verify the specification in HEART using HeaRT View. The example results are presented in Fig. 12.6.
7. Generate test cases in HEART using HeaRT View and save them as `Example Test.java` file (Fig. 12.7). The generated test cases are the same as the test cases presented in Table 12.4.
8. Fill in package and enum imports in `ExampleTest.java` and then run the prepared tests file in JUnit plugin (right click on file → run as... → JUnit test).

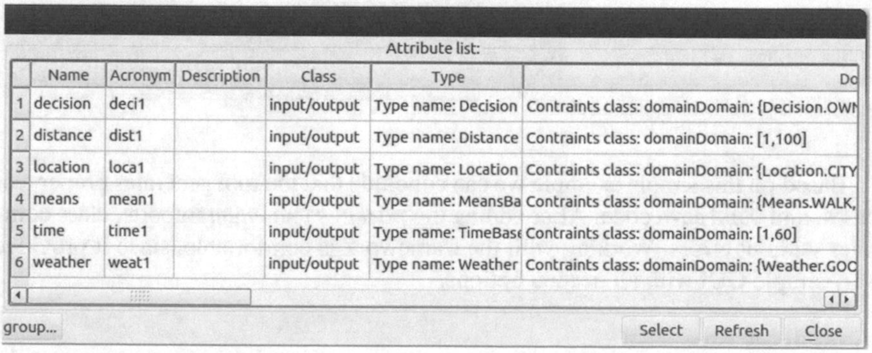

	Name	Acronym	Description	Class	Type		Do
						Attribute list:	
1	decision	deci1		input/output	Type name: Decision	Contraints class: domainDomain: {Decision.OWN	
2	distance	dist1		input/output	Type name: Distance	Contraints class: domainDomain: [1,100]	
3	location	loca1		input/output	Type name: Location	Contraints class: domainDomain: {Location.CITY	
4	means	mean1		input/output	Type name: MeansBa	Contraints class: domainDomain: {Means.WALK,	
5	time	time1		input/output	Type name: TimeBase	Contraints class: domainDomain: [1,60]	
6	weather	weat1		input/output	Type name: Weather	Contraints class: domainDomain: {Weather.GOO	

group...   Select   Refresh   Close

Fig. 12.4 List of attributes for testing

▲▼ (?) distance	▲▼ (?) time	(->) means
> 5	∈ {[1,60]}	:= Means.DRIVE ▶
∈ {[1,5]}	∈ {[1,15]}	:= Means.DRIVE ▶
∈ {[1,5]}	> 15	:= Means.WALK ▶

Table id: tab_2 -

▲▼ (?) means	▲▼ (?) location	▲▼ (->) decision
= Means.DRIVE	= Location.CITY_CENTRE	:= Decision.TAXI ▶
= Means.DRIVE	∈ {Location.OUTSKIRTS,Location.COUNTRY}	:= Decision.OWN_CAR ▶

Table id: tab_3 - selectDriveType

▲▼ (?) means	▲▼ (?) weather	▲▼ (->) decision
= Means.WALK	= Weather.BAD	:= Decision.WALK_WITH_UMBRELLA ▶
= Means.WALK	= Weather.GOOD	:= Decision.WALK ▶

Table id: tab_4 - selectWalkType

Fig. 12.5 Full specification as a XTT2 model

Fig. 12.6 Specification
verification results

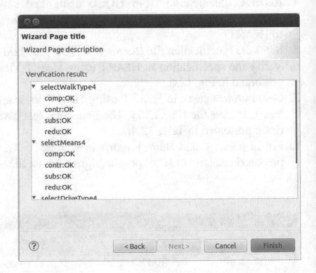

Based on this simple example we can conclude that the tool generates proper test cases, and valid Java code. After adding the *package* and *enum* imports, class compiles without errors. Working with the framework is comfortable, since it provides very simple GUI with all needed options.

```
🗎 Example.java      ⌧ Example.hml     ⚙ Example.plh [HMR]      🗎 ExampleTest.java ⌗
    /* insert package here */
  import static org.junit.Assert.*;
    import org.junit.Test;

    public class ExampleTest{
        Example objExample;

        @Test
        public void testselectWalkType1(){
            //testcase
            objExample = new Example();
            objExample.setmeans(Means.WALK);
            objExample.setweather(Weather.BAD);
            objExample.selectWalkType();
            assertTrue(objExample.getdecision() == Decision.WALK_WITH_UMBRELLA);

        }

        @Test
        public void testselectWalkType2(){
            //testcase
            objExample = new Example();
            objExample.setmeans(Means.WALK);
            objExample.setweather(Weather.GOOD);
            objExample.selectWalkType();
            assertTrue(objExample.getdecision() == Decision.WALK);

        }

        @Test
        public void testselectMeans1(){
            //testcase
            objExample = new Example();
            objExample.setdistance(6);
            objExample.settime(1);
            objExample.selectMeans();
            assertTrue(objExample.getmeans() == Means.DRIVE);
```

Fig. 12.7 Resulting JUnit code

12.5 Summary

This chapter presented a concept of a black-box unit testing technique based on XTT2
rules. Decision tables were used to simplify the design of the test cases specification
and to provide superior means of ensuring that all important test data is used and all
test cases are generated, thanks to the formalized logical model. This is a proposal
for supporting the unit testing process that will be developed in the future. Possible
extensions include the integration of this method into a *Test-Driven Development*
where the development of the specification and test cases is tightly connected to
the actual coding [3]. In a more general case testing would be considered within
the scope of the Model-Driven Development [14]. Moreover, more evaluation of
the completeness of the test specification [15] as well as formalizing the test cases
specification [16] can be provided.

This work continues the integration of the XTT2 rule formalism into software engineering processes. In the next chapter we consider a broader integration of SKE with business processes.

References

1. Sommerville, I.: Software Engineering. International computer science, 7th edn. Pearson Education Limited, Boston (2004)
2. Hunt, A., Thomas, D.: Pragmatic Unit Testing in Java with JUnit. Pragmatic Programmers. (2003)
3. Astels, D.R.: Test-Driven Development: A Practical Guide. Prentice Hall, USA (2003)
4. Tahchiev, P., Leme, F., Massol, V., Gregory, G.: JUnit in Action, 2nd edn. Manning Publications, (2010)
5. Sharma, M., Chandra, B.: Automatic generation of test suites from decision table - theory and implementation. In: Fifth International Conference on Software Engineering Advances (ICSEA). pp. 459–464 (2010)
6. Nalepa, G.J.: Proposal of business process and rules modeling with the XTT method. In Negru, V., et al. (eds.) Symbolic and numeric algorithms for scientific computing, 2007. SYNASC Ninth international symposium. September 26–29, Los Alamitos, California, Washington, Tokyo. IEEE Computer Society. IEEE, CPS Conference Publishing Service, pp. 500–506 September 2007
7. Nalepa, G.J., Ligęza, A., Kaczor, K.: Formalization and modeling of rules using the XTT2 method. Int. J. Artif. Intell. Tools **20**(6), 1107–1125 (2011)
8. Anjaneyulu, K.: Expert systems: An Introduction. Research scientist in the knowledge based computer systems group at NCST. Resonance article. (1998)
9. Ligęza, A., Nalepa, G.J.: A study of methodological issues in design and development of rule-based systems: proposal of a new approach. Wiley Interdiscip. Rev. Data Min. Knowl. Discov. **1**(2), 117–137 (2011)
10. Nalepa, G.J., Kaczor, K.: Proposal of a rule-based testing framework for the automation of the unit testing process. In: Proceedings of the 17th IEEE International Conference on Emerging Technologies and Factory Automation ETFA 2012, Kraków, Poland, 28 September 2012. (2012)
11. Nalepa, G., Bobek, S., Ligęza, A., Kaczor, K.: Algorithms for rule inference in modularized rule bases. In Bassiliades, N., Governatori, G., Paschke, A. (eds.) Rule-Based Reasoning, Programming, and Applications of Lecture Notes in Computer Science, vol. 6826, pp. 305–312, Springer, Heidelberg (2011)
12. Kaczor, K., Nalepa, G.J., Kutt, K.: Hadesclipse– integrated environment for rules (tool presentation). In Nalepa, G.J., Baumeister, J. (eds.) Proceedings of 9th Workshop on Knowledge Engineering and Software Engineering (KESE9) co-located with the 36th German Conference on Artificial Intelligence (KI2013), Koblenz, Germany, September 17, 2013. (2013)
13. Nalepa, G., Bobek, S., Ligęza, A., Kaczor, K.: HalVA – rule analysis framework for XTT2 rules. In Bassiliades, N., Governatori, G., Paschke, A. (eds.) Rule-Based Reasoning, Programming, and Applications of Lecture Notes in Computer Science, vol. 6826, pp. 337–344, Springer, Heidelberg (2011)
14. Xu, D., Xu, W., Wong, W.E.: Automated test code generation from class state models. Int. J. Softw. Eng. Knowl. Eng. **19**(4), 599–623 (2009)
15. Medders, S.C., Allen, E.B., Luke, E.A.: Using rule structure to evaluate the completeness of rule-based system testing: A case study. Int. J. Softw. Eng. Knowl. Eng. **20**(7), 975–986 (2010)
16. Liu, S., Tamai, T., Nakajima, S.: A framework for integrating formal specification, review, and testing to enhance software reliability. Int. J. Softw. Eng. Knowl. Eng. **21**(2), 259–288 (2011)

Chapter 13
Integrating Business Process Models with Rules

When it comes to practical software design, UML [1] is the standard for modeling software applications [2]. This graphical modeling language has become the dominant notation among software engineers and attempts to be a universal visual notation for software design. However, the design of complex business management systems requires much more than just UML for design. In such systems the detailed implementation of application logic is separated from business processes (the way the company works). This requires advanced modeling solutions that are comprehensible not only for software engineers, but also for business architects, knowledge engineers, regular programmers, as well as business people.

In the case of process modeling, UML is far too expressive and over-complicated to be understood by the average business user. Thus, BPMN [3] was introduced to deal with this issue. As it provides a notation emphasizing the workflow in a process and clearly describes activities of an organization at an abstract level, it became a *de facto* standard for BP modeling. Although BPMN can perfectly model a workflow in the process, the detailed logic of the process can not be specified in BPMN. The notation is not suitable for the specification of the low level logic of particular tasks in the process or even the specification of the more complex rules that control the flow. Thus, such details of the process have to be specified in other ways. Recently, the Business Rules approach [4] has been used for this purpose.

Although there is an important difference in abstraction levels of BP and BR, rules can be complementary to processes. The proposal of integration of BP with BR was put forward in Chap. 5. A formal model for the integration was provided, where the BPMN component defines the high level behavior of the system while the low level logic is defined by rules in XTT2.

In this chapter we continue that discussion on a practical level. In Sect. 13.1 we discuss challenges that need to be addressed to provide full integration, not just on the design but also the runtime level. Then in Sect. 13.2 we demonstrate how the SKE design process introduced in Chap. 9 can be applied to this goal. Furthermore, in

© Springer International Publishing AG 2018   313
G.J. Nalepa, *Modeling with Rules Using Semantic Knowledge Engineering*,
Intelligent Systems Reference Library 130,
https://doi.org/10.1007/978-3-319-66655-6_13

Sect. 13.3 we elaborate on an MDA perspective to our approach. Then in Sect. 13.4 we discuss selected metrics for the evaluation of BP complexity. Moreover, we provide the application of this approach using the PLI case study in Sect. 13.5. Finally, in Sect. 13.6 evaluation is provided. The chapter ends with a summary in Sect. 13.7.

13.1   Motivation and Challenges

Despite the ongoing research on the integration of BP and BR, there is no standardized and coherent approaches available. The existing approaches to process and rules integration are not formalized, the integration details are not well specified, and there are not many tools which provide such an integrated modeling environment. Providing an approach for modeling and executing BP with BR is therefore extremely important in order to ensure the high quality of information systems in the future.

In our approach the following challenges dealing with BP and BR are addressed:

a. the *design representation mismatch*;
b. the *model susceptibility to change*;
c. the *execution of the integrated model*.

The first challenge is related to the fact that concepts on which the rule representation is based cannot be directly modeled in process models using the BPMN notation. Such a separation of processes and rules is consistent with the BPMN 2.0 specification [3], which clarifies that BPMN is not suitable for modeling such concepts as rules. A BPMN process model should define the high level behavior of the system while the low level process logic can be described by rules. However, the specification does not define the connection between processes and rules, especially it does not clarify how process variables can be used in rules and how data can be shared between processes and rules.

Another aspect is the mismatch between the abstract analytic model of the company processes and the executable models for BPMS. In some cases, a process is not explicitly represented as an executable model [5]. Nevertheless, BPMN supports executable specification and there are guidelines how to transform an analytic BPMN model to an executable one [6]. It is also possible to design executable models directly. However, such models are generally hard to comprehend by business users [7].

The second challenge is related to changes, as nowadays processes and rules in companies change often. This is caused by changes of law, product specification, or other factors resulting from striving for innovation and competitiveness or the regulations imposed by stakeholders. Thus, the management system has to keep up with these changes. According to [8] most companies hire external consultants for constructing their models. However, the manual redesign of process models is costly, time-consuming and error-prone, and in such cases companies often fail to keep their models up to date [9]. Moreover, in the case of complex process models designed manually, it is hard to determine if they are compliant to real world processes and properly capture the right business practices [10]. Therefore, to deal with this challenge, automation or even semi-automation of model generation is usually introduced. It can lead to significant improvements, both in compliance issues and in such aspects as cycle or service times [11]. It can help in fitting rules to processes, keeping up with changes as well as preparing an executable model.

The third challenge consists in the executing of such an integrated model. If processes and rules are managed separately during modeling, this can cause some problems during execution, especially related to enforcement of rules and data manipulation. Although there are many tools that provide an execution environment for process models, in the case of models with rules the execution is not always supported. As there are no coherent requirements how such models should be integrated, this influences how they should be proceeded. Such an issue as managing several rule sets in a process is also a nontrivial task.

The three above mentioned challenges are addressed in the research presented in Chaps. 5, 6, and finally in this chapter. In fact, the *design representation mismatch challenge* was addressed in Sect. 5.2 by the introduction of a formalized General Business Logic Model. It integrates a process model with rules and can be applied to specific approaches, such as the SKE-specific Business Business Logic Model presented in Sect. 5.3. Furthermore, in Sect. 6.4 an algorithm for the automatic generation of Business Process models was presented. It allows for quick response to changes of the system specification. Moreover, the metrics for measuring model complexity allows for constant managing of the process complexity. Thus, these address the *challenge concerning model susceptibility to change*.

In this chapter we discuss the tool support for the above mentioned solutions. This includes practical design and execution tools. As the mentioned algorithm creates executable models, which can be deployed and tested in the provided execution environment for the integrated approach, this solution addresses the *execution challenge*.

13.2 Design and Execution Using the Integrated Approach

The approach presented in this section extends the presented design process of SKE towards Business Processes integrated with rules. It uses BPMN for modeling and a workflow engine that runs BPMN-based business process models for execution purposes. The proposed design process includes System Specification, Modeling and Execution phases. The overview of the process with tools and files exchanged between them can be observed in Fig. 13.1.

1. The SKE conceptual modeling phase was extended with a web application – WebARD[1] – that supports on-line system specification in terms of the ARD models.

---

[1]The application was developed at AGH UST by Artur Smaroń as a student project: http://ai.ia.agh. edu.pl/wiki/pl:dydaktyka:wshop:prv:2014:ardeditor:.

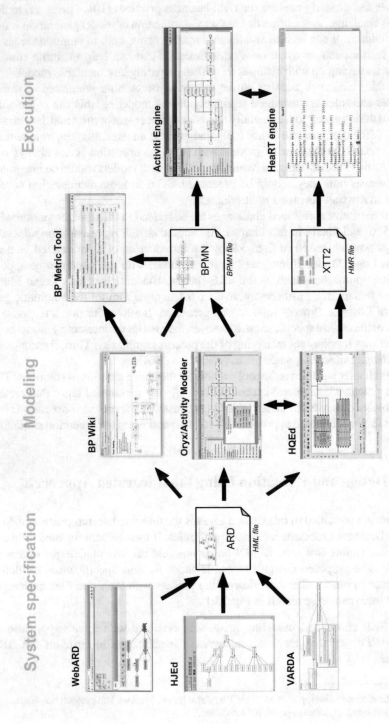

Fig. 13.1 Overview of the design approach for BP with BR

2. In the case of modeling phase the environment takes advantage of standalone BPMN supporting editors for modeling processes (such as Oryx, Activiti Modeler, jBPM Designer, or Activiti Designer) [12, 13] as well as wiki-based editor (BPWiki) [14].
3. For execution purposes, a prototype integration of a process engine (jBPM or Activiti) with the HEART inference engine was provided [13]. Moreover, the proposed toolchain also supports creation of knowledge in a collaborative wiki-based environment – BPWiki [14], as well as measuring the complexity of Business Process models [15].

The modeling and execution phases are described in the following paragraphs.

Modeling

The modeling phase in the proposed approach uses the BPMN notation for Business Processes. The semi-automatic transition from the system specification phase to the modeling phase is based on the algorithm presented in Sect. 6.4, and allows for generating BPMN process models with decision table schemas for Business Rule tasks and form attributes for the User tasks. As a proof of concept, the first attempt to support the transition algorithm from ARD to BPMN was provided using the prototype ProM plugin.[2]

For the modeling phase, an integrated tool framework for BP and BR was proposed in [12]. The framework uses one of the BPMN supporting editors for the modeling processes (Oryx, Activiti Modeler, jBPM Designer, or Activiti Designer), and the XTT2-based tools for rules. Oryx[3] and Activiti Modeler[4] are web-based editors for modeling business processes, which enables the modeling of processes using a web browser. jBPM and Activiti Designers are Eclipse plugins that allow for designing and more detailed development of process models; BPWiki [14] allows for importing models and simple management as well as collaborative work (see Sect. 15.4). The XTT2 knowledge representation is used for Business Rules, supported by the HQEd editor. This allows for visual modeling of both Business Processes and Rules.

Oryx

It is an academic open source project, allowing for modeling processes in BPMN 2.0 [16]. Its user interface was adapted for choosing a proper XTT2 decision table for a particular Business Rule task in the BPMN model [12]. The editor was selected for this research for a couple reasons. It is implemented as a web-based application and allows for the collaborative modeling of processes. It is possible to extend the application with new functions or adapt it to enhance existing functionality as it allows

---

[2]The prototype version of a plugin was developed at AGH UST by Piotr Winiarski as a student project: http://ai.ia.agh.edu.pl/wiki/pl:dydaktyka:wshop:prv:2014:ardplugin:start.

[3]See: http://code.google.com/p/oryx-editor.

[4]See: http://activiti.org/components.html.

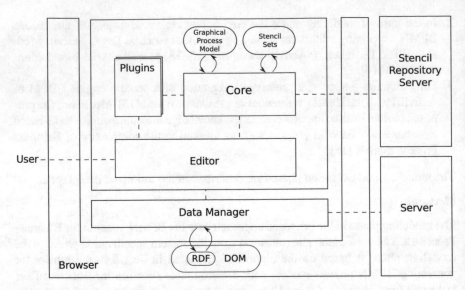

Fig. 13.2   Outline of the Oryx editor's architecture, after [16]

for implementing plug-ins (see Fig. 13.2) and the architecture of the intermediate layer is based on the concept of Representational State Transfer (REST).[5]

Using the extended Oryx platform, there are two possible ways to edit Business Rule tasks:

- editing rules using the dedicated HQEd editor for XTT2, which can be connected via the TCP/IP protocol with the editor, or
- editing rules within Oryx using a dedicated Oryx plugin.

In Fig. 13.3, a user interface supporting the integration of the BPMN Business Processes with the XTT2-based BR. for the PLI use case is presented. As HQED provides a socket-based interface for its services, the XTT2 rules can be edited in the HQEd editor connected via network with Oryx. In such a case opening a rule task in Oryx allows for XTT2 editing in HQEd. Based on the experience from the Oryx tool, a similar approach to model and edit Business Rule tasks was implemented in the following BPMN editors: jBPM Designer, Activiti Modeler and Activiti Designer [13].

BPWiki

Wiki systems [17, 18] are lightweight solutions supporting collaborative cooperation between users. The proposed environment was extended with a wiki supporting

---

[5]This means that it is a network-based application with stateless collaboration among its member components. Each request is expected to be self-contained (only the server has access to resources with unique URIs; the client sends a request to the server, and the server returns a resource representation to the client). In this way, the server does not store the information about the state, but the client is responsible for transitions between states.

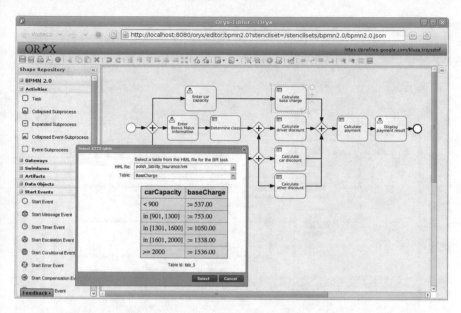

Fig. 13.3 Screenshot from the prototype Oryx GUI for XTT2, after [12]

Business Process design and development – BPWiki [14]. It provides a human-readable textual notation for BPMN that is easy to be edited manually if needed. The syntax in the BPWiki editor is highlighted and the visualization of the BPMN model in the wiki system is provided (see Fig. 13.4). The system also supports the translation of the BPMN models from their XML serialization to the human-readable textual notation. What is more important, the tool allows for decomposition of the BP model to wiki pages corresponding to particular tasks, or to wiki namespaces corresponding to specific subprocesses, as well as it provides an environment for model commenting and discussions between modelers. More details of BPwiki integrated with the LOKI platform will be provided in Chap. 15.

Complexity Metrics

Metrics have been used for many purposes in software development, such as predicting errors [19], supporting re-factoring or estimating costs of software [20] or measuring software functional size [21]. Thus, as processes are quite similar to software programs (in such respects as focusing on information processing, dynamic execution that follows a static structure and having a similar compositional structure), the business process metrics can be used for controlling quality and improving models. The complexity of the designed BP model can be measured using additional tools [15]. The provided plugin allows for measuring several complexity metrics for models, and configuring a metric based on the configured ones. As the metrics are expressed as easy to interpret numbers, they are intuitive. Thus, they provide some

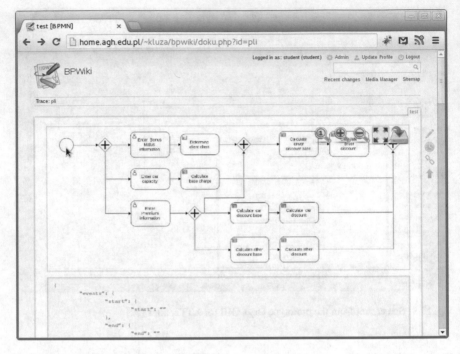

Fig. 13.4 Exemplary process model in BPWiki

basic information about the complexity of models. This allows for a quick comparison of two different models and an evaluation of their quality in the modeling phase. Selected metrics are used in evaluation section. More information on this topic will given in Sect. 13.4.

Execution

To address the execution challenge, the proposed environment uses the workflow engine which communicates with the HeaRT rule engine. The workflow engine can be either one of the BPEL [22] engines (a subset of BPMN models can be transformed to BPEL) or some native execution environment for processes, such as jBPM[6] or Activiti.[7] HEART executes the logic of particular tasks, is responsible for handling the selection and execution of rules. This approach allows for executing a fully specified BPMN model.

The BP model with the specified rules in BR tasks can be executed in the process engine, such as jBPM [23] or Activiti Engine [24], extended with suitable integration plugins [13, 25]. Both of these tools are generic process engines based on the Process Virtual Machine (PVM),[8] which can execute business processes described in BPMN.

---

[6]See: http://www.jboss.org/jbpm.

[7]See: http://www.activiti.org/.

[8]See: http://docs.jboss.com/jbpm/pvm.

The architectures are based on the WfMC's reference model [26]. As jBPM provides a process engine and has a pluggable architecture, it can be integrated with this runtime environment. Activiti is a workflow engine started by former jBPM developers and is based on their workflow system experience.

Both jBPM and Activiti provide suitable interfaces and thus they could be integrated with HeaRT. Thus, the proposed integrated architecture that supports executing the combined BP and BR model is based on these open-source workflow engines that runs a BPMN-based process model, and communicates with the HEART rule engine for running the low-level rule-based logic.

Integration of Rule and Process Engines

For specific Business Rule tasks in a process model, the workflow engine calls HEART using a socket-based interface. The HEART engine runs in a server mode and handles the communication from the jBPM/Activiti run as a client. This is done by connecting the BPMN model tasks with the corresponding XTT2 tables. In the case of jBPM, the process engine sends a message to HEART in the defined format:

1. *Rule task* with names starting from "H" are executed using HEART. Each rule task corresponds to one XTT2 table.
2. *Ruleflow groups* are associated with the list of XTT2 tables in the rule base.
3. If the model state is not specified, the HEART engine starts from its current state.
4. State attributes which are the results of the HEART inference are stored as context variables of the process instances.

Communication between these tools is possible via dedicated API. HEART introduces an integration layer for the most common programming languages, including Java, PHP and Python. HEART communication with jBPM uses a network protocol and is handled by a dedicated Java class. For the implementation the dedicated Drools Knowledge API[9] provided by jBPM (Drools) was used. This knowledge-centric API is written in Java and provides classes and interfaces for knowledge manipulation.

13.3 Model-Based Perspective

Modeling of processes with rules can be seen as the Model Driven Engineering (MDE) approach [27]. In this approach, models are considered as first-class citizens as opposite to other approaches, e.g. where documentation plays a dominant role. This section briefly presents the main concepts of Meta Object Facility (MOF). Moreover, an overview of Model Driven Architecture (MDA), an architectural framework for MDE developed by OMG is given. Finally, our approach is compared to the MDA.

---

[9]See: http://docs.jboss.org/jbpm/v5.1/javadocs.

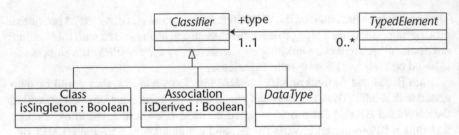

Fig. 13.5 Fragment of the MOF metamodel abstract syntax, after [35]

Meta Object Facility

Most languages are specified using a kind of a grammar, mostly in Backus Naur Form (BNF) [28]. Such a grammar describes how to develop correct expressions in a specific language. As BNF is a text-based notation, it is suitable for text-based languages. In the case of visual languages having a graphical syntax, like BPMN or UML, a visual mechanism for defining this syntax is needed [29].

A language (or more specifically a model) used to define models is called a meta-model (or a model of a model) [30]. The abstract syntax of BPMN is defined by the BPMN metamodel [3], and the metamodel is defined using MOF [31]. It is an OMG standard as well as an international standard ISO/IEC 19502:2005 [32] that defines the language for defining modeling languages. It provides a universal way of describing modeling constructs [33, 34], so MOF is defined using MOF itself. A fragment of the MOF metamodel is shown in Fig. 13.5. Its syntax is based on UML class diagrams: artifacts can be modeled as classes and their properties as attributes of the class, the relationships between artifacts can be modeled as associations between classes representing these artifacts. In order to clarify the metamodel one can also use OCL expressions [33].

OMG defined a four-layered architecture for MOF, consisting of the following conceptual layers: a meta-metamodel layer (an abstract language used to define meta-models), a metamodel layer (a language used to define models), a model layer (comprised of metadata that describe data in the information layer), and an information layer. Each element in a layer describes elements in the layer below it.

Model Driven Architecture

Model Driven Architecture (MDA) [36] is an approach developed by OMG that defines an architectural framework for MDE. MDA supports detailed specification based on models. As it separates the design specification from the specific technology platform, this allows for producing interoperable, reusable and portable software components and data models. In such a software development process, a modeler becomes a sort of programmer, and modeling replaces or displaces programming activity [35]. MDA requires a strict definition of a model, as there is a need for automatic transformations within the MDA framework. Such a definition must use a language with a well-defined syntax and semantics, suitable for an automated computer interpretation.

In the traditional software life cycle transformations from model to model, or from model to code, are done manually by a software architect or developer. As MDA models can be interpreted and processed by machines, some aspects of model transformations can be automated. MDA defines the following model abstraction layers [36]:

- *Computation Independent Business Models (CIM)* is on the highest-level of business model [29]. It shows a system from the computation independent viewpoint, not showing structure details. As it focuses on the system requirements and environment [36], it is often compared to a domain model specification and to an ontology [37] or uses some high level business process language [29].
- *Platform Independent Models (PIM)* shows a system from the computation-dependent perspective [37]. It hides the details necessary for certain platforms and defines a model independent of any implementation technology by focusing on the operation of a system [36].
- *Platform Specific Models (PSM)* shows a system from a platform specific perspective [36], focusing on the usage of this platform. It specifies a system in terms of certain implementation constructs.
- *A code model* can also be distinguished. It represents the code, usually in a high-level language [38].

As transformations between these models are essential in MDA. The MDA goal is to be able to transform a high level PIM into a PSM, and then each PSM to code. The possibility of automatic transformations allows for increasing the abstraction level of the design.

Business Logic Model in the Context of MDA

In software development, MDE [29] increases productivity by constructing models as a software specification and reusing them. Models can be designed and generated to a certain level of detail, and then complemented with some source code. MDE consists of consecutive phases of system modeling. In the approach presented here the Business Logic Model can be considered as the M2 Layer of MOF (see Fig. 13.6). It defines the BPMN and XTT2 Concepts. In practice, the model can be instantiated as Business Logic Model Instance.

Similarly to MDE, our approach consists of consecutive phases. Using the algorithm proposed in Sect. 6.4 the integrated models can be generated automatically (see transformation 1 in Fig. 13.7). Then, they can be complemented with rules (see transformation 2 in Fig. 13.7), deployed and executed in the execution environment, described in this chapter. Thus, our integrated approach fits into the MDE paradigm [27] as it focuses on creating and exploiting domain models and simplifies the design process.

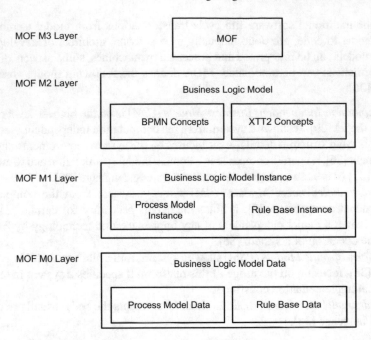

Fig. 13.6 Business Logic Model in the MOF Modeling Infrastructure

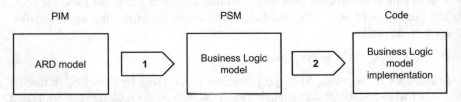

Fig. 13.7 Steps of the approach in the context of the MDE development process

13.4 Complexity Assessment

Complexity metrics for Business Processes can be used for better understanding, and controlling the quality of the models, thus improving their quality [39]. The design process of BP models can be improved by the availability of metrics that are transparent and easy to be interpreted by the designers.

In [40] four perspectives to process complexity were distinguished:

- activity complexity – affected by the number of activities a process has,
- control-flow complexity – affected by such elements and constructs as splits, joins, loops, start and end,
- data-flow complexity – concerns data structures, and the number of formal parameters of activities, and the mappings between data of activities [41],

• resource complexity – concerns different types of resources that have to be accessed during process execution.

Here, we consider mainly activity complexity and control-flow complexity.

Furthermore, several simple complexity metrics adapted from software engineering were introduced in [42]. Based on the simple metric, which counts the number of Lines of Code (LOC) of a program, they proposed three metrics: NOA – Number of activities in a process, NOAC – Number of activities and control-flow elements, and NOAJS – Number of activities, joins, and splits.

Another metric is the Information Flow Metric. They proposed a metric Interface Complexity (IC) of an activity, defined as: $IC = Length * (NoI * NoO)^2$. In the formula, the length of the activity can be calculated using traditional SE metrics such as LOC and the number of inputs (NoI) and outputs (NoO) of the activity follow can be directly calculated from the model.

Latva-Koivisto [43] introduced the Coefficient of Network Complexity (CNC) metric for BP. CNC is a widely used metric in network analysis and was proposed to measure the degree of complexity of a critical pass network. Cardoso et al. [42] proposed a simple process version of this formula:
$CNC = numberofarcs/(no.ofactivities, joins, andsplits)$.

Selected simple metrics were defined by Sánchez-González et al. [44]. They are presented in Table 13.1. Each of these metrics describes a single aspect of a process model and is easy to understand.

In [45], Mendling presented an overview of process metrics with reference to model partitionability and connector interplay. *Sequentiality* is defined by a number of connections between non-gateway nodes divided by the total number of connections. If the whole model is sequential, the sequentiality ratio is 1. It is based on the assumption that models with more gateways are more complex and have more errors. *Gateway heterogeneity (GH)* – shows how many different gateway types was used in the model. Higher heterogeneity indicates a higher probability of gateway mismatch

Table 13.1 Simple metrics defined in [44]

Metric	Description
Number of nodes	Number of activities and routing elements in a model
Diameter	Length of the longest path from a start node to an end node
Average Gateway Degree (AGD)	Average of the number of both incoming and outgoing arcs of the gateway nodes in the process model
Maximum Gateway Degree (MGD)	Maximum sum of incoming and outgoing arcs of these gateway nodes
Gateway Mismatch	Sum of gateway pairs that do not match with each other, e.g. when an AND-split is followed by an OR-join
Concurrency	Maximum number of paths in a process model that may be concurrently activate due to AND-splits and OR-splits

(gateway pairs that do not match with each other). If only one type of gateway was used, the GH is 1. As it is emphasized, BP model complexity cannot be directly determined by only one type of metric, as each of them has some limitations, e.g.:

- size metrics (such as NOA, NOAC, NOCAJS) or Diameter – bigger models can be more understandable than smaller ones if they are more sequential,
- CNC – one can find a models with the same value of CNC, which vary in comprehensibility due to different types of nodes.
- gateway degree (GH, AGH, MGD) – two models which differ in size can have the same value of gateway degree.
- sequentiality – models with the same value of sequentiality can vary in complexity due to different kind of gateways.

The metrics elaborated in this section will be used for evaluation purposes of our approach.

13.5   Practical Application of the Model

For evaluation purposes several benchmark cases were selected. Here we only discuss results for the PLI case study (Sect. 4.4). It was presented previously during the description of ARD+ model in Sect. 6.2. Here, a refined version of the case study is presented.

Description

In the PLI case the rate for the automobile liability insurance for protection against third party insurance claims is to be calculated. The price is calculated based on various reasons, which can be obtained from the insurance domain expert. The main factors in calculating the liability insurance premium are data about the vehicle: the automobile engine capacity, the automobile age, the number of seats, and a technical examination. Additionally, information such as the driver's age, the period of holding the license, the number of accidents in the last year, and the previous class of insurance, have impact on the final insurance rate.

ARD model

Here a short excerpt of the G_A^{PLI} (see Sect. 5.3) with types and domains of the attributes is presented:

$type(accidentNo) = integer, \quad \mathbb{D}_{accidentNo} = [0; inf],$
$type(clientClass) = integer, \quad \mathbb{D}_{clientClass} = [-1; 9],$
$type(carCapacity) = integer, \quad \mathbb{D}_{carCapacity} = [0; inf],$
$type(baseCharge) = integer, \quad \mathbb{D}_{baseCharge} = [0; inf],$
$type(driverAge) = integer, \quad \mathbb{D}_{driverAge} = [16; inf],$
$type(driverLicenceAge) = integer, \quad \mathbb{D}_{drLicAge} = [0; inf],$
$type(driverDiscountBase) = integer, \quad \mathbb{D}_{drLicAge} = [-inf; inf],$

$type(driverDiscount) = integer, \ \mathbb{D}_{drLicAge} = [-inf; inf],$

$type(carAge) = integer, \ \mathbb{D}_{carAge} = [0; inf],$

$type(antiqueCar) = boolean, \ \mathbb{D}_{antiqueCar} = [true; false],$

$type(seatsNo) = integer, \ \mathbb{D}_{seatsNo} = [2; 9],$

$type(technical) = boolean, \ \mathbb{D}_{technical} = [true; false],$

$type(carDiscountBase) = integer, \ \mathbb{D}_{drLicenceAge} = [-inf; inf],$

$type(carDiscount) = integer, \ \mathbb{D}_{drLicenceAge} = [-inf; inf],$

$type(installmentNo) = integer, \ \mathbb{D}_{installmentNo} = [1; 2],$

$type(insuranceCont) = boolean, \ \mathbb{D}_{insuranceCont} = [true; false],$

$type(insuranceCarsNo) = integer, \ \mathbb{D}_{insuranceCarsNo} = [0; inf],$

$type(otherInsurance) = boolean, \ \mathbb{D}_{otherInsurance} = [true; false],$

$type(insuranceHistory) = integer, \ \mathbb{D}_{insuranceHistory} = [0; inf],$

$type(otherDiscountBase) = integer, \ \mathbb{D}_{drLicAge} = [-inf; inf],$

$type(otherDiscount) = integer, \ \mathbb{D}_{drLicenceAge} = [-inf; inf],$

$type(payment) = float, \ \mathbb{D}_{payment} = [0; inf].$

Execution

First, refer to the discussion of the SKE model on page 118. Here the full BPMN model for the PLI case study containing forms and decision table schemas is presented in Fig. 13.8.

One can observe that the BR tasks in the process are connected with the decision tables (for clarity in Fig. 13.8 only decision table schemas are presented), e.g. the "Calculate other discount base" Business Rule task in the process model is connected with the decision table schema:

$$schema(t) = (\{installmentNo, insuranceCont, insuranceCarsNo\},$$
$$\{otherDiscountBase\}).$$

For the further execution purposes, the decision tables schemas were complemented with rules. The Table 13.2 shows the XTT2 decision table for the BR task "Calculate base charge". It calculates the baseCharge price based on the carCapacity (capacity of the car engine). The rules from this table are specified using the executable HMR representation, a fragment of the HMR representation for the "Calculate base charge" XTT2 decision table is presented in Listing 13.1. All decision tables with rules with the corresponding HMR representation can be found in [46].

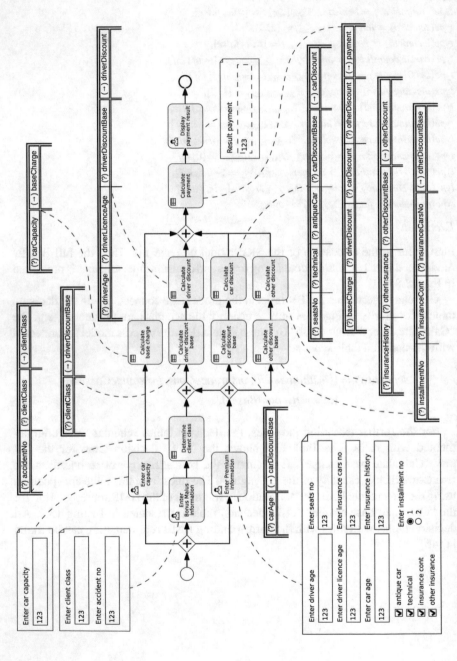

Fig. 13.8 The BPMN model for the PLI case study with forms and rules

Table 13.2 "Calculate base charge" XTT2 decision table

(?) carCapacity	(–>) baseCharge
in [901, 1300]	753.00
< 900	537.00
>= 2000	1536.00
in [1601, 2000]	1338.00
in [1301, 1600]	1050.00

```
1  xschm 'Calculate base charge': [carCapacity] ==> [baseCharge].
2  xrule 'Calculate base charge'/1: [carCapacity in [901 to 1300]] ==> [↩
       baseCharge set 753.00].
3  xrule 'Calculate base charge'/2: [carCapacity lt 900] ==> [baseCharge ↩
       set 537.00].
4  xrule 'Calculate base charge'/3: [carCapacity gt 2000] ==> [baseCharge ↩
       set 1536.00].
5  xrule 'Calculate base charge'/4: [carCapacity in [1601 to 2000]] ==> [↩
       baseCharge set 1338.00].
6  xrule 'Calculate base charge'/5: [carCapacity in [1301 to 1600]] ==> [↩
       baseCharge set 1050.00].
```

Listing 13.1 Fragment of the HMR representation for the "Calculate base charge" XTT2 decision table

For the PLI case study, the correct BPMN model with 4 forms for User tasks and 9 decision table schemas for Business Rule tasks were created. The model constitutes an executable BPMN model. The BPMN model XML code along with the corresponding rules can be found in [46].

13.6 Evaluation of the Example

Deployment

As the proposed approach is supported by the proof-of-concept environment, the models described in the previous section have been deployed and tested in the execution environment that is based on Activiti integrated with the HeaRT rule engine. In order to execute the models generated using the proposed algorithm, each decision table schema in the model was complemented with suitable rules. The code of the full implemented BPMN models (in the BPMN 2.0 XML format with Activiti extensions) and the HMR rule representation for the selected use cases can be found in [46].

Business solutions need comprehensive testing for ensuring correct and reliable performance. Although Business Process testing focuses on system and user acceptance testing [47], regression testing is used as well [48]. Regression testing helps to ensure the quality of modified models by providing unit tests for checking the behavior of models.

To validate the integrated models, unit testing was applied. In general, unit testing provides a way to keep bugs in check [49]. It is a method of validating software by providing input, if needed, and examining software behavior, usually in the form of its output.

JUnit is an open source Java library that supports unit testing [50]. It was previously discussed in Chap. 12. In JUnit, testers write assertions which specify the desired outcomes and compare them with the actual outcomes. The assertion succeeds when the desired and actual outcomes are identical. Otherwise, it fails. For the our example model:

1. 10 unit tests that check various scenarios were prepared.
2. The prepared tests were executed in the following way:

   2.1 A test sets the initial values of the attributes for the process instance.
   2.2 It completes the User tasks that requires input from a user.
   2.3 Business Rule tasks in the process are automatically executed using the HeaRT rule engine.
   2.4 The result of the test is obtained and checked with the testing value.

In Listing 13.2 a fragment of a JUnit test for testing PLI use case is presented. The *ProcessEngine* and the services, initialized in the form of a JUnit rule, are available to the test class through the getters of the `activitiRule` (see Line 6). Test input specification is provided as a hashmap (see Lines 15–18) and a new process instance is created based on the process definition with the given key with the variables to pass to the process instance (see Line 20). JUnit assertions check if the process instance was created (see Line 21). Then, the test completes all the User tasks found. As the Business Rule tasks are executed automatically, the test checks if a single User task (displaying the output) is available (see Line 36) and if the result exists (see Line 39), if it is equal to the desired value (see Line 41).

```
1  package org.activiti.designer.test;
2  import static org.junit.*;
3  public class ProcessTestPli {
4      private String filename = "pli.bpmn";
5      @Rule
6      public ActivitiRule activitiRule = new ActivitiRule←
           ();
7      @Test
8      public void test1() throws Exception {
9          Logger log = Logger.getLogger("mylog");
10         RepositoryService repositoryService = ←
               activitiRule.getRepositoryService();
11         RuntimeService runtimeService = activitiRule.←
               getRuntimeService();
12         TaskService taskService = activitiRule.←
               getTaskService();
13         repositoryService.createDeployment().←
               addInputStream("pli.bpmn20.xml",
```

```
14                          new FileInputStream(filename)).deploy();
15          Map<String, Object> variableMap = new HashMap<↩
                String, Object>();
16          variableMap.put("carCapacity", "993");
17          ...
18          variableMap.put("driverAge", "30");
19          ProcessInstance processInstance = runtimeService
20              .startProcessInstanceByKey("pli", ↩
                    variableMap);
21          assertNotNull(processInstance.getId());
22          log.info("Created process instance "
23              + "(process instance ID:" + processInstance↩
                    .getId() + ", process
24              model ID:  " + processInstance.↩
                    getProcessDefinitionId() + ")");
25
26          List<Task> allTaskList = taskService.↩
                createTaskQuery()
27                  .processInstanceId(processInstance.getId())↩
                        .list();
28          log.info(allTaskList.size() + " tasks found");
29          for (Task task : allTaskList) {
30              taskService.complete(task.getId());
31              log.info(" Task " + task.getName() + " ↩
                    completed");
32          }
33              allTaskList = taskService.createTaskQuery()
34                          .processInstanceId(↩
                                processInstance.getId()).↩
                                list();
35          log.info(allTaskList.size() + " task found");
36          assert(allTaskList.size() == 1);
37          Long payment = Long.parseLong((runtimeService
38                          .getVariable(processInstance.getId↩
                                (), "payment"), 10);
39          assertNotNull(payment);
40          log.info("Payment result is: " + payment);
41          assert(payment == 753);
42          for (Task task : allTaskList) {
43                  taskService.complete(task.getId());
44                  log.info(" Task " + task.getName() + "↩
                        completed");
45          }}}
```

Listing 13.2   A fragment of an exemplary JUnit test for testing PLI use case

Using the *log4j* logging utility the test shows some interesting runtime events on
a console. The result of the execution of the test from Listing 13.2 can be observed
in Listing 13.3. Lines 1–10 shows that a process engine was initialized. Then, the
processing resource pli.bpmn20.xml was successfully deployed (Lines 11–12),
and a process instance was created (Lines 17–18). After all input User tasks are
completed (Lines 19–26), the result of the process instance is shown (Lines 27–28).

```
 1 2014-11-20 16:14:06 org.springframework.beans.factory.↩
      xml.XmlBeanDefinitionReader loadBeanDefinitions
 2 INFO: Loading XML bean definitions from class path ↩
      resource [activiti.cfg.xml]
 3 2014-11-20 16:14:08 org.activiti.engine.impl.db.↩
      DbSqlSession executeSchemaResource
 4 INFO: performing create on engine with resource org/↩
      activiti/db/create/activiti.h2.create.engine.sql
 5 2014-11-20 16:14:08 org.activiti.engine.impl.db.↩
      DbSqlSession executeSchemaResource
 6 INFO: performing create on history with resource org/↩
      activiti/db/create/activiti.h2.create.history.sql
 7 2014-11-20 16:14:08 org.activiti.engine.impl.db.↩
      DbSqlSession executeSchemaResource
 8 INFO: performing create on identity with resource org/↩
      activiti/db/create/activiti.h2.create.identity.sql
 9 2014-11-20 16:14:08 org.activiti.engine.impl.↩
      ProcessEngineImpl <init>
10 INFO: ProcessEngine default created
11 2014-11-20 16:14:08 org.activiti.engine.impl.bpmn.↩
      deployer.BpmnDeployer deploy
12 INFO: Processing resource pli.bpmn20.xml
13 2014-11-20 16:14:08 org.activiti.engine.impl.bpmn.↩
      parser.BpmnParse parseDefinitionsAttributes
14 INFO: XMLSchema currently not supported as typeLanguage
15 2014-11-20 16:14:08 org.activiti.engine.impl.bpmn.↩
      parser.BpmnParse parseDefinitionsAttributes
16 INFO: XPath currently not supported as ↩
      expressionLanguage
17 2014-11-20 16:14:09 org.activiti.designer.test.↩
      ProcessTestPli test1
18 INFO: Created process instance (process instance ID:5, ↩
      process model ID:  pli:1:4)
19 2014-11-20 16:14:09 org.activiti.designer.test.↩
      ProcessTestPli test1
20 INFO: 3 tasks found
21 2014-11-20 16:14:09 org.activiti.designer.test.↩
      ProcessTestPli test1
22 INFO:  Task Enter car capacity  information completed
23 2014-11-20 16:14:09 org.activiti.designer.test.↩
      ProcessTestPli test1
24 INFO:  Task Enter Bonus Malus information completed
25 2014-11-20 16:14:09 org.activiti.designer.test.↩
      ProcessTestPli test1
26 INFO:  Task Enter Premium information completed
27 2014-11-20 16:14:09 org.activiti.designer.test.↩
      ProcessTestPli test1
28 INFO: Payment result is: 753
```

Listing 13.3 A result of executing the test for the PLI use case from Listing 13.2

The testing results for the PLI use case can be observed in Fig. 13.9.

For all three case studies, the BPMN model was executed and tested. This provides a validation of results as a part of model quality assurance. Model files and tests for

Fig. 13.9 The JUnit tests for the PLI model

Table 13.3 Simple metrics – number of elements

Metrics	PLI
Number of user tasks	4
Number of business rule tasks	9
Number of parallel gateways	4
Number of activities (NOA)	13
Number of activities and control flow elements (NOAC)	19
Number of activities, joins and splits (NOAJS)	17
All Elements (ALL)	41

the PLI use case and two other cases are available at the website of the HiBuProBuRul project.[10]

Measurement of Model Complexity

In order to evaluate complexity of the case in this section selected complexity metrics are calculated in Tables 13.3 and 13.4.

Table 13.5 presents other parameters related to rules.

In order to better assess the complexity of the integrated model, the following combined metrics is introduced:

---

[10]See: http://geist.agh.edu.pl/pub:projects:hibuproburul:start.

Table 13.4 Business process
metrics based on the number
of elements

Metrics	PLI
Coefficient of Network Complexity (CNC)	1.29
Interface Complexity (IC)	13
Diameter	4
Average Gateway Degree (AGD)	4
Maximum Gateway Degree (MGD)	5
Concurrency	4
Sequentiality	0.32
Gateway Heterogeneity (GH)	1

Table 13.5 Other parameters
of cases related to rules

Other parameters	PLI
Number of ARD attributes ($|A| + |C|$)	30
Number of ARD physical attributes ($|A|$)	22
Number of ARD dependencies ($|D|$)	22
Number of XTT2 decision tables ($|T_{\mathcal{X}}|$)	9
Number of XTT2 rules ($|R|$)	61

$$Complexity(\mathcal{M}_{SKE}) = \frac{|R|}{|T_{\mathcal{X}}|} * \frac{NoA}{ALL} * Concurrency$$

This metrics combines the complexity of the knowledge base (an average number of rules in the decision component) and the complexity of a process model (the ratio of the number of activities to the number of all elements) with the concurrency of the model. For the PLI case $Complexity(\mathcal{M}_{SKE}) = 8.60$. This single complexity metrics is comparable to the complexity metrics of the integrated model components.

13.7   Summary

In this chapter a practical approach to the integration of business processes and rule based systems was presented. This approach is based on the previously introduced formalized model. The model opens up an opportunity of integrating the XTT2 rules with business processes on a level of rule tasks. From the perspective of the structured rule base the model provides an explicit inference flow determined by the business process control flow. Furthermore, the design approach considered in this chapter extends the previously introduced SKE design approach. In the chapter we discuss software tools supporting the design process. We also continue the development of the previously introduced PLI case study to demonstrate the feasibility of the approach. The evaluation is provided by the means of practical implementation of the case study, as well as by the means of process complexity metrics applied to the integrated models. Finally, we consider how this approach for the integration of SKE with business process management follows the MDA principles.

In the next chapters we will discuss the application of SKE with Semantic Web technologies and extension towards Collective Knowledge Engineering.

References

1. OMG: Unified Modeling Language version 2.1.2. infrastructure specification. Technical report formal/2007-11-04, Object Management Group (November 2007). http://www.omg.org/cgi-bin/doc?formal/2007-11-04.pdf
2. Hunt, J.: Guide to the Unified Process Featuring UML, Java and Design Patterns. Springer, Berlin (2003)
3. OMG: Business Process Model and Notation (BPMN): Version 2.0 specification. Technical report formal/2011-01-03, Object Management Group (January 2011)
4. Ross, R.G.: Principles of the Business Rule Approach, 1st edn. Addison-Wesley Professional, Reading (2003)
5. Nascimento, G., Iochpe, C., Thom, L., Reichert, M.: A method for rewriting legacy systems using business process managemet technology. In: Proceedings of the 11th International Conference on Enterprise Information Systems (ICEIS), pp. 57–62 (2009)
6. Dumas, M., La Rosa, M., Mendling, J., Reijers, H.A.: Fundamentals of Business Process Management. Springer, Berlin (2013)
7. Goedertier, S., Vanthienen, J.: Declarative process modeling with business vocabulary and business rules. In: On the Move to Meaningful Internet Systems 2007: OTM 2007 Workshops, Springer (2007) 603–612
8. Rosemann, M., Schwegmann, A., Delfmann, P.: Preparation of process modeling. Process Management: A Guide for the Design of Business Processes, 2nd edn, pp. 41–90. Springer (2011)
9. Weber, B., Reichert, M., Mendling, J., Reijers, H.A.: Refactoring large process model repositories. Comput. Ind. **62**(5), 467–486 (2011)
10. Rosa, M.L., ter Hofstede, A.H.M., Wohed, P., Reijers, H.A., Mendling, J., van der Aalst, W.M.P.: Managing process model complexity via concrete syntax modifications. IEEE Trans. Ind. Inform. **7**(2), 255–265 (2011)
11. Reijers, H.A.: Design and Control of Workflow Processes: Business Process Management for the Service Industry. Springer, Berlin (2003)
12. Kluza, K., Kaczor, K., Nalepa, G.J.: Enriching business processes with rules using the Oryx BPMN editor. In Rutkowski, L., et al. (eds.): Artificial Intelligence and Soft Computing: 11th International Conference, ICAISC 2012: Zakopane, Poland, April 29–May 3, 2012. Lecture Notes in Artificial Intelligence, vol. 7268, pp. 573–581. Springer (2012)
13. Nalepa, G.J., Kluza, K., Kaczor, K.: Proposal of an inference engine architecture for business rules and processes. In Rutkowski, L., et al. (eds.): Artificial Intelligence and Soft Computing: 12th International Conference, ICAISC 2013: Zakopane, Poland, 9–13 June 2013. Lecture Notes in Artificial Intelligence, vol. 7268, pp. 453–464. Springer (2013)
14. Nalepa, G.J., Kluza, K., Ciaputa, U.: Proposal of automation of the collaborative modeling and evaluation of business processes using a semantic wiki. In: Proceedings of the 17th IEEE International Conference on Emerging Technologies and Factory Automation ETFA 2012, Kraków, Poland, 28 Sept 2012. (2012)
15. Kluza, K., Nalepa, G.J., Lisiecki, J.: Square complexity metrics for business process models. In Mach-Król, M., Pełech-Pilichowski, T. (eds.): Advances in Business ICT. Advances in Intelligent Systems and Computing, vol. 257, pp. 89–107. Springer (2014)
16. Tscheschner, W.: Oryx Dokumentation. Universitat Potsdam, Hasso Plattner Institut (2007)

17. Adrian, W.T., Bobek, S., Nalepa, G.J., Kaczor, K., Kluza, K.: How to reason by HeaRT in a semantic knowledge-based wiki. In: Proceedings of the 23rd IEEE International Conference on Tools with Artificial Intelligence, ICTAI 2011, Boca Raton, Florida, USA (November 2011) 438–441
18. Nalepa, G.J.: Collective knowledge engineering with semantic wikis. J. Univers. Comput. Sci. **16**(7), 1006–1023 (2010)
19. Wang, H., Khoshgoftaar, T.M., Hulse, J.V., Gao, K.: Metric selection for software defect prediction. Int. J. Softw. Eng. Knowl. Eng. **21**(2), 237–257 (2011)
20. Grady, R.: Successfully applying software metrics. Computer **27**(9), 18–25 (1994)
21. Monsalve, C., Abran, A., April, A.: Measuring software functional size from business process models. Int. J. Softw. Eng. Knowl. Eng. **21**(3), 311–338 (2011)
22. Sarang, P., Juric, M., Mathew, B.: Business Process Execution Language for Web Services BPEL and BPEL4WS. Packt Publishing (2006)
23. The jBPM team of JBoss Community: jBPM User Guide. 5.2.0.final edn. (Dec 2011) online: http://docs.jboss.org/jbpm/v5.2/userguide/
24. Rademakers, T., Baeyens, T., Barrez, J.: Activiti in Action: Executable Business Processes in BPMN 2.0. Manning Pubs Co Series. Manning Publications Company (2012)
25. Kaczor, K., Kluza, K., Nalepa, G.J.: Towards rule interoperability: design of Drools rule bases using the XTT2 method. Trans. Comput. Collect. Intell. XI **8065**, 155–175 (2013)
26. Hollingsworth, D.: The workflow reference model. Issue 1.1 TC00-1003, Workflow Management Coalition (Jan 1995)
27. Schmidt, D.C.: Model-driven engineering. IEEE. Computer **39**(2), 25–31 (2006)
28. Informatics: A Propaedeutic View. Elsevier Science Ltd, London (2000)
29. Kleppe, A., Warmer, J., Bast, W.: MDA Explained: The Model Driven Architecture: Practice and Promise. Addison Wesley, Reading (2003)
30. Henderson-Sellers, B., Atkinson, C., Kühne, T., Gonzalez-Perez, C.: Understanding metamodelling (October 2003)
31. OMG: Meta object facility (MOF) version 2.0, core specification. Technical report formal/2006-01-01, Object Management Group (January 2006). http://www.omg.org/cgi-bin/doc?formal/2006-01-01.pdf
32. International Organization for Standardization: Information technology – Meta Object Facility (MOF) (2005)
33. Ignizio, J.P.: An Introduction To Expert Systems. The Development and Implementation of Rule-Based Expert Systems. McGraw-Hill, Maidenheach (1991)
34. Kluza, K., Nalepa, G.J.: MOF-based metamodeling for the XTT knowledge representation. In: Tadeusiewicz, R., Ligęza, A., Mitkowski, W., Szymkat, M. (eds.) CMS'09: Computer Methods and Systems: 7th conference, 26–27 November 2009, pp. 93–98. Poland, Cracow, AGH University of Science and Technology, Cracow, Oprogramowanie Naukowo-Techniczne, Kraków (2009)
35. Frankel, D.S.: Model Driven Architecture: Applying MDA to Enterprise Computing. Wiley Publishing, Indianapolis (2003)
36. Miller, J., Mukerji, J.: MDA Guide Version 1.0.1. OMG. (2003)
37. Gasevic, D., Djuric, D., Devedzic, V.: Model Driven Architecture and Ontology Development. Springer, Berlin (2006)
38. Pilone, D., Pitman, N.: UML 2.0 in a Nutshell. O'Reilly (2005)
39. Kluza, K., Nalepa, G.J.: Proposal of square metrics for measuring business process model complexity. In Ganzha, M., Maciaszek, L.A., Paprzycki, M., (eds.): In: Proceedings of the Federated Conference on Computer Science and Information Systems – FedCSIS 2012, pp. 919–922. Wroclaw, Poland, 9-12 September 2012 (2012)
40. Cardoso, J.: Control-flow complexity measurement of processes and weyuker's properties. In: 6th International Enformatika Conference. Transactions on Enformatika, Systems Sciences and Engineering, vol. 8. Budapest, Hungary, 26– 28 October (2005)

41. Cardoso, J.: About the data-flow complexity of web processes. In: Proceedings from the 6th International Workshop on Business Process Modeling, Development, and Support: Business Processes and Support Systems: Design for Flexibility. In: The 17th Conference on Advanced Information Systems Engineering (CAiSE'05), pp. 67–74. Porto, Portugal, 13–17 June 2005 (2005)

42. Cardoso, J., Mendling, J., Neumann, G., Reijers, H.A.: A discourse on complexity of process models. In Eder, J., Dustdar, S., et al. (eds.) In: Proceedings of the 2006 international conference on Business Process Management Workshops, Vienna, Austria. BPM'06, pp. 117–128. Springer-Verlag, Berlin, Heidelberg (2006)

43. Latva-Koivisto, A.M.: Finding a complexity for business process models. Technical report, Helsinki University of Technology (Feb 2001)

44. Sánchez-González, L., García, F., Mendling, J., Ruiz, F., Piattini, M.: Prediction of business process model quality based on structural metrics. In: Proceedings of the 29th international conference on Conceptual modeling, Vancouver, Canada. ER'10, pp. 458–463. Springer-Verlag, Berlin, Heidelberg (2010)

45. Mendling, J.: Metrics for business process models. In: Metrics for Process Models. Lecture Notes in Business Information Processing, vol. 6, pp. 103–133. Springer, Berlin, Heidelberg (2009)

46. Kluza, K.: Methods for Modeling and Integration of Business Processes with Rules. Ph.D. thesis, AGH University of Science and Technology (March 2015) Supervisor: Grzegorz J. Nalepa

47. Li, Z.J., Sun, W.: BPEL-unit: JUnit for BPEL processes. Service-Oriented Computing – ICSOC 2006, pp. 415–426. Springer, Berlin (2006)

48. Liu, H., Li, Z., Zhu, J., Tan, H.: Business Process Regression Testing. Service-Oriented Computing ICSOC 2007. Springer, Berlin (2007)

49. Louridas, P.: Junit: unit testing and coiling in tandem. Software, IEEE 22(4), 12–15 (2005)

50. Tahchiev, P., Leme, F., Massol, V., Gregory, G.: JUnit in Action, 2nd edn. Manning Publications (2010)

Chapter 14
Rule-Based Systems and Semantic Web

This chapter discusses the practical application of the SKE approach in the context
of Semantic Web technologies. One of the objectives of the Semantic Web was to
deliver rule-based reasoning on top of the ontological layer. However, this is non-
trivial task as we discussed in Sect. 3.3. In this chapter we present an original solution
to a heterogeneous integration of forward chaining rules (that can be found in classic
rule-based shells) with Description logic. We investigate the possible application of
the XTT2 rules.

The motivation for our research is presented in Sect. 14.1. Starting from
preliminary research [1, 2] on the possible applications of established rule technologies in the Semantic Web, further results were achieved and the DAAL (Description And Attributive Logic) formalism was proposed. It provides the integration of
the ALSV(FD)-based rule solution with Description Logics. It was first introduced
in [3] and discussed in [4]. We present it in Sect. 14.2. Furthermore, the PELLET-
HEART framework enables practical runtime integration of the ontology reasoner
PELLET with the HEART engine. The framework was originally introduced in [5]
and is discussed in Sect. 14.3. In Sect. 14.4 the use of DAAL and PELLET-HEART
is demonstrated using the BOOKSTORE example. Finally, in Sect. 14.5 the chapter is
summarized.

14.1 Integrating SKE with the Semantic Web

Motivation

Description Logics [6] provide an effective formal foundation for the Semantic Web
ontologies described with OWL. They allow for simple inference tasks, e.g. corresponding to concept classification. Rules are the next layer in the Semantic Web

© Springer International Publishing AG 2018   339
G.J. Nalepa, *Modeling with Rules Using Semantic Knowledge Engineering*,
Intelligent Systems Reference Library 130,
https://doi.org/10.1007/978-3-319-66655-6_14

stack that should be provided on top of the ontological layer in order to make the Semantic Web applications use powerful reasoning. However, the solution is not straightforward [7].

From this perspective important challenges include:

1. *Knowledge representation formalism* for the design of rules, integrated with Description Logics describing ontologies,
2. *Rule inference framework* that combines classic RBS with facts stored in ontologies, and
3. *Knowledge engineering platforms* allowing users to author the semantic knowledge using the two of the above.

The proposed solution to the *first challenge* is to allow the use of the XTT2 rules to work in the context of the Semantic Web. This would open up the possibility to use the SKE design tools to design Semantic Web rules, as well as use the existing verification solutions. To achieve this goal, the logical foundations of XTT2 and ontologies were compared in [3, 4].

Based on the analysis of knowledge representation with DL and ALSV(FD), a language called DAAL was formulated in [4]. It provides the foundation for integration of SKE tools with DL-based applications. Another research objective is to run HEART engine in ontology-based systems. This is a generic approach where rules can be designed visually using the XTT2 representation. Simultaneously the rule-based reasoning could exploit facts from OWL ontologies where inference tasks use DL.

As a solution to the *second challenge* a practical framework called PELLET-HEART combining HEART rule engine with a DL reasoner is proposed. The primary goal of this prototype is to run HEART inference over ontologies. The system has a hybrid architecture: HEART is a control component, responsible for rule handling, selection and execution. In the rule format used by HEART rule preconditions may include complex formulas based on ALSV(FD). These formulas describe relations among system attributes, which are mapped onto DL descriptions. For handling the relations between concepts and individuals, the Pellet[1] DL reasoner is used.

To overcome the *third challenge* a new semantic wiki system platform called LOKI is proposed. The principal idea consists in representing the semantic annotations in a formalized way, and simultaneously enriching them with an expressive rule-based knowledge representation. Semantic annotations should be based on standard solutions, such as RDF, and RDFS and possibly OWL for ontologies of concepts. The knowledge base is coupled with the contents of the wiki. Moreover, the basic wiki engine is integrated with an inference engine. Practical implementation of these concepts provide compatibility with important existing semantic wiki implementations. Work on LOKI was further extended into a collective knowledge engineering platform described in more detail in Chap. 15.

---

[1]See http://clarkparsia.com/pellet.

Overview of Description Logics

Description Logics are a family of knowledge representation languages discussed in detail by Baader et al. in [6]. Historically related to semantic networks, conceptual graphs, and frame languages [8], they describe the world of interest by means of *concepts*, *individuals* and *roles*. However, contrary to their predecessors, they provide formal semantics and thus allow for automated reasoning. Basic DL take advantage of their relation to predicate calculus. On one hand, they adopt its semantics, which makes them more expressive than propositional logic. On the other hand, by restricting the syntax to formulae with at most two variables, they remain decidable and more human-readable. These features made Description Logics a popular formalism used for designing ontologies for the Semantic Web.

There exist a number of DL languages that are defined and distinguished by concept descriptions they allow, which influences language expressiveness. Obviously, the more expressive the language is, the more complex the reasoning. Expressive languages, such as $\mathcal{SHOIN}(\mathbf{D})$ on which OWL DL is based, or $\mathcal{SROIQ}(\mathbf{D})$ for OWL2DL, remain challenging in terms of computational complexity.

The vocabulary in DL consists of *concepts*, which denote sets of individuals, and *roles*, which denote the binary relations between individuals. Elementary descriptions in DL are *atomic concepts* (A) and *atomic roles* (R). More complex descriptions can be built inductively from those using *concept constructors*. Respective DL languages are distinguished by the constructors they provide. A minimal language of interest is AL (*Attributive Language*) [6].

In order to define a formal semantics, an *interpretation* $\mathcal{I} = (\Delta^{\mathcal{I}}, \cdot^{\mathcal{I}})$ is considered. It consists of the *domain of interpretation* which is a non-empty set and an *interpretation function* that assigns a set $A^{\mathcal{I}} \subseteq \Delta^{\mathcal{I}}$ to every atomic concept A, and for every atomic role R a binary relation $R^{\mathcal{I}} \subseteq \Delta^{\mathcal{I}} \times \Delta^{\mathcal{I}}$. The interpretation function is extended over concept descriptions [6].

Description Logics provide tools to build a knowledge base and to reason about it. The knowledge base consists of two parts: *TBox* ($\mathcal{T}$) and *ABox* ($\mathcal{A}$).

TBox provides terminology and contains taxonomy expressed in the form of set of *axioms*. They define concepts, specify relations between them and introduce set constraints. Therefore, TBox stores knowledge about sets of individuals in the world of interest. Formally, a terminology $\mathcal{T}$ is a finite set of terminological axioms. If C and D denote concept names, and R and S role names, then the terminological axioms may be in forms: $C \sqsubseteq D$ ($R \sqsubseteq S$) or $C \equiv D$ ($R \equiv S$). Equalities having an atomic concept on the left-hand side are called *definitions*. Equalities express necessary and sufficient conditions. *Specialization* statements i.e. axioms of the form $C \sqsubseteq D$, specify constraints (necessary conditions) only. The interpretation *satisfies* an axiom $C \sqsubseteq D$ if: $C^{\mathcal{I}} \subseteq D^{\mathcal{I}}$, and satisfies a concept definition $C \equiv D$ if: $C^{\mathcal{I}} = D^{\mathcal{I}}$. If the interpretation satisfies all definitions and all axioms in $\mathcal{T}$, it satisfies the terminology $\mathcal{T}$ and is called a *model* of $\mathcal{T}$.

ABox contains explicit assertions about individuals in the conceived world. They represent extensional knowledge about the domain of interest. Statements in the ABox may be: *concept assertions*, e.g. $C(a)$ or *role assertions*, $R(b, c)$. An interpretation $\mathcal{I}$ maps each individual name to an element in the domain. With regard to terminology $\mathcal{T}$, the interpretation satisfies a concept assertion $C(a)$ if $a^{\mathcal{I}} \in C^{\mathcal{I}}$, and a role assertion $R(b, c)$ iff $(b^{\mathcal{I}}, c^{\mathcal{I}}) \in R^{\mathcal{I}}$. If it satisfies all assertions in ABox $\mathcal{A}$, then it satisfies $\mathcal{A}$ and $\mathcal{I}$ is a model of $\mathcal{A}$.

Although the terminology and the world description share the same model-theoretic semantics, it is convenient to distinguish these two parts while designing a knowledge base or stating particular inference tasks.

Inference in DL can be separated for reasoning tasks for TBox and ABox. With regards to terminology $\mathcal{T}$, one can pose a question if a concept is *satisfiable*, if one concept *subsumes* another, if two concepts are *equivalent* or *disjoint*. A concept C is satisfiable with respect to $\mathcal{T}$ if there exists a model $\mathcal{I}$ of $\mathcal{T}$ such that $C^{\mathcal{I}}$ is not empty. A concept C is subsumed by a concept D w.r.t. $\mathcal{T}$ if $C^{\mathcal{I}} \subseteq D^{\mathcal{I}}$ for every model $\mathcal{I}$ of $\mathcal{T}$. Concepts C and D are equivalent w.r.t. $\mathcal{T}$ if $C^{\mathcal{I}} = D^{\mathcal{I}}$ for every model $\mathcal{I}$ of $\mathcal{T}$. Finally, two concepts C and D are disjoint w.r.t. $\mathcal{T}$ if $C^{\mathcal{I}} \cap D^{\mathcal{I}} = \varnothing$ for every model $\mathcal{I}$ of $\mathcal{T}$. *Satisfiability* and *subsumption checking* are the main reasoning tasks for TBox; all others can be reduced to them, and either one can be reduced to the other.

For ABox, there are four main inference tasks: *consistency checking, instance checking, realization* and *retrieval*. An ABox $\mathcal{A}$ is consistent w.r.t. a TBox $\mathcal{T}$, if there is an interpretation that is a model of both $\mathcal{A}$ and $\mathcal{T}$. We say that an ABox is consistent, if it is consistent w.r.t. the empty TBox. Realization tasks consist in finding the most specific concept for a given individual. Retrieval returns individuals which are instances of a given concept. All these tasks can be reduced to *consistency checking* of ABox w.r.t. TBox.

The next section contains the discussion of how ALSV(FD) relates to DL.

14.2   DAAL Rule Language

Description Logics allow for a complex descriptions of objects in the universe of discourse and the relations between them. The static part of a system is expressed in TBox part of a DL Knowledge Base. The information about individuals is represented by means of facts asserted in ABox, which in DL is limited in terms of its syntax and semantics. Only a simple concept and role assertions are allowed, which together with the knowledge expressed in TBox lay the ground for inferencing.

The main goal of ALSV(FD) is to provide an expressive notation for dynamic system state transitions in Rule-Based Systems. The knowledge specification with ALSV(FD) is composed of: state specification with facts and transition specification with formulas building decision rules.

The language of DL consists of *concepts*, *roles* and *constants* (*individuals*). The meaning of the symbols is defined by an *interpretation function*, which assigns a set of objects to each concept, a binary relation to each role, and an object in the universe of discourse to each individual.

In the ALSV(FD) language the following symbols are used: $\mathbb{A}$ – a finite set of attribute names, $\mathbb{A} = \{A_1, A_2, \ldots, A_n\}$, and $\mathbb{D}$ – a set of possible attribute values (their *domains*), $\mathbb{D} = \mathbb{D}_1 \cup \mathbb{D}_2 \cup \cdots \cup \mathbb{D}_n$. The semantics of ALSV(FD) is based on the interpretation of the symbols of the alphabet (see Sect. 4.1).

Both logics describe the universe of discourse by identifying certain entities. In ALSV(FD) they are called attributes, in DL – concepts. They have their *domains* to which the interpretation maps the attribute and concept name symbols. Every attribute in ALSV(FD) has a domain, which constraints its values. In DL this kind of specification is done by means of TBox axioms. In order to be able to express a finite domain in DL, a *set of* constructors (denoted by $\mathcal{O}$) is needed.[2] Once the attributes (in ALSV(FD)) and concepts (in DL) are defined, they can be used in the specification of rules. Legal ALSV(FD) formulae specify the constraints that an attribute value has to match in order for a rule to be fired. Attribute values define the state of the system under consideration. A statement in ALSV(FD) that an attribute A_i holds a certain value d_i may be interpreted as a DL statement that a certain object d_i is an instance of the concept A_i. This is valid for both simple and generalized attributes, as explained in Sect. 14.2.

After this analysis, a proposal of a new language integrating DL with ALSV(FD) is introduced next, after the paper [4].

Language Overview

To address the challenge of providing a rule language for DL, a hybrid framework for integrating Attributive Logic and Description Logic is proposed. A language called DAAL (Description And Attributive Logic) is introduced. It is syntactically based on Description Logics, but tries to capture ALSV(FD) semantics and thus enables expressing ALSV(FD) models in DL. The ideas of the DAAL framework can be summarized as follows:

- In the universe of discourse, entities which correspond to attributes in ALSV(FD) and concepts in DL are identified.
- The domains of the entities are defined in DAAL in the form of TBox definitions.
- The formulae in rules are written in DAAL in the form of DL TBox axioms.
- The state is modeled in DAAL as a DL ABox.
- The actions taken as consequences of rule execution generates new states of the system, encoded in DAAL as new ABoxes.

DAAL uses the $\mathcal{AL}$ DL language. In the subsequent subsections the introduced definitions use DL concepts and syntax to form the DAAL formulae.

A novel idea in the DAAL framework is the existence of a *static* TBox with domain definitions, *Temporary Rule TBoxes* and *Temporary ABoxes*. *Temporary Rule TBoxes* express the preconditions of the system rules. During the execution of the reasoning process they are loaded into and unloaded from a reasoner. Therefore, they are not a static part of the system ontology. *Temporary ABoxes* correspond to system states. As the system changes its state, new ABoxes replace the previous ones.

---

[2]See the *DL Complexity Navigator* at http://www.cs.man.ac.uk/~ezolin/dl.

Syntax and Semantics

The language syntax using the DL symbols is defined below.

Vocabulary of DAAL language consists of:

$$\mathbf{A}, \mathbf{B}, \mathbf{C}, \quad \text{concept names} \tag{14.1}$$

$$\bot, \quad \text{bottom concept} \tag{14.2}$$

$$\mathbf{a}, \mathbf{b}, \mathbf{c}, \quad \text{individuals} \tag{14.3}$$

$$\equiv, \neg, \sqcap, \sqsubseteq, \wedge, \rightarrow, \quad \text{operators} \tag{14.4}$$

$$(,), \{, \}, ., \quad \text{delimiters} \tag{14.5}$$

Let us consider concept descriptions $\mathbf{C}, \mathbf{D}$, and instances $\mathbf{c}_1, \mathbf{c}_2, \ldots, \mathbf{c}_n$. Admissible concept descriptions in DAAL:

$$\mathbf{C}, \tag{14.6}$$

$$\neg\mathbf{C} \tag{14.7}$$

$$\{\mathbf{c}_1, \mathbf{c}_2, \ldots, \mathbf{c}_n\} \tag{14.8}$$

$$\mathbf{C} \sqcap \mathbf{D} \tag{14.9}$$

Formulae in DAAL are of two sorts: *terminological axioms* and *concept assertions*. Admissible formulas in DAAL:

$$\mathbf{C} \equiv \mathbf{D}, \mathbf{C} \sqsubseteq \mathbf{D} \quad \text{terminological axiom} \tag{14.10}$$

$$\mathbf{C}(\mathbf{a}), \quad \text{concept assertion} \tag{14.11}$$

A rule in DAAL is of the following forms:

$$(\phi_1 \wedge \cdots \wedge \phi_n) \rightarrow (\theta_1 \wedge \cdots \wedge \theta_n) \tag{14.12}$$

where $\phi_i, i = 1, \ldots, k$ are terminological axioms, and $\theta_i, i = 1, \ldots, n$ are concept assertions.

The semantics of DAAL formulae is defined by the interpretation $\mathcal{I} = (\Delta^{\mathcal{I}}, \cdot^{\mathcal{I}})$, where $\Delta^{\mathcal{I}} = \mathbb{D}$ (see Sect. 4.1). For a given system, the interpretation function maps the DAAL symbols into the domain of interpretation defined by the sets used in ALSV(FD) description of the system. For every ALSV(FD) set $\mathbb{S}$, DAAL concept $\mathbf{S}^{\mathcal{I}} = \mathbb{S}$, for instance:

- for an attribute domain: $\mathbf{D}_i^{\mathcal{I}} = \mathbb{D}_i$
- for a set of values: $\mathbf{V}_i^{\mathcal{I}} = \mathbb{V}_i$

Individuals are mapped onto the elements of the respective sets, e.g. $(\mathbf{d}_1, \mathbf{d}_2, \ldots, \mathbf{d}_k)^{\mathcal{I}} = (\mathbf{d}_1^{\mathcal{I}}, \mathbf{d}_2^{\mathcal{I}}, \ldots, \mathbf{d}_k^{\mathcal{I}}) = (d_1, d_2, \ldots, d_k)$.

The logical operator $\wedge$ denotes a *conjunction* with a classical meaning. The symbol $\rightarrow$ separates the rule preconditions and conclusion. Its interpretation follows the one from Logic Programming.

Table 14.1 Formulae in ALSV(FD) and terminological axioms in DAAL, domain definitions

Attributive logic		DAAL	
Attribute name	Attribute domain	Concept name	Concept constructors
A_i	$\mathbb{D}_i$ $\mathbb{D}_i = \{d_{i_1}, d_{i_2}, \ldots, d_{i_n}\}$	$\mathbf{A_i}$	$\mathbf{A_i} \equiv \mathbf{D_i}$ $\mathbf{D_i} \equiv \{\mathbf{d_{i_1}}, \mathbf{d_{i_2}}, \ldots, \mathbf{d_{i_n}}\}$

Table 14.2 State specification in ALSV(FD) and respective assertions in DAAL

Attributive logic		DAAL
Attribute type	Formula	Assertion
Simple attribute	$A_i := d_i$	$\mathbf{A_i(d_i)}$
Generalized attribute	$A_i := V_i$ $V_i = \{v_{i_1}, v_{i_2}, \ldots, v_{i_n}\}$	$\mathbf{A(v_{i_1}).A_i(v_{i_2}).\ldots, A_i(v_{i_n})}.$

Conceptual Modeling in DAAL

In DAAL, the ALSV(FD) attributes are modeled as DL concepts (corresponding to OWL classes). They are subclasses of the general OWL Attribute class for attributes. Let the domain $\mathbb{D}_i$ of an attribute A_i be finite. Then, the transition shown in Table 14.1 holds.

The actual state of a system is modeled as a set of DL assertions. To express a value of a simple attribute a single assertion is used. For generalized attributes there are as many assertions as values the attribute takes. A formula $A_i := d$ denotes that the attribute A_i *is assigned a value d* at the certain moment. In DAAL this can be represented as an assertion in ABox, namely: $\mathbf{A_i(d_i)}$. In the case of generalized attributes, there is no direct counterpart in the DL for an ALSV(FD) formula: $A_i = V_i$, where V_i is a set. However, the same meaning can be acquired in another way. Based on the assumption that V_i is a finite set of the form: $V_i = \{v_{i_1}, v_{i_2}, \ldots, v_{i_n}\}$ one adds *all* of the following assertions to the DAAL ABox: $\mathbf{A_i(v_{i_1}).A_i(v_{i_2}).\ldots A_i(v_{i_n})}$. State specification in ALSV(FD) and DL is shown in Table 14.2.

Specification of Rules

The formulae used in rule preconditions specify the constraints of the attribute values. They constitute a schema to which a state of a system in a certain moment of time is matched. The DAAL approach to the mapping from ALSV(FD) to DL consists in a translation of the ALSV(FD) formulas into TBox-like DL axioms.

For simple attributes, the following transition holds: the formula: $A_i = d$ is logically equivalent to $A_i \in \{d\}$, so we express it in DAAL as: $\mathbf{A_i} \equiv \{\mathbf{d}\}$ (instances of concept $\mathbf{A_i}$ belongs to the set $\{d\}$). Another formula $A_i \in V_i$ constraint the set of possible values to the set V_i. This corresponds to the DAAL axiom: $\mathbf{A_i} \equiv \mathbf{V_i}$.

For generalized attributes one cannot express all the ALSV(FD) formulae in the DL TBox. This is due to the fact that the constraints in the DL TBox apply to *individuals* and not sets of objects. For example, one can say that all the instances of

Table 14.3 Simple attributes formulae in ALSV(FD) and respective axioms in DAAL

ALSV(FD)	DAAL
Formula	Axiom
$A_i = d$	$\mathbf{A_i} \equiv \{\mathbf{d}\}$
$A_i \in V_i$	$\mathbf{A_i} \equiv \mathbf{V_i}$
$A_i \neq d$	$\mathbf{A_i} \equiv \neg\{\mathbf{d}\}$
$A_i \notin V_i$	$\mathbf{A_i} \equiv \neg\mathbf{V_i}$
$A_i \in \varnothing$	$\mathbf{A_i} \equiv \bot$

Table 14.4 Generalized attributes formulae in ALSV(FD) and respective axioms in DAAL

ALSV(FD)	DAAL
Formula	Axiom
$A_i \subseteq V_i$	$\mathbf{A_i} \sqsubseteq \mathbf{V_i}$
$A_i \supseteq V_i$	$\mathbf{V_i} \sqsubseteq \mathbf{A_i}$
$A_i \sim V_i$	$\neg(\mathbf{A_i} \sqcap \mathbf{V_i} \equiv \bot)$
$A_i \nsim V_i$	$\mathbf{A_i} \sqcap \mathbf{V_i} \equiv \bot$

$\mathbf{A_i}$ are also instances of $\mathbf{V_i}$ and the DL axiom $\mathbf{A_i} \sqsubseteq \mathbf{V_i}$ corresponds to the ALSV(FD) formula $A_i \subseteq V_i$. However, it is impossible to specify that all elements of V_i appear in the ABox. Hence, the ALSV(FD) formula $A_i = V_i$ cannot be expressed in DL. An axiom $\mathbf{A_i} \equiv \mathbf{V_i}$ only restricts the values of $\mathbf{A_i}$ so that $\forall x \mathbf{A_i}(x) \leftrightarrow \mathbf{V_i}(x)$, but it cannot force the concept $\mathbf{A_i}$ to *take all the values* from $\mathbf{V_i}$. This is a consequence of the Open World Assumption. If some values are missing in the explicit assertions, it does not mean that the assertions do not hold. To sum up, the ALSV(FD) formulae for simple attributes can be represented as terminological axioms in DL as shown in Tables 14.3 and 14.4.

Inference Scenario

The axioms for rule representation introduced in the preceding section are called *temporary TBoxes*. They constitute a schema to which particular system states are matched. At a given moment of time, the state of the system is represented as a conjunction of ABox formulae. In order to check if rule preconditions are satisfied appropriate inference tasks have to be performed. The inference rules in ALSV(FD) were presented in Sect. 4.2. For DAAL the corresponding task is *consistency checking*. For each rule a *consistency check* of the state assertions with regards to the rule preconditions is performed. If the consistency holds, the rule can be fired. Rule firing generates a new ABox which represents the new state of the system.

The architecture of this hybrid reasoning framework consists of HEART, a control component, and a DL reasoner. It is introduced in the next section.

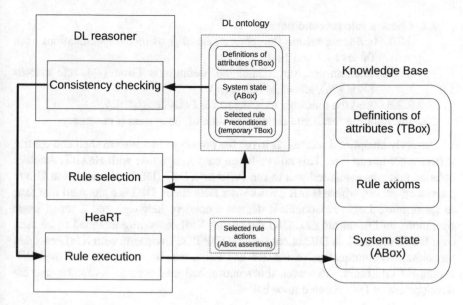

Fig. 14.1 Hybrid system combining HEART and Pellet, after [5]

14.3 Hybrid Reasoning with PELLET-HEART

A practical architecture of a hybrid reasoning framework is realized by combining the dedicated XTT2 inference engine – HEART, a control component, and a DL reasoner – Pellet. The architecture can be observed in Fig. 14.1. The inference process is controlled by the HEART tool. It takes care of rule handling, including context-based rule selection and execution. Pellet's task is to check the rule preconditions. More specifically, it checks the consistency of the ontology built from the state of the system and the particular preconditions of the rules. In a loop appropriate rules are loaded into the DL reasoner together with the actual state of the system. Each rule is temporarily joined with an existing TBox in which the definitions of the concepts are stored. The state is a *temporary ABox*. The DL reasoner checks the consistency of the ontology resulting from the TBox and ABox representing the system at a given time. If the ontology is consistent, then the rule can be fired. The rule axioms are then unloaded and the loop continues. Rules are able to change the knowledge base of the system. Adding and removing facts is allowed. These operations generate a new ABox which represents the new state of the system. The reasoning with PELLET-HEART includes the following steps:

1. Build a TBox T_1 with definitions of types and attributes; build additional statements: `owl:allDifferent` in OWL for individuals.
2. Run a user-defined inference mode:

    2.1. In each state build an ABox A representing this state.

    2.2. Check a rule preconditions:

        2.2.1. build rule axioms (temporary TBox T_2), using the specifications from Tables 14.3 and 14.4,

        2.2.2. build an ontology containing: definitions TBox (T_1), rule axioms TBox (T_2), and state ABox (A),

        2.2.3. send the ontology to Pellet to verify its consistency.

    2.3. Interpret the result and change the state of the system if necessary.

Currently an optimal solution is to run the prototype in a system shell and control it from a command line. This allows for an easy integration with HEART. Another solution that was evaluated was to run Pellet as a DIG (DL Implementation Group interface)[3] server, where HEART works as a DIG client. DIG is a standard interface to Description Logic reasoners. It defines a concept language and a set of basic operations on DL ontologies. DIG defines an XML encoding intended to be used over HTTP protocol. A DIG client sends HTTP POST requests with XML-encoded messages. Communication of HEART and Pellet was implemented with the SWI-Prolog HTTP library.[4] However, this solution had mediocre performance, and the development of DIG seemed to be halted.

14.4 DAAL in Practice

The most expressive reasoning is possible once the HEART engine communicates with the ontology reasoner Pellet. In this case combining HEART and Pellet reasoning capabilities is possible. In order to do this, translation to DAAL language is needed. Let us re-consider the XTT2 rule base for the BOOKSTORE recommendation system, first introduced in Sect. 4.1.

Translation to DAAL

In this section a formalized description of the case in both ALSV(FD) and DAAL is given. Based on this, the PELLET-HEART implementation of the case is presented next.

    Description in ALSV(FD) Within the system the following attributes are identified:

$$\mathbb{A} = \{fav\_genres, age, age\_filter, recently\_read, rec\_read\_filter,$$
$$known\_language, rec\_book\}.$$

with corresponding domains:

$$\mathbb{D} = \mathbb{D}_{fav\_genres} \cup \mathbb{D}_{age} \cup \mathbb{D}_{age\_filter} \cup \mathbb{D}_{recently\_read} \cup \mathbb{D}_{rec\_read\_filter} \cup$$
$$\mathbb{D}_{known\_language} \cup \qquad \mathbb{D}_{rec\_book}.$$

---

[3]See http://dl.kr.org/dig.

[4]See http://www.swi-prolog.org/pldoc/package/http.html.

defined as:

$\mathbb{D}_{fav\_genres} = \{$horror, handbook, fantasy, science, historical, poetry$\}$,
$\mathbb{D}_{age} = \{1 - 99\}$,
$\mathbb{D}_{age\_filter} = \{$young_horrors, young_poetry, adult_horrors, adult_poetry,
     all_science, adult_handbooks, handbooks, horrors$\}$,
$\mathbb{D}_{recently\_read} = \mathbb{D}_{rec\_read\_filter} = \mathbb{D}_{rec\_book} = \{$'At the mountains of madness', 'It',
     'Insomnia', 'Bag of bones', 'Betty Crocker Christmas Cookbook',
     'Desperation', 'Logical Foundations for RBS', 'The Call of Cthulhu',
     'The Christmas Table', 'The Raven and Other Poems'$\}$,
$\mathbb{D}_{known\_language} = \{$english, polish, german, french, spanish, greek, italian, hebrew$\}$.

Selected rules are presented below:

$R1 : age < 18 \wedge fav_genres = $ horror $\longrightarrow age\_filter :=$ young_horrors
$R2 : age = \_ \wedge fav\_genres \in \{$science$\} \longrightarrow age\_filter :=$ all_science
$R3 : recently\_read \in \{$'Desperation'$\} \longrightarrow rec\_read\_filter := \{$'Insomnia', 'It'$\}$
$R4 : age\_filter \in \{$all_science$\} \wedge rec\_read\_filter = \_$
    $\wedge known\_language \in \{$polish, english$\} \longrightarrow rec\_book :=$
    'Logical Foundations for RBS'
$R5 : age\_filter \in \{$young_horrors, adult_horrors$\} \wedge rec\_read\_filter = any$
    $\wedge known\_language = any \longrightarrow rec\_book :=$ 'The Call of Cthulhu'

Description in DAAL Within the same system, when described in DAAL language, the following concepts are distinguished:

FavGenres, **Age**, **AgeFilter**, **RecentlyRead** **RecReadFilter**, **KnownLanguage**, **RecBook**.

The definitions of the concepts are as follow:
FavGenres $\equiv \{$horror, handbook, fantasy, science, historical, poetry$\}$,
Age $\equiv \{1, 2, \ldots, 99\}$,
AgeFilter $\equiv \{$young_horrors, young_poetry, adult_horrors, adult_poetry,
        all_science,adult_handbooks, handbooks, horrors$\}$,
RecentlyRead $\equiv$ **RecentlyReadFilter** $\equiv$ **RecommendedBook** $\equiv$
    $\{$'At the mountains of madness', 'Bag of bones', 'Betty Crocker Christmas
    Cookbook', 'Desperation', 'Insomnia', 'It', 'Logical Foundations for RBS',
    'The Call of Cthulhu', 'The Christmas Table', 'The Raven and Other Poems'$\}$,
KnownLanguage $\equiv \{$english, polish, german, french, spanish, greek, italian,
    hebrew$\}$.
Additional concept **Young** is defined as follows:
Young $\equiv \{1, 2, 3, 4, 5, 6, 7, 8, 9, 10, 11, 12, 13, 14, 15, 16, 17\}$
Using the transition specified in Table 14.3 the rules are rewritten as follow:
$R1 : ($**Age** $\equiv$ **Young**$) \wedge ($**FavGenres** $\equiv \{$horror$\}) \rightarrow$ **AgeFilter**(young_horrors).
$R2 : ($**FavGenres** $\equiv \{$science$\}) \rightarrow$ **AgeFilter**(all_science).

$R3$: (**RecentlyRead** $\equiv$ {'Desperation'}) $\rightarrow$ **RecReadFilter**('Insomnia').
RecReadFilter('It').
$R4$: (**AgeFilter** $\equiv$ {all_science}) $\wedge$ (**KnownLanguages** $\equiv$ {polish, english})
$\longrightarrow$ **RecBook**('Logical Foundations for RBS').
$R5$: (**AgeFilter** $\equiv$ {young_horrors, adult_horrors}) $\longrightarrow$ **RecBook**
('The call of Cthulhu').

Inference process In a given moment t_n the state is represented as a *temporary ABox $_n$*.[5] The inference process is as follows:

1. The state $\text{ABox}_g jn$ and the preconditions of rule R1 are loaded into the DL reasoner.
2. The DL reasoner performs the consistency check of the state with respect to rule preconditions.
3. Because the ontology built from the state assertions and R1 precondition formulas is consistent ($Month(jan)$ is consistent w.r.t. $Month \equiv SummerMonths$ ($SummerMonths \equiv \{dec, jan, feb\}$)) the rule is fired.
4. The conclusion of the rule generates a new assertion in ABox. The new ABox_1 replaces the old one (ABox_0). ABox_1 is as follows:
   $Month(jan).Day(mon).Hour(11).Season(summer)$.
5. The state ABox_1 and the preconditions of rule R2 are loaded into the DL reasoner.
6. The DL reasoner performs the consistency check of the state with respect to rule preconditions.
7. Because this time the ontology built from the state assertions and R2 precondition formulas again is consistent $Day(mon)$ is consistent w.r.t. $Day \equiv WorkingDays$ ($WorkingDays \equiv \{mon, tue, wed, thr, fri\}$) the rule is fired.
8. The conclusion of the rule generates a new assertion in ABox. The new ABox_2 replaces the old one (ABox_1). ABox_2 is as follows:
   $Month(jan).Day(mon).Hour(11).Season(summer).Today(workday)$.
9. The reasoning continues with new ABox state and the next rules.

Reasoning with DAAL *and* PELLET-HEART

Hybrid reasoning in PELLET-HEART offers the possibility of combining flexible rule formulations with ontology reasoning tasks. Therefore, it extends both HEART reasoning capabilities and ontology reasoning. The reasoning process is explained in Sect. 14.3. Now, consider the TBox definitions in HMR:

```
xtype [name: genres, base: symbolic,
       domain: [horror,science,handbook,fantasy,poetry,historical]].
xattr [name: fav_genres,
       abbrev: fav1, class: general, type: genres, comm: in].
```

---

[5]While we use terms "moments" (time points), as well as "temporary" we do not refer to or address any *temporal aspects*. In fact, these terms are used only technically, to denote a *sequence* of ABoxes.

They are translated into OWL (XML serialization) as follows

```xml
<!-- Definition of the attribute: fav_genres -->
<owl:Class rdf:ID="fav_genres">
  <rdfs:subClassOf>
    <owl:Class>
      <owl:oneOf rdf:parseType="Collection">
        <rdf:Description rdf:about="#horror" />
        <rdf:Description rdf:about="#science" />
        <rdf:Description rdf:about="#handbook" />
        <rdf:Description rdf:about="#fantasy" />
        <rdf:Description rdf:about="#poetry" />
        <rdf:Description rdf:about="#historical" />
      </owl:oneOf>
    </owl:Class>
  </rdfs:subClassOf>
</owl:Class>
```

Regular HEART invocation is executed, but with an additional parameter:

```
?- gox(gjn, [recommend_books], gdi, pellet).
State "gjn":
[age, 35],[recently_read, the_call_of_cthulhu], [knows_language,
    [english, polish, hebrew]], [favourite_genre, science],
    [favourite_author, [witold_gombrowicz, h_p_lovecraft]],
[age_filter, recently_read, recommend_books]
...
```

In a given moment rules chosen by HEART need to be send to Pellet in order to determine their satisfiability. For instance, the rule preconditions:

```
xrule age_filter/7:
      [age in [any], fav_genres in [science]]
    ==>
      [age_filter set [all_science]].
```

are translated into OWL as follows:

```xml
<!-- Preconditions of the rule:age_filter/7 -->
<!-- The precondition in HMR:age in [any] -- skipped -->
<!-- The precondition in HMR: fav_genres in [science] -->
<owl:Class rdf:ID="{fav_genres}">
  <owl:equivalentClass>
    <owl:Class>
      <owl:oneOf rdf:parseType="Collection">
        <rdf:Description rdf:about="#science" />
```

```
        </owl:oneOf>
      </owl:Class>
    </owl:equivalentClass>
  </owl:Class>
```

In order to ensure the Unique Name Assumption the OWL `owl:AllDifferent` construct must be used for all the attributes which are currently checked:

```
<owl:AllDifferent>
  <owl:distinctMembers rdf:parseType="Collection">
    <fav_genres rdf:about="#horror" />
    <fav_genres rdf:about="#science" />
    <fav_genres rdf:about="#handbook" />
    <fav_genres rdf:about="#fantasy" />
    <fav_genres rdf:about="#poetry" />
    <fav_genres rdf:about="#historical" />
  </owl:distinctMembers>
</owl:AllDifferent>
```

The state in HMR:

```
xstat gjn: [age, 35].
xstat gjn: [recently_read,the_call_of_cthulhu].
xstat gjn: [favourite_genre, science].
```

is translated into OWL as follows:

```
<age rdf:about="#35" />
<recently_read rdf:about="#The call of Cthulhu" />
<fav_genres rdf:about="#science" />
```

Once the knowledge base is built, it is sent to Pellet along with a consistency check request. In this example (state `gjn`, rule `age_filter/7`) the answer is:

```
?- check_consistency_detailed('kb.owl',pellet).
Consistent: Yes
```

In the opposite case, with rule conditions unsatisfied, the response would be:

```
?- check_consistency_detailed('kb.owl',pellet).
Consistent: No
Reason: No specific explanation was generated.
Generic explanation: An individual is sameAs and
  differentFrom another individual at the same time
```

Combining HEART with Pellet opens up the possibility of independent modeling of the domain ontology and the rule base. Here, for clarity, only a small proof of the concept example was shown.

14.5 Summary

The DAAL language presented in this chapter considers a practical rule representation using DL. This representation employs both TBox an ABox axioms. TBox axioms are used in rule preconditions whereas ABox statements appear in rule conclusions. This approach is different from SWRL [9] and its subsets. In SWRL rule preconditions may include concepts and role assertions (ABox statements). Such a representation is also possible for translation from ALSV(FD) to DL but is beyond the scope of this work. DL rules [10] represent rules with a certain tree structure. The approach presented here does not impose any restrictions on the rule structure. Describing states of a dynamic system using DL constructs implies the problem of updating the state description. The idea of updating ABox over time was investigated in [11] and appropriate DL languages have been defined. In DAAL the updated ABoxes are treated by HEART as a separate component, so there is no direct requirement for ABox updates support. For more recent work on integration of rules with OWL2 see [12–14].

PELLET-HEART was a prototype solution to demonstrate and investigate the practical use of DAAL. Its main drawback was the communication that has an impact on performance. The functionality did not include any user interface that helps in creating knowledge, and querying the system. In fact it worked well as proof of concept. However it was not further developed. On the other hand, the focus of the applied work was put on LOKI as a practical example of a hybrid semantic wiki architecture. It will be presented in more detail in the next chapter on the Collective Knowledge Engineering.

References

1. Furmańska, W.T., Nalepa, G.J.: Review of selected semantic web technologies. Technical report CSLTR 6/2009, AGH University of Science and Technology (2009)
2. Furmańska, W.T., Nalepa, G.J.: Nowe metody reprezentacji reguł dla sieci semantycznej. In: Grzech, A., et al. (eds.) Inżynieria Wiedzy i Systemy Ekspertowe. Problemy Współczesnej Nauki, Teoria i Zastosowania. Informatyka, Warszawa, Akademicka Oficyna Wydawnicza EXIT, pp. 265–275 (2009)
3. Nalepa, G.J., Furmańska, W.T.: New challenges in computational collective intelligence. In: 1st International Conference on Computational Collective Intelligence - Semantic Web, Social Networks and Multiagent Systems. Studies in Computational Intelligence, pp. 15–26. Springer (2009)
4. Nalepa, G.J., Furmańska, W.T.: Integration proposal for description logic and attributive logic - towards semantic web rules. In: Nguyen, N.T., Kowalczyk, R. (eds.) Transactions on Com-

putational Collective Intelligence II. Lecture Notes in Computer Science, vol. 6450, pp. 1–23. Springer, Berlin (2010)

5. Nalepa, G.J., Furmańska, W.T.: Pellet-HeaRT – proposal of an architecture for ontology systems with rules. In: Dillmann, R., et al. (eds.) KI 2010: Advances in Artificial Intelligence: 33rd Annual German Conference on AI: Karlsruhe, Germany, 21–24 September 2010. Lecture Notes in Artificial Intelligence, vol. 6359, pp. 143–150. Springer, Berlin (2010)

6. Baader, F., Calvanese, D., McGuinness, D.L., Nardi, D., Patel-Schneider, P.F. (eds) The Description Logic Handbook: Theory, Implementation, and Applications. Cambridge University Press, Cambridge (2003)

7. Adrian, W.T., Nalepa, G.J., Kaczor, K., Noga, M.: Overview of selected approaches to rule representation on the semantic web. Technical report CSLTR 2/2010, AGH University of Science and Technology (2010)

8. van Harmelen, F., Lifschitz, V., Porter, B. (eds.) Handbook of Knowledge Representation. Elsevier Science, Amsterdam (2007)

9. Horrocks, I., Patel-Schneider, P.F., Boley, H., Tabet, S., Grosof, B., Dean, M.: SWRL: a semantic web rule language combining OWL and RuleML, W3C member submission 21 May 2004. Technical report, W3C (2004)

10. Krötzsch, M., Rudolph, S., Hitzler, P.: ELP: tractable rules for OWL 2. In: 7th International Semantic Web Conference (ISWC2008) (2008)

11. Drescher, C., Liu, H., Baader, F., Guhlemann, S., Petersohn, U., Steinke, P., Thielscher, M.: Putting ABox updates into action. In: Ghilardi, S., Sebastiani, R. (eds.) The Seventh International Symposium on Frontiers of Combining Systems (FroCoS-2009). Lecture Notes in Computer Science, vol. 5749, pp. 149–164. Springer (2009)

12. Knorr, M., Hitzler, P., Maier, F.: Reconciling OWL and non-monotonic rules for the semantic web. In: Raedt, L.D., Bessière, C., Dubois, D., Doherty, P., Frasconi, P., Heintz, F., Lucas, P.J.F. (eds.) ECAI 2012 - 20th European Conference on Artificial Intelligence. Including Prestigious Applications of Artificial Intelligence (PAIS-2012) System Demonstrations Track, Montpellier, France, 27–31 August 2012. Frontiers in Artificial Intelligence and Applications, vol. 242, pp. 474–479. IOS Press (2012)

13. Martínez, D.C., Hitzler, P.: Extending description logic rules. In: Simperl, E., Cimiano, P., Polleres, A., Corcho, Ó., Presutti, V. (eds.) The Semantic Web: Research and Applications - 9th Extended Semantic Web Conference, ESWC 2012, Heraklion, Crete, Greece, 27–31 May 2012. Proceedings. Lecture Notes in Computer Science, vol. 7295, pp. 345–359. Springer (2012)

14. Mutharaju, R., Mateti, P., Hitzler, P.: Towards a rule based distributed OWL reasoning framework. In: Tamma, V.A.M., Dragoni, M., Gonçalves, R., Lawrynowicz, A. (eds.) Ontology Engineering - 12th International Experiences and Directions Workshop on OWL, OWLED 2015, Co-located with ISWC 2015, Bethlehem, PA, USA, 9-10 October 2015, Revised Selected Papers. Lecture Notes in Computer Science, vol. 9557, pp. 87–92. Springer (2015)

Chapter 15
Collaborative Knowledge Engineering with Wikis

Important phases of the knowledge engineering process were discussed in Chap. 2. The whole process becomes much more complex when we consider the participation of many individual knowledge engineers. This is especially important during the acquisition and modeling phases. These individuals might participate in the process in roughly the same time using distributed KE tools. They might also take part in a gradual development of knowledge bases. Moreover, for the research considered in this chapter we want to distinguish between several cases of participation. By *cooperative* we understand the case where many individuals work on the same knowledge base, but they have their individual (possibly conflicting) goals. By *collective* we interpret the case where the workload is somehow partitioned between different, possibly equal participants. The most interesting case is the *collaborative* one where there is a clear mutual engagement of participants in a coordinated effort to solve the KE problem together. This of course requires some special KE tools. In our opinion they can be provided based on the wiki technology.

A wiki can be described as a collection of linked webpages. Wikis were created to provide a conceptually simple tool for collaborative knowledge sharing and social communication. Important features of a wiki include: remote editing using a basic web browser, simplified text-based syntax for describing context (wikitext), rollback mechanism, thanks to built in versioning, diversified linking mechanism (internal, interwiki, and external links), and access control mechanisms. A comprehensive comparison of different wiki systems can be found online.[1]

While wiki systems provide an abstract representation of the content they store, as well as standard searching capabilities, they lack facilities that help in expressing the semantics of the content. This is especially important in the case of collaborative systems, where number of users work together. Thus, wikis became one of the main applications and testing areas for the Semantic Web technologies. One of the most

---

[1]See http://www.wikimatrix.org.

© Springer International Publishing AG 2018
G.J. Nalepa, *Modeling with Rules Using Semantic Knowledge Engineering*,
Intelligent Systems Reference Library 130,
https://doi.org/10.1007/978-3-319-66655-6_15

important technologies in this area are *semantic wikis* that extend basic wikis with knowledge representation features.

In this chapter we discuss original results in the area of *Collaborative Knowledge Engineering* (CKE). We begin with the presentation of the LOKI platform in Sect. 15.1. It is an original semantic wiki solution that combines the ideas of SKE with semantic wikis to provide an original KE platform for CKE. In Sect. 15.2 we present a practical example of the use of the platform. We extended LOKI with business users in mind. To this goal we provided extended capabilities of SBVR authoring in the form of the SBVRwiki plugin described in Sect. 15.3. Furthermore, the second extension called BPwiki allows for authoring of the BPMN models in LOKI as described in Sect. 15.4. We summarize the chapter in Sect. 15.5.

15.1   Semantic Knowledge Engineering in LOKI

Semantic Wikis

These systems enrich standard wikis with the semantic information expressed by a number of mechanisms. Some basic questions every semantic wiki needs to address are according to Oren et al. [1]: (1) how to annotate content, (2) how to formally represent content, and (3) how to navigate content. Multiple implementations of semantic wikis were developed, including IkeWiki [2], OntoWiki [3], SemanticMediaWiki [4], SemperWiki [5], SweetWiki [6], and AceWiki [7].

In general, in these systems the standard wikitext is extended with semantic annotations. These include relations (RDF triples) and categories (here RDFS is needed). It is possible to query the semantic knowledge thus providing dynamic wiki pages (e.g. with the use of SPARQL). Some systems also allow for building an OWL ontology of the domain to which the content of the wiki is related. This extension introduces not just new content engineering possibilities, but also semantic search and an analysis of the content.

A survey of semantic wiki systems is available online.[2] Some of them are in the development stage, others have been discontinued. IkeWiki [2, 8] was one of the first semantic wikis. The system offered semantic annotations with RDF and OWL support. It introduced simple ontology editing in the wiki with certain visualization techniques. OntoWiki [3] provided improved visualization of the ontological content and advanced collaborative editing features. SemperWiki [5] used advanced semantic annotations with explicit differentiation of documents and concepts. SweetWiki [6] was based on the CORESE engine (RDF engine based on Conceptual Graphs) and used an ontology-based model for wiki organization.

When it comes to active implementations, one of the most popular was SMW (Semantic MediaWiki) [4]. It is built on top of the MediaWiki engine, and extends

---

[2]See http://semanticweb.org/wiki/Semantic_Wiki_State_Of_The_Art.

it with lightweight semantic annotations and simple querying facilities. AceWiki [7] uses ACE (Attempto Controlled English) for natural language processing in the wiki. An FP7 project Kiwi[3] aimed at providing a collaborative knowledge management based on semantic wikis.

It can be observed that from a knowledge engineering point of view, expressing semantics is not enough. In fact, a knowledge-based system should provide both effective knowledge representation and processing methods. In order to extend semantic wikis to knowledge-based systems, ideas to use rule-based reasoning and problem-solving knowledge have been introduced. An example of such a system is the *KnowWE* semantic wiki [9, 10]. The system allows for introducing knowledge expressed with decision rules and trees related to the domain ontology.

Motivation for LOKI

LOKI (Logic-based wiki)[4] uses the logic programming paradigm to represent knowledge in the wiki, including semantic annotations and rules [11, 12]. The main design principles are to provide an expressive underlying logical representation for semantic annotations and rules, allow for strong reasoning support, and preserve compatibility with existing wikis, e.g. SMW.

The design has been based on the use of Prolog for programming and knowledge representation. LOKI provides certain important features, namely:

- Prolog representation for the Semantic annotations of SMW,
- RDF and OWL support in Prolog,
- integration of the Prolog engine and a wiki engine,
- support for an expressive knowledge representation with XTT2 rules, and
- integration with the PELLET- HEART engine.

The architecture of this solution is described next.

LOKI *Architecture*

Considering the features mentioned above, the following LOKI architecture was given (Fig. 15.1), [11, 12]. The wikitext from the regular wiki contains basic semantic annotations. Additionally, it contains a Prolog code, HMR rules, and ontological data. These are extracted by the LOKI engine and combined into a LOKI knowledge base. The main engine is integrated with the HEART interpreter coupled with Pellet. It also supports querying the knowledge base using both generic Prolog queries, as well as SPARQL. This architecture was implemented with a proof-of-concept prototype called PLWIKI presented in [13].

The main features of the system are described in the following subsections.

Semantic MediaWiki Support

LOKI provides a compatibility layer for SMW. There are three main methods of semantic annotations in SMW that are supported by LOKI. These are:

---

[3] See http://www.kiwi-project.eu.
[4] See http://loki.ia.agh.edu.pl.

Fig. 15.1 LOKI architecture

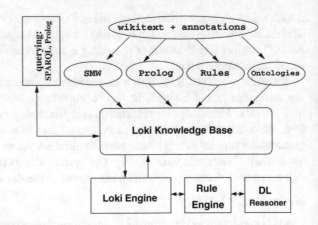

- Categories – a simple form of annotation that allows users to classify pages. To state that an article (Wiki page) belongs to the Category Cities one has to write [[Category:Cities]] within an article.
- Relations – there is a possibility to describe relationships between two Wiki pages by assigning annotations to existing links. For example, there is a relation capital_of between Warsaw and Poland. To express this, one has to edit the page Warsaw and add [[capital_of::Poland]] within the page content.
- Attributes – allow users to specify relationships of Wiki pages to things that are not Wiki pages. For example, one can state that Wiki page Warsaw was created on April 26 2011 by writing [[created:=April 26 2011]].

In SMW, annotations are usually not shown at the place where they are inserted. Category links appear only at the bottom of a page, relations are displayed like normal links, and attributes just show the given value. A factbox at the bottom of each page enables users to view all extracted annotations, but the main text remains undisturbed. Users can search for articles using a simple query language developed based on the wiki syntax.

For example to display the name of the city which is the capital of Poland, the following query is used:

```
{{#ask: [[category:city]] [[capital of::Poland]]}}.
```

The query functionality of Semantic MediaWiki can be used to embed dynamic content into pages, which is a major advantage over traditional wikis. Several other forms of queries can be found in the SMW online documentation.

LOKI allows users to describe categories, relations and attributes as in SMW. They are represented by appropriate Prolog terms. Examples are given below, with the SMW syntax first, followed by the corresponding Prolog representation.

```
[[category:cities]] Warsaw is in Poland.
    wiki_category('subject_page_uri','cities').
Warsaw is [[capital_of::Poland]].
    wiki_relation('subject_page_uri','capital_of','Poland').
[[created:=April 26 2011]].
    wiki_attribute('subject_page_uri','created','April 26 2011').
```

LOKI also provides a direct low-level support for RDF and OWL.

RDF and OWL Support

Plain RDF annotations are supported and separated from the explicit SMW annotations mentioned above. The RDF annotation can be embedded directly in the XML serialization then it is parsed by the corresponding Prolog library, and turned to the internal representation, that can also be used directly. Using the *semweb/rdf_db* library SWI-Prolog represents RDF triples simply as:

```
    rdf(?Subject, ?Predicate, ?Object).
```

So mapping the above example would result in:

```
    rdf('Warsaw',capital_of,'Poland').
```

RDFS is also supported by the *semweb/rdfs* library, e.g.:

```
    rdfs_individual_of('Warsaw',cities).
```

SPARQL queries are handled by the *semweb/sparql_client*. The SWI-Prolog RDF storage is highly optimized. It can be integrated with the provided RDFS and OWL layers, as well as with the *ClioPatria* platform[5] that provides SPARQL queries support. SWI-Prolog supports OWL using Thea [14].

Reasoning in the Wiki

Two approaches to reasoning are provided: pure Prolog, and a rule-based one. Thanks to the availability of the full Prolog engine in the wiki, the inference options are rich. Prolog uses backward chaining with program clauses. However, it is very easy to implement meta-interpreters for forward chaining. Compound queries can be easily created and executed as Prolog predicates. A simple clause finding recently created pages might be as follows:

```
recent_pages(Today,Page) :-
    wiki_attribute(Page,created,date(D,'May',2011)),
    I is Today - D,
    I < 7.
```

---

[5]See http://e-culture.multimedian.nl/software/ClioPatria.shtml.

This generic approach provides a lot of flexibility. However, it requires the knowledge of the Prolog language. This is why, a higher level rule-based knowledge representation is also provided for LOKI.

The following scenarios for embedding rules in the wiki are considered, including: (1) a single collection of rules (possibly defining a decision table) for a single wiki page, and (2) rules working in the same context present in multiple pages in the same namespace. Rules are designed with the use of XTT2, serialized to HMR and embedded in the wiki pages with related content. Before the inference process starts, rules are extracted by the LOKI parser and concatenated into a single HMR file corresponding to a wikipage or namespace.

The discussed knowledge representation features were implemented with a prototype called PLWIKI.

Prototype

A prototype implementation of the LOKI was called PLWIKI (Prolog-based wiki) and was presented in [11, 13]. The main goal of the system design is to provide a generic and flexible semantic wiki solution. There are numerous wiki engines available. Most of them are similar with respect to the main concepts and features. The LOKI idea is to use a ready and extensible wiki engine, that could be possibly extended with knowledge representation and processing capabilities. Instead of modifying an existing wiki engine or implementing a new one, a development of an extension to the DokuWiki[6] system was selected. The basic idea was to build a layered knowledge wiki architecture, where the expressive Prolog representation is used at the lowest knowledge level. This representation is embedded within the wiki text as an optional extension. On top of it a number of layers are provided. These include meta-data descriptions with RDF and semantic annotations with RDFS and OWL.

The PLWIKI architecture can be observed in Fig. 15.2. The stack is based on a runtime including the Unix environment with the Unix filesystem, the Apache web server and the PHP stack. Using this runtime the standard DokuWiki installation is run. PLWIKI functionality is implemented with the use of a plugin allowing for enriching the wikitext with Prolog clauses, as well as running the SWI-Prolog interpreter. It is also possible to extend the wikitext with explicit semantic information encoded with the use of RDF and possibly the OWL representation. This layer uses the Semantic Web library provided by SWI-Prolog.

DokuWiki supports a flexible plugin system, providing several kinds of plugins. These include: *Syntax Plugins*, extending the wikitext syntax; *Action Plugins*, redefining selected core wiki operations; and *Renderer Plugins*, allowing users to create new export modes (possibly replacing the standard XHTML renderer). The current version of PLWIKI implements both the *Syntax* and *Renderer* functionality. Text-based wikipages are fed to a lexical analyzer (Lexer) which identifies the special wiki markup. The standard DokuWiki markup is extended by a special `<pl>...</pl>` markup that contains Prolog clauses. The stream of tokens is then passed to the Helper that transforms it to renderer instructions, parsed by the Parser.

---

[6]See http://www.dokuwiki.org.

Fig. 15.2 PLWIKI
architecture, after [13]

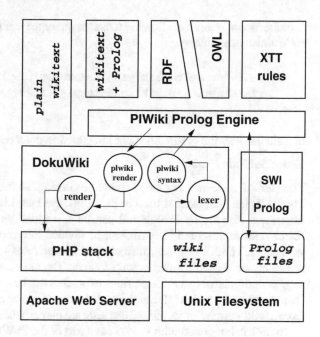

The final stage is the Renderer, responsible for creating a client-visible output. In this stage the second part of the plugin is used for running the Prolog interpreter.

The functionality of the PLWIKI *Syntax Plugin* includes parsing the Prolog code embedded in the wikitext and generating the knowledge base composed of files containing the Prolog code (where each wikipage has a corresponding file). The PLWIKI *Renderer Plugin* is responsible for executing the Prolog interpreter with a given goal, and rendering the results via the standard DokuWiki mechanism.

As mentioned previously, PLWIKI can directly interpret the SMW syntax. Moreover, it allows for embedding any Prolog code, providing more expressive knowledge. Both facts and goal may be specified. To specify the goal (query) for the interpreter the following syntax is used:

```
<pl goal="country(X),write(X),nl,fail">
     country(france).
     country(spain).
</pl>
```

It is possible to specify a given *scope* of the query (using wiki namespaces):

```
<pl goal="country(X),write(X),nl,fail"
     scope="prolog:examples">
</pl>
```

A bidirectional interface, allowing a user to query the wiki contents from the Prolog code is also available, e.g.:

```
<pl goal="consult('lib/plugins/prolog/plwiki.pl'),
    wikiconsult('plwiki/pluginapi'),list.">
</pl>
```

In a similar way the XTT2 rules are integrated as demonstrated next.

Embedded Rule Engine

The HEART inference engine is written in Prolog, so it can be run using PLWIKI. The HMR language that is used to represent rule-based knowledge in HMR is also interpreted directly by Prolog and can be embedded on wiki pages as well. The XTT2 rules represented in HMR can be divided into modules spread over several wiki pages. The XTT2 types, attributes, tables and rules – all of them can be defined separately on multiple areas in wiki system. The area can be understood as a wiki page or entire namespace. Later, rules to process can be chosen by defining contexts (namespaces) for which inference should be run. It can be either the entire knowledge from a wiki system, or its parts from selected namespaces.

HEART is integrated with PLWIKI as a part of the PLWIKI plugin. HMR language can be embedded on wiki pages with `<pl>` `</pl>` tag, as shown below:

```
<pl>
xtype [name: genres,
       base: symbolic,
       domain: [horror,science,handbook,fantasy,poetry,historical]
       ].
xattr [name: fav_genres,
       abbrev: fav1,
       class: general,
       type: genres,
       comm: in
       ].
xschm age_filter: [age,fav_genres] ==> [age_filter].
xrule age_filter/1:
      [age lt 18.000,
       fav_genres eq horror]
    ==>
      [age_filter set [young_horrors]]
    :recommend_books.
</pl>
```

To run a reasoning process the `<pl scope="" goal="">` tag has to be used. If the goal is a valid HEART command for running inference process, then the reasoning is performed by the engine, and result calculated, and rendered on a wiki page.

To run the inference process in HEART rule engine a `gox` command is used. It takes three parameters values of input attributes, rules to be processed, and reasoning mode. Values of input attributes using a named state. The HMR state element stores values of attributes values. An example of running inference in PLWIKI is shown below: `<pl scope="*" goal="gox(init,[result_table],gdi">`.

The meaning of the example is: run the Goal-Driven inference using the `result_table` as a goal table and taking values of input attributes from the state called `init`. The `scope` parameter in `<pl>` tag is optional and it specifies a namespace from which types, attributes, tables and rules should be taken as an input for the reasoning process. If not specified, the entire knowledge in the wiki is processed. HEART is integrated with PLWIKI using two modules. The first module is responsible for rendering wiki pages and extracting the HMR code. The second one is embedded within the PLWIKI engine and it is responsible for performing inference based on the HMR model passed to it by the PLWIKI engine. The process of rendering a wiki page looks as follows:

- The Wiki engine parses the wiki page and extracts the HMR code and reasoning queries (goals),
- Depending on the scope defined in the goal, PLWIKI merges the HMR code from the wiki pages that falls into the given scope and passes it to the HEART inference engine,
- HEART performs reasoning and returns results to the PLWIKI engine,
- PLWIKI renders a complete wiki page with a previously parsed regular text and an answer to a given goal produced by HEART.

The explicit use of the ontological data is not considered mandatory now (XTT2 rules can work on data explicitly defined in the wiki pages). However, when attributes defined in an ontology are used, then the engine can invoke the DL reasoner using the previously mentioned PELLET- HEART framework which provides DL reasoning.

These features are presented on an extended example in the next section.

15.2 Case Study Implementation in LOKI

In this section, we continue the development of the exemplary case called the BOOKSTORE recommendation system, first introduced in Sect. 4.1. The system goal is to identify books potentially interesting to users, based on the books properties, user data and preferences. The system was modeled in LOKI. A modularized rule base for the system was developed in HQED.

Implementation in LOKI

The basic version of this example was implemented as a LOKI benchmark case [15]. In the system there are five main types of pages:

- `genre` in `bookstore:genre` namespace,
- `publisher` in `bookstore:publisher` namespace,
- `author` in `bookstore:author` namespace,
- `book` in `bookstore:book` namespace,

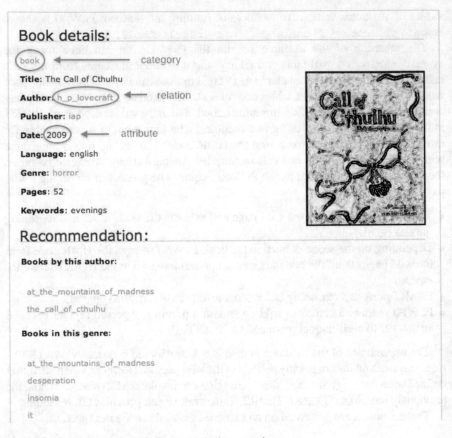

Book details:

book ← category

Title: The Call of Cthulhu

Author: h_p_lovecraft ← relation

Publisher: iap

Date: 2009 ← attribute

Language: english

Genre: horror

Pages: 52

Keywords: evenings

Recommendation:

Books by this author:

at_the_mountains_of_madness

the_call_of_cthulhu

Books in this genre:

at_the_mountains_of_madness

desperation

insomia

it

Fig. 15.3 Bookstore example in LOKI: semantic annotations

- user in bookstore:user namespace.

The namespaces contain information about authors, publishers, users and books. The most important namespace bookstore:book contains information about the books. A exemplary page bookstore:book:the_call_of_cthulhu page source is as follows (page view is presented in Fig. 15.3):

```
====== Book details:  ======

[[category:book]]

**Title**:   [[title:=The Call of Cthulhu]]
**Author**: [[author::bookstore:author:h_p_lovecraft]]
**Publisher**: [[publisher::bookstore:publisher:iap]]
**Date**:   [[date:=2009]]
**Language**:   [[language:=english]]
```

Author: betty_crocker

Publisher: wiley_publishing

Date: 2006

Language: english

Genre: handbook

Pages: 352

Keywords: christmas, cooking

Recommendation:

Books by this author:

betty_crocker_christmas_cookbook

Books in this genre:

betty_crocker_christmas_cookbook

logical_foundations_for_rule_based_systems

the_christmas_table

Books by this publisher:

betty_crocker_christmas_cookbook

Export to RDF/XML

Export to RDF

Fig. 15.4 Bookstore example in LOKI: RDF/XML export

```
**Genre**: [[genre::bookstore:genre:horror]]
**Pages**: 52
**Keywords**:  [[keyword:=evenings]]
```

The following Prolog code is associated with this page is (Fig. 15.4):

```
wiki_category('bookstore:book:the_call_of_cthulhu','book').
wiki_attribute('bookstore:book:the_call_of_cthulhu','title',
        'The Call of Cthulhu').
wiki_relation('bookstore:book:the_call_of_cthulhu','author',
        ':bookstore:author:h_p_lovecraft').
wiki_relation('bookstore:book:the_call_of_cthulhu','publisher'
        ,'bookstore:publisher:iap').
wiki_attribute('bookstore:book:the_call_of_cthulhu','date','2009').
wiki_attribute('bookstore:book:the_call_of_cthulhu','language',
```

```
          'english').
wiki_relation('bookstore:book:the_call_of_cthulhu','genre',
          'bookstore:genre:horror').
```

The page also contains suggestions for some related items.

```
====== Recommendation: ======
**Books by this author**:  {{#ask: [[category:book]]
          [[author::bookstore:author:h_p_lovecraft]] }}
**Books in this genre**: {{#ask: [[category:book]]
          [[genre::bookstore:genre:horror]] }}
**Books by this publisher**: {{#ask: [[category:book]]
          [[publisher::bookstore:publisher:iap]] }}
```

The suggestions are the results of the reasoning process based on semantic annotations. In fact, it narrows down to querying the wiki system for pages with desired properties. This flexible mechanism enhances searching within the system. If a new book is added to the system, it will automatically be captured by the recommendation mechanism. The only requirement is to define an author, a publisher, and a genre of this new book.

More complex recommendations may be defined thanks to the possibility of combining SMW markup with Prolog code. For instance, in Christmas some customer recommendations may be adequate:

```
<pl cache="true">
    custom_recommendations(X)  :-
        wiki_attribute(X,'keyword','christmas').
</pl>
```

And on the page with book details:

```
<pl goal="custom_recommendations(X),write(X),nl,fail" scope="*"></pl>
```

Custom recommendations can be easily modified, for example:

```
<pl cache="true">
    custom_recommendations(X)  :-
        wiki_attribute(X,'keyword','easter').
</pl>
```

The possibility of combining the SMW markup with the Prolog code is one of the main advantages of LOKI.

Rule-Based Reasoning with HEART

Simple classification and flexible Prolog rules may be further developed and enhanced. Thanks to the possibility of embedding HMR rules within the wiki text and

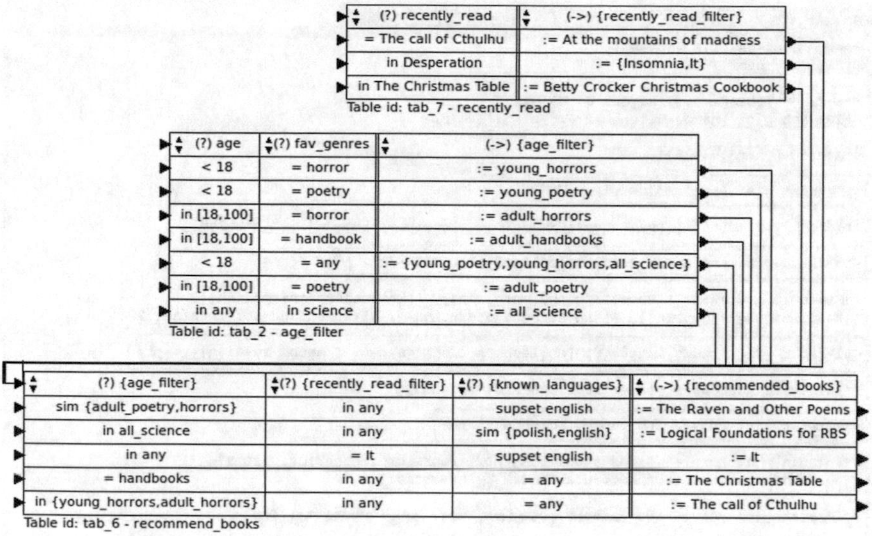

(?) recently_read	(->) {recently_read_filter}
= The call of Cthulhu	:= At the mountains of madness
in Desperation	:= {Insomnia,It}
in The Christmas Table	:= Betty Crocker Christmas Cookbook

Table id: tab_7 - recently_read

(?) age	(?) fav_genres	(->) {age_filter}
< 18	= horror	:= young_horrors
< 18	= poetry	:= young_poetry
in [18,100]	= horror	:= adult_horrors
in [18,100]	= handbook	:= adult_handbooks
< 18	= any	:= {young_poetry,young_horrors,all_science}
in [18,100]	= poetry	:= adult_poetry
in any	in science	:= all_science

Table id: tab_2 - age_filter

(?) {age_filter}	(?) {recently_read_filter}	(?) {known_languages}	(->) {recommended_books}
sim {adult_poetry,horrors}	in any	supset english	:= The Raven and Other Poems
in all_science	in any	sim {polish,english}	:= Logical Foundations for RBS
in any	= It	supset english	:= It
= handbooks	in any	= any	:= The Christmas Table
in {young_horrors,adult_horrors}	in any	= any	:= The call of Cthulhu

Table id: tab_6 - recommend_books

Fig. 15.5 Rule base for the BOOKSTORE system

invoking HEART from the system, the development of a complex rule base is possible. Let us consider the following XTT2 rule base for the bookstore recommendation system designed in HQED (see Fig. 15.5):

Selected rules generated by HQED may be embedded in the wiki within the dedicated tags. Examples are as follows:

```
<pl>
xrule age_filter/1:
    [age lt 18.000, fav_genres eq horror]
    ==>
    [age_filter set [young_horrors]]
    :recommend_books.

xrule recently_read/2:
    [recently_read in ['Desperation']]
    ==>
    [recently_read_filter set ['Insomnia','It']]
    :recommend_books/2.

xrule recommend_books/1:
    [age_filter in [all_science], recently_read_filter in [any],
     known_languages in [polish,english]]
    ==>
    [recommended_books set ['Logical Foundations for RBS']].
</pl>
```

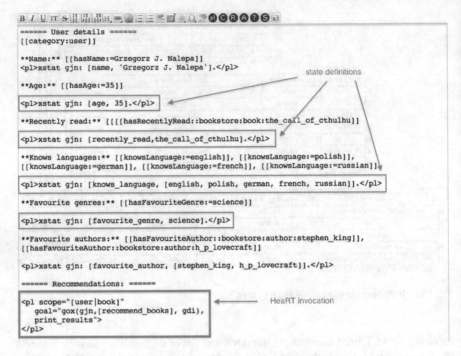

Fig. 15.6 Bookstore example in PLWIKI: wikitext with embedded HMR

If the system is modeled in wiki, then each page may represent a different state of the system. Actual values of the attributes used in rule preconditions are specified on the user page (see Fig. 15.6) e.g.:

```
<pl>
xstat gjn: [age, 35].
xstat gjn: [recently_read,the_call_of_cthulhu].
xstat gjn: [favourite_genre, science].
</pl>
```

Once the state and the rules are defined, more complex rule-based recommendations are possible:

```
<pl scope="[user|book]"
    goal="gox(gjn,[recommend_books], gdi), print_results">
</pl>
```

Starting from basic KE capabilities we extended LOKI with business users in mind. To this goal we provided an opportunity of SBVR authoring in the form of the SBVRwiki plugin for LOKI described next.

15.3 Collaborative Rule Authoring with SBVR

Motivation

SBVR (Semantics of Business Vocabulary and Business Rules) [16] is a mature standard for capturing expressive business rules [17], see Sect. 3.1 for previous discussion. It is also suitable to model the semantics of BR, including vocabularies in a formalized way. Furthermore, it can be perceived as a useful tool in the communication of business analytics with business people. Finally, the set of vocabularies and rules described with the use of SBVR can be an important part of requirements specification from the classic software engineering methodologies. However, an effective use of the SBVR notation is non trivial, as it requires certain KE skills. Moreover, practical software tools are needed to support business analytics in the rule acquisition process. Such tools should allow for syntax checking, and automatic hinting, as well as a preliminary evaluation of the resulting set of rules on the semantic level.

Currently, there are only a few specialized tools that offer proper SBVR authoring. In fact, this is one of the limiting factors in the wider adoption of the notation. RuleXpress[7] is a tool in which a user can define terms, facts and rules using natural language. It does not support SBVR natively but is compliant with and allows a user to import the SBVR definitions of concepts and rules. Moreover, it provides a mechanism of rule quality checking using simple lexical validation. Although RuleXpress provides an additional web-based interface, it allows only for browsing the content of a knowledge base and does not support editing functionality. SBeaVeR[8] is a plugin for the Eclipse integrated development environment. The tool supports defining terms, facts and business rules in Structured English, provides also a syntax highlighting feature as well as it allows for syntax verification. As it is implemented as an Eclipse IDE plugin, it is addressed rather to software engineers than to an average enterprise employee. SBeaVeR does not provide any web-based interface for collaborative content editing. SBVR Lab 2.0[9] is a web application used to edit concepts and business rules using SBVR that provides syntax highlighting, simple verification and visualization features. However, the tool has several disadvantages, it does not support exporting of the created terms and rules to other formats or a local file. Moreover, all the specified elements are stored in one place and it is not possible to separate term glossary from facts or rules, as well as the application does not support dividing a rule set into subsets or categories. Thus, in the case of large, real world examples, rules are not transparent. Moreover, because of the online verification, the application slows down so much so that typing new rules or searching for a particular data becomes a time consuming task.

In order to improve the situation a new tool called *SBVRwiki* was developed [18]. It is an online collaborative solution that allows for distributed and incremental rule authoring for all participating parties. SBVRwiki is integrated with LOKI and uses

---

[7] See: http://www.rulearts.com/RuleXpress.

[8] See: http://sbeaver.sourceforge.net.

[9] See: http://www.sbvr.co.

the DokuWiki back-end for storage and unlimited version control, as well as user authentication. It supports the creation of vocabularies, terms and rules in a transparent, user friendly fashion. Furthermore, it provides a visualization and evaluation mechanisms for created rules.

The main features of the SBVRwiki plugin can be summarized as follows:

1. Creation of a new SBVR project composed of vocabularies, facts, and rules using a set of predefined templates,
2. Authoring project using structured vocabularies, with identified categories,
3. SBVR syntax verification and highlighting in text documents, as well as syntax hinting,
4. Visualization of vocabularies and rules as UML class diagrams to boost the transparency of the knowledge base,
5. File export in the form of SBVR XMI,
6. Knowledge interchange with the existing PlWiki platform,
7. Integration with the BPwiki plugin for building combined specification of business rules and processes.
8. Full support for the SBVR syntax, including at least binary facts,
9. Ease of use including templates for creating new sets of facts and rules, and
10. Constant assistance during the editing of the SBVR statements, including the elimination of common errors, the use of undefined concepts, duplicated entries, etc.

We discuss next the specific implementation of the plugin.

Implementation

SBVRwiki implements two main plugin components for syntax and actions. The *SBVRwiki Action Plugin* is responsible for the file export in the XMI (XML) format. Moreover, it handles the user interface events, and extends the built-in DokuWiki editor with number hooks that implement shortcuts for common SBVR constructs.

The process of creating a new SBVR projects is supported by a set of simple built in wizards that guide a user. The project starts with the definition of concepts, using them facts can be defined, and rules can be authored. Each of these categories is stored as a separate namespace in the wiki. The Lexer module detects all the defined tokens which allows not only for proper syntax highlighting, but also for detecting the use of undefined concepts. Full interaction of the user with the plugin can be observed in Fig. 15.7.

The *SBVRwiki Syntax Plugin* is used to enter SBVR expressions as wiki text. To make it possible, a special wiki markup < sbvr > is introduced. Using it, a user can enter legal SBVR expressions. The plugin offers rich syntax highlighting, as presented in Fig. 15.9. Moreover, vocabularies can be visualized with the dynamic translation to UML class diagrams. The diagrams are then rendered by the wiki using the PlantUML tool,[10] see Fig. 15.8.

---

[10]See http://plantuml.sf.net.

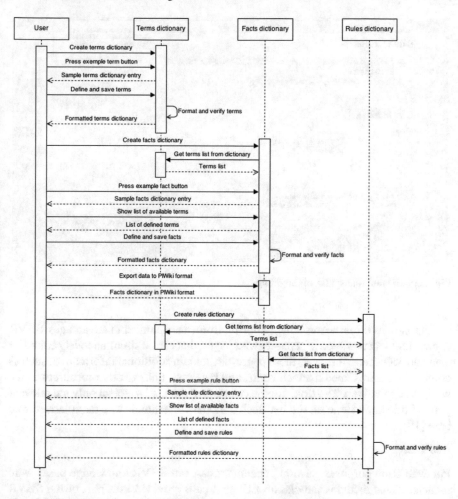

Fig. 15.7 User interaction with SBVRwiki

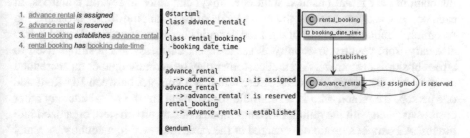

Fig. 15.8 Diagram visualization with PlantUML

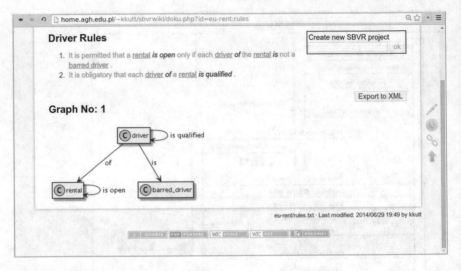

Fig. 15.9 EU rent rule syntax highlighting

The use of wiki as the implementation platform has a number of advantages. SBVR expressions can be stored in separate wiki pages, that can be simultaneously edited by a number of users. Moreover, these pages can contain additional information, such as comments, figures, media attachments, and hypertext links to other resources in the wiki and on the Web. The Loki engine can be programmed to select only the relevant parts of this knowledge on the fly. Such a model corresponds to a modularized rule base [19].

Use Case

For evaluation purposes, several benchmark cases of SBVR knowledge bases were modeled. This includes the classic EU Rent case provided as a part of the SBVR specification [16] and published as a separate document [20]. EU-Rent is a (fictional) international car rental business with operating companies in several countries. In each country it offers broadly the same kinds of cars, ranging from "economy" to "premium" although the mix of car models varies between countries. Rental prices also vary from country to country. It seeks repeat business, and positions itself to attract business customers. A rental customer may be an individual or an accredited member of a corporate customer. A car rental is a contract between EU-Rent and one person, the renter, who is responsible for payment for the rental and any other costs associated with the rental. Different models of cars are offered, organized into groups. All cars in a group are charged at the same rates within a country. A rental booking specifies: the car group required; the start and end dates/times of the rental; the EU-Rent branch from which the rental is to start. A visualization of the parts of the vocabulary modeled by the wiki can be observed in Figs. 15.9, 15.10.

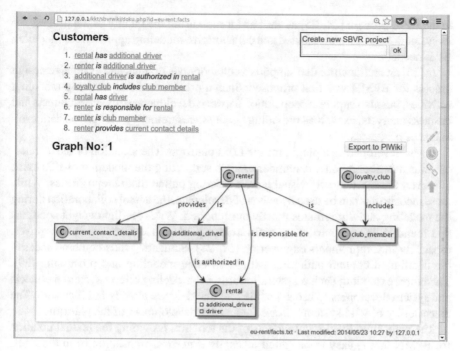

Fig. 15.10 EU rent fact visualization

As applications of SKE also include integration of BR with BP, as described in Chap. 13, we extended LOKI in this direction too. In the next section we demonstrate the collaborative authoring of BP models in BPMN.

15.4 Collaborative Modeling of BPMN Models

Motivation

We introduced BPMN in Sect. 3.2 as a leading notation for modeling Business Processes. It captures processes describing activities of the organization, especially at an abstract design level, and provides a notation emphasizing the control flow. Some tasks or subprocesses can be used to describe either particular kinds of work to complete or some subgoals of the process. Thanks to such a modularization of the BP model, subprocesses can partly help to deal with the problem of model complexity [21]. In this spirit we provided the integration proposal in Chap. 13.

There are claims that BPM should emphasize a gradual, continuous improvement of BP rather than a thorough re-engineering [22]. BP models are commonly used to gather requirements from the early stages of a project and can be a valuable source of information [23]. There is also a pressure for cooperation between people engaged in the BPM process [24, 25] as well as for measuring the quality of the process for its evaluation. Thus, a distributed and collaborative modeling approach for modeling BP is needed.

In [26] an architecture that supports a collaborative, gradual and evaluative design process for BPMN was first proposed. Such a process involves not only modeling activities, but also supports cooperation between developers, software architects and business analysts, as well as providing quality measurement tools for constant evaluation of processes.

It was developed as a plugin for the LOKI platform. The selection of the technology was a deliberate one. A number of tools supporting the modeling of BPs exist. However, these are mostly visual editors offering online model repositories. While these repositories can be used by teams of developers, the actual collaboration during the modeling activity remains mostly unsupported. Wikis are lightweight solutions that found their way into many software companies as well as small enterprises, especially in requirements engineering [27, 28]. Semantic wikis combine accessible distributed content authoring with knowledge modeling and processing [29]. Therefore, extending such a system towards BP modeling allows system architects and system developers to use a flexible tool that they are already familiar with. The extensibility of wiki systems allows for an easy development of the platform.

The main functional requirements of the tool are: providing the textual notation for BPMN that is easy to be edited manually if needed, translation from the XML serialization of the BPMN model to the textual notation, rendering the visualization of the BPMN model in the wiki system in (soft) real time, decomposition of the BP model to wiki pages corresponding to particular tasks, or decomposition of the BP model to wiki namespaces corresponding to specific subprocesses, providing an environment for model commenting and discussions between designers, and enabling the integration of BP evaluation procedures in the wiki.

Moreover, the most important nonfunctional requirements include: providing the repository with the previous model versions with traceability of the source code, supporting management on different scale processes, providing access lists for various types of users, providing the possibility of non-conflict cooperation between different kinds of users.

Implementation

The *BPWiki* functionality is implemented as a set of DokuWiki plugins to be integrated with LOKI. The plugins provide both the *parser* and the *renderer* modes. The first one parses the DokuWiki webpage and interprets its syntax, then the renderer provides a visualization for the parsed data.

Such a solution has several important advantages. Diagrams stored in BPWiki can be easily compared using the SVN/Git version control plugin. When diagrams are stored in the wiki using this syntax it is possible to utilize the wiki version control to easily compare changes in diagrams. The proposed syntax is also modular to a degree and allows for generating Process Diagrams including Sub-Processes stored

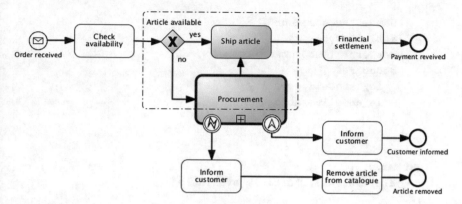

Fig. 15.11 BPMN model for the *Order Fulfillment Process*

separately. Since the focus was on the Business Process Diagrams, only some of the BPMN elements are covered.

Case Studies

In Fig. 15.11, an exemplary model (based on the example from the "BPMN 2.0 by examples" document [30]) can be observed. It depicts a process of the order fulfillment process, which starts after receiving an order message and continues to check whether the ordered article is available or not. An available article is shipped to the customer followed by a financial settlement (a collapsed subprocess). If an article is not available, it has to be procured. The shape of this collapsed subprocess is thickly bordered which means that it is a call activity. It is a wrapper for a globally defined task or, like in this case, a subprocess. The procurement subprocess has two events attached to a boundary. This allows for handling events that can spontaneously occur during the execution of a task or subprocess. In this case, the delivery can be belated or undeliverable, and a customer should be properly informed.

As an example, the textual representation of the shaded fragment of the model presented in Fig. 15.11 is as follows:

```
{
  events: {
    ...
  },
  activities: {
    ...
    at_shipArticle: "Ship article",
    as_finSettlement: {
      name: "Financial settlement",
      markers: [subprocess]
    },
    as_procurement: {
```

```
        name: "Procurement",
        activityType: call,
        markers: [subprocess],
        boundary: [
          ie_lateDelivery,
          ie_undeliverable
        ]
      }
    },
    gateways: {
      g_articleAvail: "Article available"
    },
    flow: {
      f1: ...
      f3: [
        g_articleAvail,
        at_shipArticle,
        "yes"
      ],
      ...
      f6: [
        g_articleAvail,
        as_procurement,
        "no"
      ],
      f7: [ as_procurement, at_shipArticle ],
    }
}
```

The model is decomposed into subpages and namespaces corresponding to subtasks and subprocesses respectively. Every wiki page provides space for discussion and comments. What is more important is the possibility of integrating BP evaluation and rule-based scoring modules. A similar approach has been also recently considered (in a different domain however) [31].

Furthermore, the modeling of the PLI case study can be observed in Fig. 15.12. The structure of the process can be seen in the upper part. The resulting textual representation is in the bottom. The whole process was modularized into separate wiki pages that can be edited collaboratively.

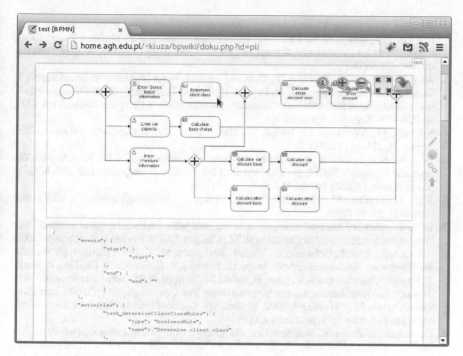

Fig. 15.12 BPMN model of PLI in BPwiki

15.5   Summary

This chapter was devoted to the discussion of the use of a semantic wiki-based solution for KE. LOKI supports the collaborative knowledge engineering process where many knowledge engineers can cooperate. Besides basic KE we also demonstrated two important extensions related to business domain that allow for SBVR and BPMN authoring. LOKI is an example of an important application of SKE in business systems domain. Our future work includes features regarding the monitoring, and boosting of collaboration between users, as well as a semantic changelog for more informative change tracking.

In the next two chapters we move away from business domain to more hardware related applications of SKE.

References

1. Oren, E., Delbru, R., Möller, K., Völkel, M., Handschuh, S.: Annotation and navigation in semantic wikis. In Völkel, M., Schaffert, S. (eds.) SemWiki. CEUR Workshop Proceedings, vol. 206 (2006). www.CEUR-WS.org

2. Schaffert, S.: Ikewiki: A semantic wiki for collaborative knowledge management. In: WET-ICE '06: Proceedings of the 15th IEEE International Workshops on Enabling Technologies: Infrastructure for Collaborative Enterprises, pp. 388–396. IEEE Computer Society, Washington, DC, USA (2006)

3. Auer, S., Dietzold, S., Riechert, T.: Ontowiki - a tool for social, semantic collaboration. In Cruz, I.F., Decker, S., Allemang, D., Preist, C., Schwabe, D., Mika, P., Uschold, M., Aroyo, L. (eds.) International Semantic Web Conference. Lecture Notes in Computer Science, vol. 4273, pp. 736–749. Springer, Berlin (2006)

4. Krötzsch, M., Vrandecic, D., Völkel, M., Haller, H., Studer, R.: Semantic wikipedia. Web Semant. **5**, 251–261 (2007)

5. Oren, E.: Semperwiki: a semantic personal wiki. In: Proceedings of 1st Workshop on The Semantic Desktop - Next Generation Personal Information Management and Collaboration Infrastructure, Galway, Ireland (2005)

6. Buffa, M., Gandon, F.L., Erétéo, G., Sander, P., Faron, C.: Sweetwiki: a semantic wiki. J. Web Sem. **6**(1), 84–97 (2008)

7. Kuhn, T.: AceWiki: a natural and expressive semantic Wiki. In: Proceedings of Semantic Web User Interaction at CHI 2008: Exploring HCI Challenges, CEUR Workshop Proceedings (2008)

8. Schaffert, S., Eder, J., Grünwald, S., Kurz, T., Radulescu, M.: Kiwi - A platform for semantic social software (demonstration). In: Aroyo, L., Traverso, P., Ciravegna, F., Cimiano, P., Heath, T., Hyvönen, E., Mizoguchi, R., Oren, E., Sabou, M., Simperl, E.P.B. (eds.) The Semantic Web: Research and Applications, 6th European Semantic Web Conference, ESWC 2009, Heraklion, Crete, Greece, May 31–June 4, 2009, Proceedings. Lecture Notes in Computer Science, vol. 5554, pp. 888–892. Springer, Berlin (2009)

9. Baumeister, J., Puppe, F.: Web-based knowledge engineering using knowledge wikis. In: Proceedings of the AAAI 2008 Spring Symposium on "Symbiotic Relationships between Semantic Web and Knowledge Engineering", pp. 1–13. Stanford University, USA (2008)

10. Baumeister, J., Reutelshoefer, J., Puppe, F.: Knowwe: A semantic wiki for knowledge engineering. Appl. Intell. 1–22 (2011). https://doi.org/10.1007/s10489-010-0224-5

11. Nalepa, G.J.: Collective knowledge engineering with semantic wikis. J. Univ. Comput. Sci. **16**(7), 1006–1023 (2010)

12. Nalepa, G.J.: Loki – semantic wiki with logical knowledge representation. In Nguyen, N.T. (ed.) Transactions on Computational Collective Intelligence III. Lecture Notes in Computer Science, vol. 6560, pp. 96–114. Springer, Berlin (2011)

13. Nalepa, G.J.: PlWiki – a generic semantic wiki architecture. In Nguyen, N.T., Kowalczyk, R., Chen, S.M. (eds.) Computational Collective Intelligence. Semantic Web, Social Networks and Multiagent Systems, First International Conference, ICCCI 2009, Wroclaw, Poland, October 5–7, 2009. Proceedings. Lecture Notes in Computer Science, vol. 5796, pp. 345–356. Springer, Berlin (2009)

14. Vassiliadis, V., Wielemaker, J., Mungall, C.: Processing OWL2 ontologies using thea: An application of logic programming. In Hoekstra, R., Patel-Schneider, P.F. (eds.) Proceedings of the 5th International Workshop on OWL: Experiences and Directions (OWLED 2009), Chantilly, VA, United States, October 23–24, 2009. CEUR Workshop Proceedings, vol. 529 (2009). www.CEUR-WS.org

15. Kotra, M.: Design of a prototype knowledge wiki system based on Prolog. Master's thesis, AGH University of Science and Technology in Kraków (2009)

16. OMG: Semantics of business vocabulary and business rules (SBVR). Technical report dtc/06-03-02, Object Management Group (2006)

17. Ross, R.G.: Principles of the Business Rule Approach. 1 edn. Addison-Wesley Professional (2003)

18. Nalepa, G.J., Kluza, K., Kaczor, K.: Sbvrwiki a web-based tool for authoring of business rules. In: Rutkowski, L., et al. (eds.) Artificial Intelligence and Soft Computing: 14th International Conference, ICAISC 2015. Lecture Notes in Artificial Intelligence, pp. 703–713. Springer, Zakopane, Poland (2015)

19. Nalepa, G., Bobek, S., Ligęza, A., Kaczor, K.: Algorithms for rule inference in modularized rule bases. In: Bassiliades, N., Governatori, G., Paschke, A. (eds.) Rule-Based Reasoning, Programming, and Applications. Lecture Notes in Computer Science, vol. 6826, pp. 305–312. Springer, Berlin (2011)
20. OMG: SBVR Annex G - EU-Rent Example. Technical Report, Object Management Group (2013)
21. Nalepa, G.J., Kluza, K., Ernst, S.: Modeling and analysis of business processes with business rules. In: Beckmann, J. (ed.) Business Process Modeling: Software Engineering, Analysis and Applications. Business Issues, Competition and Entrepreneurship, pp. 135–156. Nova Science Publishers (2011)
22. Lee, R., Dale, B.: Business process management: a review and evaluation. Bus. Process Manag. J. **4**(3), 214–225 (1998)
23. Monsalve, C., Abran, A., April, A.: Measuring software functional size from business process models. Int. J. Softw. Eng. Knowl. Eng. **21**(3), 311–338 (2011)
24. Caballé, S., Daradoumis, T., Xhafa, F., Conesa, J.: Enhancing knowledge management in online collaborative learning. Int. J. Softw. Eng. Knowl. Eng. **20**(4), 485–497 (2010)
25. Niehaves, B., Plattfaut, R.: Collaborative business process management: status quo and quo vadis. Bus. Process Manag. J. **17**(3), 384–402 (2011)
26. Nalepa, G.J., Kluza, K., Ciaputa, U.: Proposal of automation of the collaborative modeling and evaluation of business processes using a semantic wiki. In: Proceedings of the 17th IEEE International Conference on Emerging Technologies and Factory Automation ETFA 2012, Kraków, Poland (2012)
27. Abeti, L., Ciancarini, P., Moretti, R.: Wiki-based requirements management for business process reengineering. In: ICSE Workshop on Wikis for Software Engineering, 2009. WIKIS4SE '09, IEEE, pp. 14–24 (2009)
28. Dengler, F., Vrandečić, D., Simperl, E.: Comparison of wiki-based process modeling systems. In: Proceedings of the 11th International Conference on Knowledge Management and Knowledge Technologies. i-KNOW '11, pp. 30:1–30:4. ACM, New York, NY, USA (2011)
29. Hoenderboom, B., Liang, P.: A survey of semantic wikis for requirements engineering. Technical Report RUG-SEARCH-09-B01, SEARCH, University of Groningen, The Netherlands (2009)
30. OMG: BPMN 2.0 by Example. Technical Report dtc/2010-06-02, Object Management Group (2010)
31. Doukas, C., Maglogiannis, I.: Advanced classification and rules-based evaluation of motion, visual and biosignal data for patient fall incident detection. Int. J. Artif. Intell. Tools **19**(2), 175–191 (2010)

Chapter 16
Designing Robot Control Logic with Rules

Building intelligent robots has always been an important area of both pursuit and research in Artificial Intelligence [1], and applied engineering. Creating such robots requires skills from different domains, including a deep knowledge of materials and mechanics, as well as control theory, artificial intelligence, computer science, and even psychology and linguistics, when we take human-machine communication into account. However, these days the field has become much more accessible to non-experts, thanks to a number of ready robotics solutions. In recent years, a new technological release called Mindstorms from the LEGO company improved this situation even further.

In this chapter we present the application of the SKE methods, to support the design of control logic for basic mobile robots implemented with LEGO Mindstorms. This work addresses the second generation of the LEGO hardware, also known as the NXT. We begin with a basic description of this platform in Sect. 16.1. A dedicated programming solution based on the Prolog language is described in Sect. 16.2. On top of the PLNXT platform the HEART rule engine is integrated as described in Sect. 16.3. This allows for the use of XTT2 for the control of NXT. Examples of such control cases are presented in Sect. 16.4. The Chapter ends with a brief summary in Sect. 16.5.

16.1 Robot Prototyping with Mindstorms NXT

Introduction

LEGO Mindstorms NXT is a universal robotics platform, that offers advanced robot construction possibilities, as well as sophisticated programming solutions [2]. The new version of Mindstorms is becoming a standard robotics platform for both teaching and rapid prototyping of robots. Numerous programming solutions for NXT

© Springer International Publishing AG 2018
G.J. Nalepa, *Modeling with Rules Using Semantic Knowledge Engineering*,
Intelligent Systems Reference Library 130,
https://doi.org/10.1007/978-3-319-66655-6_16

exist, including the LEGO environment, LeJOS, Bricx/NQC and others. However, they fail to provide a clean high-level declarative logic programming solution for NXT. Programming robots, especially mobile ones, is a complex task, involving some typical AI problems, such as knowledge representation and processing, planning, etc. These areas are much more accessible with the use of a logic programming solutions, compared to classic, low-level imperative languages. While numerous programming solutions exist, they fail to provide a high-level declarative programming solution for NXT.

The main objective of this chapter is to present an application of the SKE approach to the controlling of mobile robots. Thanks to its openness, LEGO Mindstorms NXT was selected as a prototyping hardware platform. A new Prolog-based API for controlling Mindstorms NXT is introduced in Sect. 16.2. The API uses a multilayer architecture, composed of a behavioral, sensomotoric, and connection layer. This platform can be used as a generic solution for programming the NXT in Prolog [3]. It also serves as a foundation for a higher-level visual rule-based programming with the XTT2 method. Rules are translated to HMR code which is executed by the HEART engine using a set of custom callbacks to call PLNXT. The engine uses a middleware for controlling a mobile robot in real-time.

Mindstorms NXT

LEGO Mindstorms NXT is the second generation of programmable robotics kit released by LEGO, it is a successor to the LEGO Robotics Invention System (also known as LEGO RCX). Since the initial release the product received a lot of attention from the academic world. This is not surprising, since LEGO actively supports efforts of employing its products in a multitude of high schools and universities across the globe. For example the NXT kit was created by a partnership with MIT Media Laboratory. Most importantly, the platform proved to be a perfect solution for easy and rapid prototyping of both hardware and software robotic designs.

Over the years LEGO has released several variations of the Mindstorms NXT kit, including the NXT 2.0 released in 2009.[1] Having said that, there are several core elements of the set that can be almost always found in the box: Brick, Servo motors, Sensors, Cables, LEGO Bricks, and a Test pad. The *Intelligent Brick* is an integrated embedded computer platform [4].

- a 32-bit ARM7 microprocessor with 64KB of RAM and 256KB of ROM (about a half of the space is used by a firmware and the other half is used to store user applications)
- a monochrome LCD display (100x64 pixels)
- four hardware buttons and a hidden reset button
- a speaker capable of playing sound files at sampling rates up to 8kHz
- three RJ12 modular connectors used as output ports with servo motors
- three RJ12 low-speed ports capable of I^2C communication used mainly with sensors

---

[1]The most recent kit at the time when this research was performed. However, when this book was being prepared, a newer version was already available from LEGO.

Fig. 16.1 LEGO
Mindstorms NXT brick
hardware, after [5]

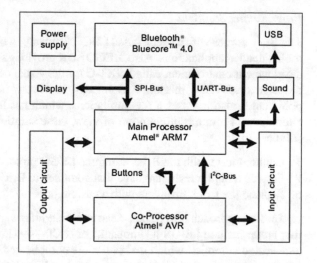

- one RJ12 high-speed port capable of both I^2C and RS-485 communication
- USB 2.0 port with a maximal transfer rate limited to 12Mbits
- a Bluetooth 2.0 EDR radio (supports only the Serial Port Profile, $\sim$200KB/s)

The AVR microcontroller handles servomotors, while the remaining functionality is handled by the ARM7 CPU see Fig. 16.1 for the Brick internals.

With its dedicated output ports, the NXT brick can support exactly three *servo motors*. Motors can provide feedback to the Brick thanks to the built-in tachometer. This is very useful for measuring speed and distance, in addition it also allows to control the motors with one degree precision.

Up to four *sensors* can be simultaneously connected to a NXT brick. Sensors provide communication with the environment. The ones provided by Lego in NXT kits are a touch sensor, an ultrasonic distance sensor, a sound sensor, a light sensor (replaced by a color sensor in the version 2.0). LEGO designed more sensors as additional accessories (sold separately), including: a compass, an accelerometer, an infrared seeker, a gyroscope, and an infrared link. Also, since the interface specification was officially published by LEGO, there are many sensors from third parties available.

To connect the sensors and the servo motors with a brick standard *cables* with 4P4C are used. Standard LEGO Technic *bricks* are used to build a robot. Moreover, a large paper board called a *test pad* is available with a scale, color palette, high contrast paths and other elements helpful for testing many NXT designs, especially the ones utilizing light sensors. The LEGO Mindstorms NXT *communication protocol* provides two options for connecting the brick with a PC. One is based on a USB connection, and the other on the Bluetooth link. Using these, a number of programming solutions are available.

Programming Solutions

The Brick comes with a preinstalled LEGO firmware, which is capable of executing multi threaded applications. Also, LEGO SDK provides a very convenient integrated development environment called NXT-G for the visual design of robot control logic based on a LabVIEW platform.[2] Being easy to use and intuitive it is a good starting point, but it also imposes a few limitations, which has led to the creation of many alternatives. From a runtime point of view, these solutions can be categorized into solutions that:

1. Communicate with the Brick using the LEGO protocol [4],
2. Provide a higher level language that compiles to Brick bytecode,
3. Replace the Brick firmware with a custom one.

The first approach is a simple, clean and straightforward one. The examples of the first group include LeJOS iCommand,[3] or NXT++.[4] The second approach requires a dedicated compiler, which makes it more complicated. In the second group there exists number of solutions including NXC,[5] or RoboLab.[6] The third solution is the most complicated one, since it requires developing a dedicated embedded operating system. This type of solution is provided by the Java-based LeJOS.[7]

Another flexible approach to robot programming is to use a high-level declarative language such as Prolog instead of low-level C-like, or Java-based programming. Besides basic programming languages, NXT robot programming can be supported on a higher logical level, offering a visual logic representation. The prime example is the default LEGO environment. In these cases the control logic is represented with the use of flowcharts representing the control algorithm. However, this is mainly a procedural representation, not a declarative one.

16.2   PLNXT Library

Using knowledge representation methods from the classic AI, such as the decision rules, rules, and tables could improve NXT programming options. XTT2 offers a generic rule-based visual programming solution, combining the power of decision tables and decision trees. XTT2 is implemented with the use of a Prolog-based inference engine. Providing a Prolog-based API for Mindstorms NXT allows to develop control logic for NXT robots with the use of the XTT2 method.

---

[2]See http://www.ni.com/labview.
[3]See http://lejos.sourceforge.net
[4]See http://nxtpp.sourceforge.net/
[5]See http://bricxcc.sourceforge.net/nxc
[6]See http://www.ceeo.tufts.edu/robolabatceeo/
[7]See http://lejos.sourceforge.net

Based on the review of existing solutions presented above, the requirements of a new Prolog API for NXT was formulated [3]. The main requirements are:

- support for all functions of the standard NXT components, that is sensors and motors,
- a cross-platform solution, for both Windows and GNU/Linux environments,
- integration with the visual rule-based logic design with XTT2.

The complete solution is ultimately composed of:

- PLNXT, a middleware executed on a PC, controlling an NXT-based robot, the control is performed with the use of the Bluetooth or USB cable connection,
- a set of custom callbacks for HEART bridging it with PLNXT,
- HEART running a control logic in HMRfor a mobile robot.

A more detailed design of the PLNXT API is presented next.

API Design

Considering the requirements the following API architecture has been designed. It is composed of three main layers as observed in Fig. 16.2:

- behavioral layer – providing higher-level functions, e.g. a drive.
- sensomotoric layer – allowing the exchange of information with sensors and motors,
- communication layer – providing low-level communication with the robot.

The behavioral layer (*nxt_movement*) exposes to the programmer some-high level functions and services. It provides abstract robot control functions, such as *go*, or

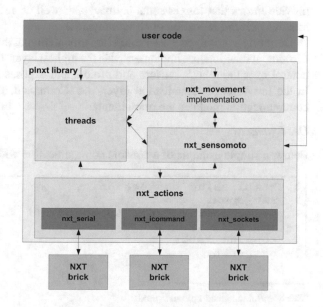

Fig. 16.2 PlNXT layered architecture

turn. Ultimately a full navigation support for different robot types can be provided. However, different robot configurations require different control logic (e.g. to move the robot).

The sensomotoric (*nxt_sensomoto*) layer controls the components of the Mindstorms NXT set motors, all the sensors, as well as Brick functions. This layer can be used to directly read the sensors, as well as program the motors. It can be used by a programmer to enhance high-level behavioral functions.

The goal of the communication layer is to execute the actions of the sensomotoric layer and communicate with the NXT Brick. Currently in this layer several modules are present, providing different means of communication:

- a pure Prolog module, using a serial port communication, and the NXT protocol commands,
- a hybrid solution based on the Java-based iCommand library,
- a hybrid socket-based solution, using the NXT++ library, that communicates with the robot.

All of these actually wrap the *Mindstorms NXT Communication Protocol* [4]. The first solution is the most straight forward one, with standard ISO Prolog stream predicates used to control the serial port. In the second case the Prolog communication module is integrated with iCommand with the use of the SWI Java to Prolog interface called JPL. In the third case, a simple server written in C++ exposes NXT communication with a TCP socket. The Prolog communication module connects to the server and controls the robot through a TCP connection. This opens up the possibility of a *remote* control, where the controlling logic is run on another machine, or even machines. Besides some basic send/receive functions the library has to provide certain services. These are event and time-based callbacks implemented. Therefore the library has to provide *timers* that trigger some callbacks, as well as *event-driven* callbacks. This requires *parallel* execution of certain threads.

Currently a prototype SWI-Prolog implementation of the API is available online.[8] Movement functions are implemented in the highest layer. The mid layer provides full control over the robot's sensors and motors and exposes timer and event services. In the low-level communication layer, the iCommand, DirectSerial, and NXT++ communication modules are implemented.

Use Examples

Below, a simple example of a console communication with a robot is provided:

```
1  % Establishing connection.
2  ?- nxt_open.
3  % Moving forward at a speed of 300 degrees/second.
4  ?- nxt_go(300).
5  % Stopping robot.
6  ?- nxt_stop.
```

---

[8]See http://ai.ia.agh.edu.pl/wiki/plnxt:

```
7  % Moving 80 cm forward at a speed of 400 degrees/↵
      second.
8  ?- nxt_go_cm(400,80).
9  % Reading touch sensor.
10 ?- nxt_touch(Value).
11 % Reading light sensor.
12 ?- nxt_light(Value).
13 % Turning the light sensor diode on.
14 ?- nxt_light_LED(activate).
15 % Reading ultrasonic sensor.
16 ?- nxt_ultrasonic(Value).
17 % Rotating 360 degrees to the right at a speed of ↵
      350 degrees/second.
18 ?- nxt_rotate(350,360).
19 % Playing tone at frequency 500 Hz for 2000 ms.
20 ?- nxt_play_tone(500,2000).
21 ?- nxt_close.
```

A simple example of a complete algorithm is provided below [6]. The robot drives straight until it encounters an obstacle (it can detect it from the 10 cm distance). Then, it turns to evade it. The robot stops when a button is pressed. It can be observed, that the declarative Prolog syntax simplifies the design of the control logic.

```
1  :- dynamic(stop/0).
2  stop(false).
3
4  start :-
5      nxt_open,
6      move_loop.
7
8  move_loop :-
9      trigger_create(_,rotate_condition,rotate_action)↵
          ,
10     trigger_create(_,stop_condition,stop_action),
11     repeat,
12         nxt_go(100),
13         sleep(1),
14     stop(true),
15     finalize.
16
17 rotate_condition :-
18     nxt_ultrasonic(Distance,force),
19     Distance =< 10.
20
21 rotate_action :-
22     nxt_stop,
23     nxt_rotate(100,80),
24     trigger_create(_,rotate_condition,rotate_action)↵
          .
25
26 stop_condition :-
27     nxt_touch(Touch, force),
```

```
28      Touch = 1.
29
30 stop_action :-
31      retractall(stop),
32      assertz(stop(false)).
33
34 finalize :-
35      trigger_killall,
36      nxt_stop,
37      nxt_close.
```

The Prolog API has been successfully tested on a number of simple control algorithms. The complete solution combining PLNXT with HEART is discussed in the
following section.

16.3 Rule-based Control with XTT2 and HEART

The principal idea advocated in this chapter is to design a control logic for a mobile
robot using the XTT2 representation, and execute it with HEART. In Fig. 16.3 we
present a layered architecture combining PLNXT with HEART that allows for a

Fig. 16.3 HeaRT/PlNXT
runtime architecture

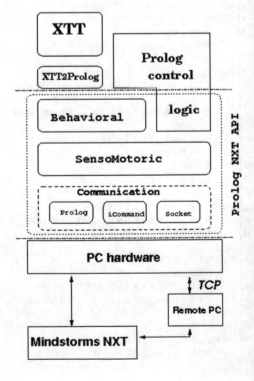

high level of flexibility. The APIs presented in the figure were created gradually by different authors, yet they still work flawlessly together. Moreover, they can be easily modified and enhanced with new features as long as the backward compatibility of the APIs is preserved.

To develop the control logic for the robot we follow the SKE design approach. First of all, a conceptual ARD+ diagram is designed in order to define the robot's attributes. Next a rule-based XTT2 model, based on the conceptual project, is created with the HQEd visual editor [7]. Then, it can be exported to an executable HMR code. Finally, an HMR file can be executed with HEART, using one of the available inference modes (algorithms presented in the next section utilize the Token-Driven Inference). The engine will continue to execute the model in a loop as long as it is traversable.

Communication with the external environment, like reading sensor states or setting motor speeds, is performed with a set of predefined synchronous callbacks and actions. This technique enables HEART to talk to the PLNXT library, which connects with the NXT bricks via USB and/or Bluetooth. Also, there is no hindrance to make HEART utilize multiple completely different libraries simultaneously, which makes this mechanism a truly powerful tool.

Most of the PLNXT predicates that are suitable for usage in XTT2 models were directly (in terms of names) translated into HeaRT callbacks and actions. As a consequence, anybody who is already familiar with PLNXT will be very comfortable with them. All of the linked predicates come from *Movement* and *Sensomoto* layers of the library. Below are a few samples of the implemented callbacks (xcall) and actions (xactn).

```
1  xcall  plnxt_motor  :[Motor,Speed]
2         >>> (alsv_values_get(current, Motor, MotorVal),
3             nxt_motor(MotorVal,SpeedVal,force),
4             alsv_new_val(Speed, SpeedVal)).
5  xactn  plnxt_motor  :[Motor,Speed]
6         >>> (alsv_values_get(current, Motor, MotorVal),
7             alsv_values_get(current, Speed, SpeedVal),
8             nxt_motor(MotorVal,SpeedVal,force)).
9  xcall  plnxt_light  :[Brightness]
10         >>> (nxt_light(Val, force),
11             alsv_new_val(Brightness, Val)).
12  xactn  plnxt_light_LED  :[Setting]
13         >>> (alsv_values_get(current, Setting, Val),
14             nxt_light_LED(Val, force)).
```

The next section contains an example of a PLNXT + XTT2 robotic control system in order to demonstrate the ease of use and capabilities of this platform.

16.4 Examples of Control Algorithms

Line Following Robot

The first model is a very popular design. It is a line following robot with two servo motors and one light sensor, essentially a basic version of the Lego's Tribot design. The robot's main objective is to follow a black line on the NXT's stock test pad. However, with just one light sensor the easiest way to actually implement the line following behavior is to follow an edge of the line. Since the line width is about 2 cm and the robot is approximately 14 cm wide, this little simplification is barely noticeable in practice.

At the highest abstraction level, we want the robot to follow a line, hence it seems reasonable to define its first attributes as Movements dependent on PathTrajectories, Fig. 16.4.

In this design the robot perceives the path trajectories with only one light sensor, consequently LightSensor attribute is a finalization of PathTrajectories. Going deeper, straight to the physical layer, a light sensor can provide the robot with a scalar value representing the luminosity of the surface below it and therefore enable it to detect its location against the line. Thus, we split and finalize LightSensor with the over_the_path attribute dependent on brightness (the left side of Fig. 16.5). Similarly, we can split Movements into two motor attributes, LeftMotor and RightMotor (the right side of Fig. 16.5). NXT servo motors are capable of sending feedback from a built-in rotation sensor, but for this simple example they can be treated just as a simple physical output attribute and finalized as left_motor_speed and right_motor_speed. Ultimately, we need to add to the design the last property linking over_the_path attribute with left_motor_speed and right_motor_speed. At the most physical level this creates the final model illustrated by Fig. 16.6.

Figure 16.7 illustrates the XTT model of the robot's logic, it consists of only two small tables and six straightforward rules. The first table, Light, translates a brightness level (an integer variable), fetched from the light sensor, to the over_the_path parameter (a symbolic variable). The second table is the heart of the algorithm, it defines three rules for three different actions:

- if the sensor is over the edge go forward with a constant wheel rotational speed of 100 degrees per second,
- if the sensor is over the black line turn right with reduced speed,
- if the sensor is over the white background turn left with reduced speed.

During the turning action the speed of one of the motors is dynamically changed by dividing it by 2, as a consequence (PlNXT rounds speed to an integer value) it

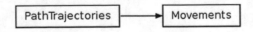

Fig. 16.4 Third conceptual level of the line following robot

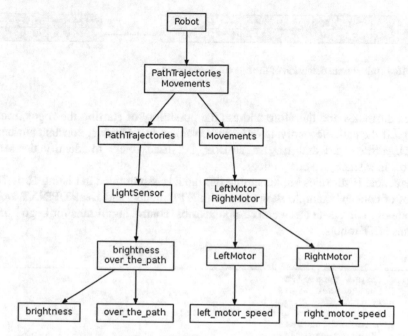

Fig. 16.5 History of the design of the line following robot

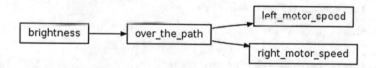

Fig. 16.6 Last conceptual level of the line following robot

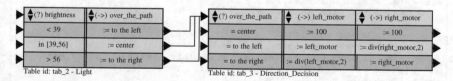

Fig. 16.7 XTT logic model of the line following robot

might eventually reach 0 while the speed of the other engine stays intact. This would cause the robot to stop in place and rotate until the path is found. Balanced speed is essential, too a high value might cause the robot to drive off the winding line. On the other hand, a speed that is too low could make the robot too slow. Both 100 degrees per second and the 2 factor for division were chosen through an experiment with NXT's the test pad.

In this example the XTT model was extended with an initial state definition (Fig. 16.8) for two reasons. Firstly, to add `left_motor` and `right_motor`

Label	(->) right_motor_id	(->) left_motor_id	(->) right_motor	(->) left_motor		(->) left_motor	(->) right_motor
1	:= 'C'	:= 'A'	:= 100	:= 100	------------	:= null	:= null

State: init   State:

Fig. 16.8 Initial state definition of the line following robot

speed attributes and therefore address the possibility of starting the robot from a place off the path. Secondly, in order to implement two auxiliary constant attributes `left_motor_id` and `right_motor_id`, used solely to identify the servo motors in `plnxt_motor` actions.

Practical HMR rules implementing this model are presented in Listing 16.1. This proof of concept example shows that with XTT models and HeaRT/PlNXT architecture one can create clean and comprehensible control algorithms for Lego Mindstorms NXT robots.

```
1  xrule 'Light Sensor'/1:
2      [brightness lt 39]
3    ==>
4      [over_the_path set 'to the left']
5    :'Direction_Decision'.
6  xrule 'Light Sensor'/2:
7      [brightness in [39 to 56]]
8    ==>
9      [over_the_path set center]
10   :'Direction_Decision'.
11 xrule 'Light Sensor'/3:
12     [brightness gt 56]
13   ==>
14     [over_the_path set 'to the right']
15   :'Direction_Decision'.
16
17 xrule 'Direction_Decision'/1:
18     [over_the_path eq center]
19   ==>
20     [left_motor set 100,
21      right_motor set 100]
22      **> [[plnxt_motor,[left_motor_id,left_motor]], [plnxt_motor ←
          ,[right_motor_id,right_motor]], [fire_model,[]]].
23 xrule 'Direction_Decision'/2:
24     [over_the_path eq 'to the left']
25   ==>
26     [left_motor set left_motor,
27      right_motor set (right_motor/2)]
28      **> [[plnxt_motor,[left_motor_id,left_motor]], [plnxt_motor ←
          ,[right_motor_id,right_motor]], [fire_model,[]]].
29 xrule 'Direction_Decision'/3:
30     [over_the_path eq 'to the right']
31   ==>
32     [left_motor set (left_motor/2),
33      right_motor set right_motor]
34      **> [[plnxt_motor,[left_motor_id,left_motor]], [plnxt_motor ←
          ,[right_motor_id,right_motor]], [fire_model,[]]].
```

Listing 16.1 Line follower rules in HMR.

Fig. 16.9 Third conceptual level of the evader robot

Evader Robot

This section describes an *evader robot* which should implement three kinds of behaviors. First of all, it is supposed to patrol its environment with random trajectories. The environment is an area limited by walls and diversified with various obstacles, all of which may damage the robot in case of an impact, therefore they have to be avoided. Lastly, when the robot hears a loud sound it should run away, which simply means moving straight with an increased speed in any random direction. Those three behaviors are of different significance, as a consequence the *Patrol* behavior can be interrupted by either *Run Away* or *Avoid Wall* actions (behaviors). The Run Away behavior can be canceled only by Avoid Wall action, the Avoid Wall behavior can not be cancel-led by any other action.

Robot's construction is fairly simple, it utilizes Lego's Tribot design, two independent motors controlling movement speed and direction (each connected to one wheel). The evader robot uses two different sensors to interact with its environment, both of which are encapsulated within the Sensors attribute in the conceptual model. Since behaviors required from this robot are quite complex there is also an additional conceptual attribute called Behaviors, representing the logic responsible for evaluating an appropriate action based on current sensors readings (Fig. 16.9).

Physical attributes of the robot are:

- distance (to an obstacle or a wall),
- loudness (an overall volume level in robot's environment),
- avoid_wall (behavior of avoiding walls and obstacles),
- run_away (running away behavior),
- patrol (patrolling behavior),
- turn (how fast the robot is turning, if at all),
- speed (robot's speed).

Figure 16.11 illustrates the last level of the conceptual design, how loudness and distance influence all the behavior attributes and how turn and speed are controlled. The history of the transformations that were performed to evaluate the conceptual model all the way down to a physical level are shown in Fig. 16.10

The robot's logic model (Fig. 16.12) can be divided into two key parts. The first one reads data from sensors (tables Loudness Decision and Distance Decision) and then decides what action to take next (table Behavior). The second part is responsible for executing an action chosen by the predecessor:

- **Avoid Wall** This behavior can be considered as the most self-preservational, the main goal here is to avoid hitting the walls and the obstacles (all of which have to be detectable by the ultrasonic sensor). This is achieved by either slowing down if an obstacle is still far from the robot or moving backwards if it is very close to the

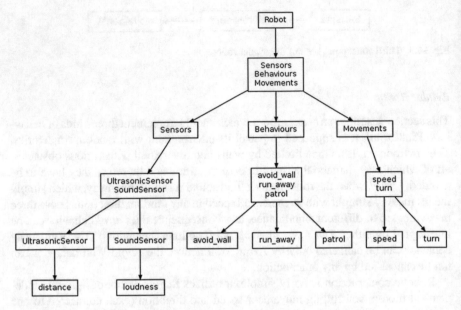

Fig. 16.10 History of the conceptual design transformations of the evader robot

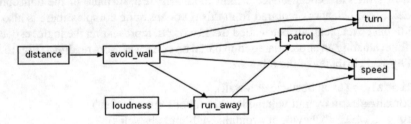

Fig. 16.11 Last conceptual ARD level of the evader robot

robot (table `Adjust Speed`). In both cases the turn is randomly adjusted (table `Wall Avoidance`), there is a 50% chance that the robot will start to turn right, and similarly 50% for turning left.

- **Run Away** The second most self-preservational behavior is `run_away`, triggered by sensing a loud sound. Since Lego NXT's sound sensor is omnidirectional and a source of a loud sound can not be easily established, the implementation of this behavior is highly simplified. It comes down to one table (`Run Away Speed`) with only one rule for setting the robot's speed. The `run_away` behavior can interrupt an execution of the `patrol` action, but its execution can be canceled by the `avoid_wall` action.
- **Patrol** When the robot's sensors do not percept anything unusual `patrol` action will be executed. It consists of two different tables, one for modifying speed (table `Patrol Speed Modification`) and the other one for modifying direction

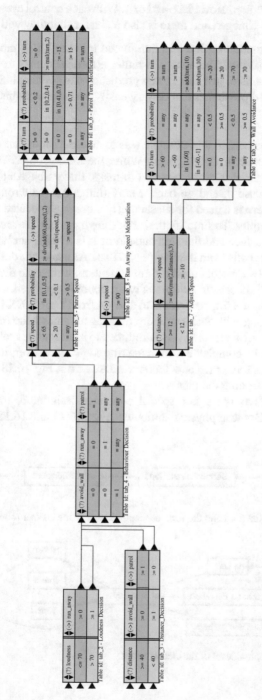

Fig. 16.12 XTT2 logic model of the evader robot

(table `Patrol Turn Modification`). With each control cycle both parameters are randomly changed and there is also a chance that they will untouched.

In this example randomization was introduced into the XTT2 model in order to add human-like behaviors. This task was rather effortless thanks to HEART's callbacks/actions mechanism, just one Prolog predicate was sufficient. Similarly many other Prolog standard predicates can be very easily adopted with this architecture.

Cleaner Robot

The main motivation behind this example was to create a robot that would perform an elaborated sequence of tasks (goals). While the master objective is to clean a certain area from litter, it requires several sub-tasks: find garbage, approach it, grab it, take it away, release it, and go back. Many things can go wrong during these operations hence there is a need for substantially more sophisticated algorithm than in the previous examples. To simplify the robot's environment, the area to be cleaned is a convex white surface and the area outside of it is a black surface. A garage is represented by small balls from the Lego's NXT set. As usual, the design is based on Lego's Tribot with two motors used for movement, an ultrasonic distance sensor, a touch sensor and a light sensor. Additionally the robot is equipped with pincers (for holding garbage), these are also a standard add-on from Lego's NXT set.

`Movements` along with `Sensors` are once again among the main conceptual attributes. A new `State` attribute was introduced to represent this robot's numerous potential states and its complex decision making logic. Moreover, it is further split into `SensorTranslations` and `InternalStates`, Fig. 16.13 is a cross level diagram briefly illustrating the idea.

`SensorTranslations` is a special container designed to translate sensors readings to simple Boolean physical attributes (Figs. 16.14 and 16.15):

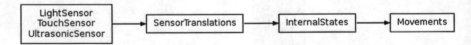

Fig. 16.13 Design of the third and the forth conceptual level of the cleaner robot

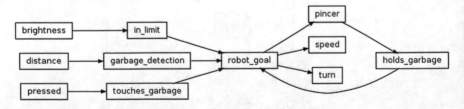

Fig. 16.14 Last conceptual level of the cleaner robot

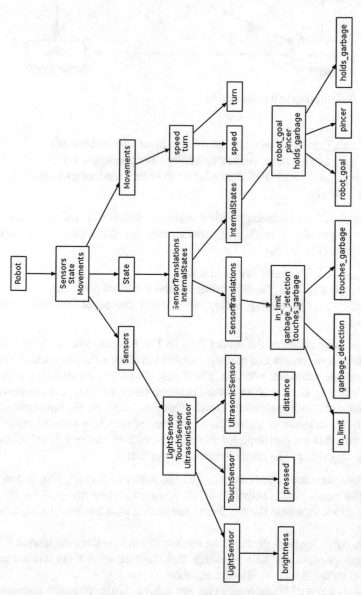

Fig. 16.15 History of the conceptual design transformations of the cleaner robot

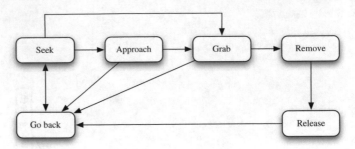

Fig. 16.16   State diagram of the cleaner robot

- `in_limit` (indicates whether the robot is inside the white area)
- `garbage_detection` (indicates whether any garbage was)
- `touches_garbage` (indicates whether the robot touches garbage, but not necessarily holds it)

`InternalStates` includes physical attributes which can not be directly calculated from single sensor readings and usually they are also dependent on the robot's previous states (Fig. 16.16):

- `robot_goal` (currently executed task),
- `pincer` (sets robot's pincer to either opened or closed position),
- `holds_garbage` (a flag indicating whether the robot currently holds any garbage).

The first four tables of the model (Fig. 16.17), as expected, are responsible for reading data from sensors and making sure that the pincers remain opened and ready to grab garbage whenever possible. The Goal Selection table is a core of the algorithm, it is in control of switching between tasks, its decisions are based on a current task and data from sensors. Goal Execution table is a hub, created solely to simplify the network of connections between the Goal Selection table and tables responsible for performing different sub-tasks, of course it is not an essential part of the algorithm. The six different sub-tasks are:

Seek Before the robot can remove a garbage it has to actually find it and that is what the seek task is responsible for. While executing this goal the robot will randomly move within the white area and scan it with the front-facing ultrasonic sensor.

Approach After spotting garbage the robot will start to advance toward it with a constant speed, this is the approach task. Like all other tasks it is initiated and stopped by the Goal Selection table.

Grab The grab goal is responsible for two things, firstly, when the ultrasonic sensor's reading indicates that the robot is very close to garbage it will significantly reduce the robot's speed. Secondly, when the touch sensor's reading indicates that the robot actually can grab the garbage it will set the speed to zero and close the pincers.

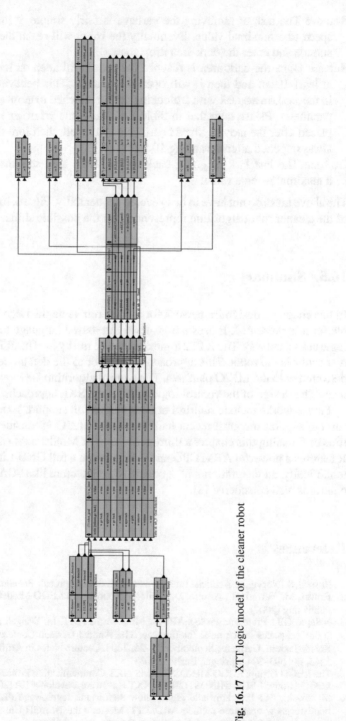

Fig. 16.17 XTT logic model of the cleaner robot

Remove The task of removing the garbage is fairly simple, it just sets the robot's speed to a maximal value. Eventually the robot will reach the end of the white surface and enter the dark area drop zone.

Release Once the dark area is reached the robot will keep on moving forward for at least 10 cm and then it will open the pincers. This behavior is implemented in the `release` task with a blocking action attached to `move_straight_cm` parameter. Please note that in table `Release` the `pincer:=opened` cell is placed after the `move_straight_cm:=10` cell, therefore the pincers will be always opened after advancing 10cm.

Go back The last task of going back to the white area sets the robot's speed to a maximal reverse value.

The above tasks do not have to be executed sequentially, Fig. 16.16 is a state diagram of the cleaner robot algorithm representing all the possible shifts.

16.5 Summary

In this chapter a design framework for mobile robots on the Lego Mindstorms NXT platform is presented. It uses a high-level rule-based language to represent control logic in a visual way. The XTT2 logic can then be run by the HEART inference engine, that controls the robot. This approach is superior to the default design environment delivered with the LEGO platform. The control algorithm is represented on a logical level. The design of the control logic follows the SKE approach.

Future works include multirobot communication support. Extension of the platform to support the most recent hardware from LEGO is considered. In fact, at the time of finishing this chapter a third generation of Mindstorms was made available. It features a powerful ARM CPU and is able to run a full GNU/Linux operating system. Finally, an integration with a cognitive architecture like SOAR for a high-level control is also considered [8].

References

1. Russell, S., Norvig, P.: Artificial Intelligence: A Modern Approach. 3rd edn. Prentice-Hall (2009)
2. Ferrari, M., Ferrari, G., Astolfo, D.: Building Robots with LEGO Mindstorms NXT. Syngress Publishing (2007)
3. Nalepa, G.J.: Prototype Prolog API for Mindstorms NXT. In: Dengel, A.R., et al. (eds.) KI 2008: Advances in Artificial Intelligence: 31st Annual German Conference on AI, KI 2008: Kaiserslautern, Germany, September 23–26, 2008. Lecture Notes in Artificial Intelligence, vol. 5243, pp. 393–394. Springer, Berlin (2008)
4. The LEGO Group: LEGO MINDSTORMS NXT Communication Protocol. 1.00 edn (2006)
5. LEGO Group: LEGO MINDSTORMS NXT Hardware Developer Kit (2006)
6. Makowski, M.: Projektowanie algorytmòw sterowania regułowego dla robotòw mobilnych mindstorms w oparciu o metodę ARD/XTT. Master's thesis, AGH University of Science and Technology, AGH Institute of Automatics, Cracow, Poland (2009)

7. Nalepa, G.J., Ligęza, A.: HeKatE methodology, hybrid engineering of intelligent systems. Int.
   J. Applied Math. Comput. Sci. **20**(1), 35–53 (2010)
8. Laird, J.E., Congdon, C.B.: The Soar User's Manual, Version 8.6.3. University of Michigan
   (2006)

Chapter 17
Rules in Mobile Context-Aware Systems

In Sect. 3.4 we identified the domain of context-aware systems (CAS) as one of the recent applications of rules. Then in Chap. 7 we discussed extensions of the XTT2 method to handle uncertain knowledge. The research on these extensions and development of software tools supporting them was motivated by applications in mobile context-aware systems. In fact, building systems that acquire, process and reason with context data is a major challenge, especially on mobile platforms. Mobile environments in which such systems operate are characterized by high dynamics. The environment changes very fast due to user mobility, but also the objectives of the system itself evolves, as the user changes his or her needs and preferences. Therefore, constant updates of knowledge models are one of the primary requirements for the mobile context-aware systems. Additionally, the nature of the sensor-based systems implies that the data required for the reasoning is not always available nor certain at the time when it is needed. In fact, such a characteristic makes it a case for a challenging big data application. Furthermore, mobile platforms can impose additional constraints, e.g. related to the privacy of data, but also resource limitations, etc.

In this chapter we discuss selected practical results of the KNOWME project.[1] We demonstrate the use of the formal model from Chap. 7. In Sect. 17.1 we distinguish three phases that every context-aware system should pass during the development and later while operating on the mobile device. These phases are the *acquisition phase*, the *modeling phase* and the *processing phase*. Furthermore, we argue, that in mobile context-aware systems, the additional *feedback loop phase* should be provided to allow for the constant adaptability of the system. Then in Sect. 17.2 an overview of the KNOWME architecture is given. Next, the knowledge modeling aspects are presented in Sect. 17.3. The use of the KNOWME toolset is given in Sect. 17.4. An evaluation of the approach is provided in Sect. 17.6. The revised development approach of KNOWME is discussed in Sect. 17.5. The work is summarized in Sect. 17.7.

---

[1] See http://geist.re/pub:projects:knowme:start.

© Springer International Publishing AG 2018
G.J. Nalepa, *Modeling with Rules Using Semantic Knowledge Engineering*,
Intelligent Systems Reference Library 130,
https://doi.org/10.1007/978-3-319-66655-6_17

17.1   Challenges for Context-Aware Systems on Mobile Platforms

The notion of context has been important in the conceptualization of computer systems for many years. In this work mostly the Dey's [1] definition will be used, that describes context as *any information that can be used to characterize the situation of an entity*. For the sake of clarity the entity is defined as a mobile user or device, while *information that characterizes its situation* is any information that can be directly obtained from the mobile device sensors (so called low-level context), or that can be inferred based on this data (high-level context).

Obtaining the low-level contextual information is performed in the first phase of building context-aware systems, which is defined as the *acquisition phase*. This phase is responsible for the delivery of low-level contextual information to the system both for the purpose of building a model and as an input for the processing phase. The successive phases of system development are respectively: the *acquisition phase*, when data is gathered, the *modeling phase*, during which a model of a system behavior is created, and the *processing phase*, during which the previously created model is executed by the inference mechanism [2]. The classic approach for building context-aware systems assumes that the context is obtained in the acquisition phase, modeled (automatically or by knowledge engineer) in the modeling phase, and finally executed in the processing phase. Such an *three-phased approach*, when applied to mobile context-aware systems, exposes serious drawbacks related with the nature of the environment and the system dynamics. In fact, this approach needs to be redefined to meet the requirements of mobile CAS.

An analysis of the literature allowed us [3–5] to formulate four main requirements that should be met by mobile context-aware system in order to assure its high quality and to cope with such drawbacks [6, 7]. These four requirements are, that a mobile CAS should:

1. Intelligibility – allow the user to understand and modify its performance.
2. Robustness – be adaptable to changing user habits or environment conditions, and be able to handle uncertain and incomplete data.
3. Privacy – assure the user that his or her sensitive data are secured and not accessible by to a third party.
4. Efficiency – be efficient both in terms of resource efficiency and high responsiveness.

Although the requirements are rather general, it can be shown that not all the phases of building a classic context-aware system equally refers to them.

Figure 17.1 shows the trade-off between the different context modeling approaches and context acquisition layer architectures with respect to the 4R requirements. The upper left triangle concerns the modeling approaches, the lower right triangle concerns context acquisition and processing architectures. Dotted areas reflects gaps which can be filled in order to meet more requirements. The processing phase is characterized by the superposition of these two. For example the context processing

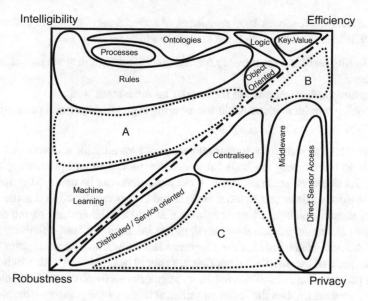

Intelligibility   Efficiency

Robustness   Privacy

Fig. 17.1 Trade-off between different modeling and architectural approaches with respect to the mobile CAS requirements, after [5]

phase that will use rules (upper left corner) and direct sensor access architecture (lower right corner) will be characterized by high intelligibility and privacy, but rather low efficiency and robustness. The combination of the approaches that will allow for meeting all the four requirements is a non trivial task, and requires a lot of modifications to existing methods for modeling and acquiring context.

The nature of a mobile environment in which such systems operate, implies important assumptions regarding their development. Most of the existing solutions were crafted for the purpose of stationary context-aware systems, which assume that the model of the environment and the user preferences are well defined a priori and do not change over time. In mobile systems this assumption does not hold, exposing the *evolutionary* nature of the models. They are influenced by constantly streaming data, which additionally is neither certain nor always available. Therefore, such raw contextual data needs to be appropriately collected, processed and applied to the model iteratively.

The dynamic nature of the mobile environment was the main reason why classic three-phased approaches was not able to support all of the four requirements. This was mainly caused by the fact that: (1) existing methods for building context-aware systems are not crafted for a dynamic, mobile environment which are characterized by thigh volume, velocity and veracity of data (3V), and that (2) the development of a mobile context-aware system is a continuous process which objective is to constantly adapt and improve the system during its operation, so that it fits changing user habits, preferences or environment conditions. Therefore, the traditional approach has to be enriched with the feedback loop consisting of sub-phases allowing to address these issues. The role of each of the three sub-phases is as follows:

1. *Collecting feedback* responsible for improving intelligibility via mediation techniques,
2. *Adapting models* responsible for improving robustness, and
3. *Adjusting providers* responsible for assuring high efficiency and preserving user privacy.

However, to successfully implement these phases additional extensions have to be made to the existing methods for acquiring, modeling and processing context. Figure 17.1 depicts three areas where these extensions can be made. These areas are defined by dotted lines, and marked as A, B and C. It is worth noting that the optimal solution should combine all the methods that join all the corners of the diagram. Therefore, it should provide a way to combine intelligibility and efficiency of rules with robustness of machine learning algorithms. From an architectural point of view, the robustness of the distributed approaches should be combined with efficiency and privacy preservation capabilities for direct sensor access to middleware architectures. This observation implies the following interpretation of the possible extensions:

- A – introduce uncertainty handling mechanisms for rules to improve robustness and allow for integration with machine learning methods,
- B – equip the acquisition layer with machine learning methods in order to learn sensor usage habits and improve energy efficiency by context-based adjustment of sampling rates, and
- C – improve robustness understood as adaptability of the system by including the user in the process of reasoning in order to collect feedback, resolve ambiguous context, modify the model.

All of these extensions can be narrowed to three areas of research: (1) uncertainty handling (concerns areas A, B), (2) adaptability (concerns areas A, B, C) and (3) intelligibility through mediation (concerns areas C). These areas are the primary focus of the research presented in [5], which main goal was to provide tools and methods for building mobile context-aware systems, that address the (4R) requirements defined at the beginning of this section. We argue that this is possible by providing extensions to the existing methods by filling in the gaps depicted in Fig. 17.1. These extensions were provided on the basis of rule-based representation. The choice of rules as the primary representation method was dictated by the fact that they were proven to be one of the most efficient and intelligible way of encoding contextual knowledge [8].

17.2   Overview of the KNOWME Architecture

Providing adaptability for a mobile context-aware system is crucial to assure that it works along with user expectations. The mobile environment changes fast, therefore constant model updates and modifications are one of the primary requirements for the mobile context-aware systems. In stationary environments, the system tuning was performed on the initial phase of the system deployment either by the expert or by the user, and did not change over time. Additionally, the stationary solutions were designed upon an assumption that all the contextual data needed for the reasoning is always available and certain. Finally, the classic approaches for building context-aware systems [9] did not include the user in the process reasoning, which limits the degree of the user's understanding and trust in the system (so-called *intelligibility* [10]). These drawbacks are even more exposed in mobile context-aware systems, where the data volume, velocity and veracity is significant.

Architecture

Figure 17.2 shows an outline of the *Adaptable Model, View, Context-Based Controller* architecture (**AMVCBC**) for context-aware applications which forms the backbone of the KNOWME project. It can be defined as an extension of a standard Model View Controller software architectural pattern [11], that includes context and adaptability as a part of the model. Our research concerns mostly two components of the architecture: the *adaptable model* and the *context-based controller*, leaving the *view* for future work. The *adaptable model* layer was designed to be responsible for discovery and adaptation to user long-term preferences and habits (profiles), but also should provide mechanisms allowing to react on dynamically changing environmental conditions. The *context-based controller* layer provides mechanisms for context-based mediation between user, and other system components that will allow to resolve vagueness and incompleteness of background knowledge data. This layer should also provide an input for an adaptable model layer that will support the adaptability of the system by taking into consideration user feedback and other mediators (probabilistic, ontology-based, etc.).

Context Data Management

In KNOWME we identified three main areas of context data management: (1) handling a large volume of contextual data during the acquisition phase, (2) building and adapting models of context-aware systems from uncertain data, and (3) reasoning based on these models under soft real-time constraints. These three areas were differently addressed by the two core components of the AMVCBC architecture, presented in Fig. 17.2. In particular, the *adaptable model* has to cope with all the three issues, providing the following mechanisms for dealing with them:

- Optimizing the acquisition of contextual information. This includes Hoeffding decision trees [12] and discovery from uncertain data streams and rule-based modeling language for the statistical analysis of historical data (concerns issues (1) and (2)).
- Delivering tools and algorithms for efficient modeling and reasoning in uncertain knowledge bases. In particular this concerns the development of efficient inference engine and algorithms for handling uncertainties caused by a missing or ambiguous context (concerns issue (3)).

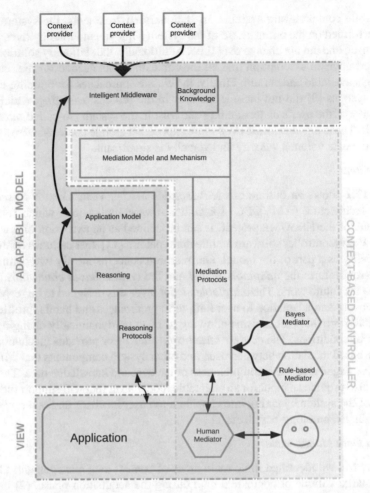

Fig. 17.2 *Adaptable Model, View, Context-Based-Controller* (AMVCBC) Architecture [5]

The *context-based controller* aims at providing mechanisms for dealing with issues (1) and (2). It assures the robustness of the mobile context-aware systems via quick uncertainty resolution in a form of implicit mediation techniques based on the so-called questions forests approach [13]. Question forests allow for improvement of the handling of ambiguous context by providing methods for modeling environment in which the system operates. This model is later used to query the user for additional information that can be used by the system to deal with missing or ambiguous data. In KNOWME we also developed methods for visualizing and initial preprocessing of large volumes of contextual data.

Optimization of Acquisition and Storage of Context

In a mobile context-aware system the acquisition phase should be considered on two levels: (1) during the system development and modeling, (2) during the system operation. In the first case, the acquisition phase is used to collect data for offline analysis which allows the developer to gain insight into the data, distinguish information that may be relevant for solving a particular problem, and to better understand the nature of the data. In the second case, the acquisition phase is considered as a part of the operational process of the system deployed on the mobile device.

Nevertheless, in both cases, the acquisition phase needs to provide an acquisition layer, that will allow for seamless integration with other system components such as inference engine, simulation framework, or context-analysis framework. What is more, the layer should be extensible enough to allow for the implementation of an adaptability module, that would intelligently tune the acquisition process in order to improve the energy efficiency.

AWARE is a mobile instrumentation middleware designed for the purpose of building context-aware applications, collecting data, and studying human behavior [14].[2] The architecture of AWARE consists of two main components: client and server. Client runs on an Android device and is responsible for collecting and data to the server. The server side of the AWARE framework is responsible for storing contextual data of multiple clients and provides remote communication between clients via the MQTT server.[3] Although AWARE was developed for the purpose of the analysis of contextual data it lacks tools that support this task. The functionality of AWARE is limited to collecting and storing of data. However, due to its extensibility and simple integration with other systems, it can be efficiently used as a middleware for more complex systems. The client side of AWARE was designed to support extensibility in a form of *plug-ins*. Therefore, the amount of contextual information that is provided by this framework is unbounded and still growing.

The client is integrated with context-aware applications with the generic Android approach based on the so-called content providers and broadcast receivers. The client side is responsible for logging readings from selected sensors. This data can be directly handled by streaming algorithms, stored in mobile device memory for further analysis, or transferred to the AWARE server for more sophisticated analysis, or research purposes. Every piece of information that AWARE logs contains it as follows: timestamp of the reading, the reading itself and the certainty associated with it. This allows for taking of full advantage from methods for uncertainty handling and adaptability support described later on.

In our work, the AWARE server is used as a repository of context only for the purpose of simplification of testing and evaluating research objectives. According to the privacy requirements, the target system is assumed to work entirely on a mobile device, without any transfer of data to an external server.

---

[2]See: http://www.awareframework.com.

[3]MQTT (formerly MQ Telemetry Transport) is a publish-subscribe based lightweight messaging protocol for use on top of the TCP/IP protocol. For more details see http://mqtt.org.

AWARE has a very simple mechanism for adjusting sampling rates of different context providers, allowing to switch between several predefined values, like *low*, *medium*, *high*, etc. The sampling rate has a direct impact on the energy efficiency of the system, therefore it has to be chosen wisely. The next section discusses the usage of an automated methods for dynamic adjustments of sampling rates of context providers in the AWARE framework.

Optimization with Learning Middleware

The idea of the *learning middleware component* is introduced in Fig. 17.2. It is shared between all models stored within the inference engine, acting as a *knowledge cache*. It is responsible for exchanging information between the sensors and the inference engine. Therefore, it minimizes the number of required requests to the sensors, improving the power efficiency of the entire system.

The prototype of learning middleware, that makes use of a machine learning approach to discover sensor usage patterns in order to minimize energy usage, was described in details in [15]. The enhanced version, of the algorithm, based on uncertain Hoeffding trees, was presented in [5]. As the results of the experiments shown, the usage of learning middleware can save up to 30% of battery power, compared to other solutions. The middleware learns sensor usage patterns and adapts to it by minimizing queries to the sensor layer when it is less likely that the sensor will provide important data. It automatically generates a model of usage habits from historical data and based on that model, it adjusts the sampling rates for the sensors appropriately. It improves the power efficiency of the system, since sampling rates are not fixed but learned from the usage patterns. On the other hand it may help in increasing responsiveness of the system, as the learned model allows for predicting not only future sensor activity but also context-aware application needs. Hence, it is possible to get the desired context in advance, before the application actually requests it. It can be especially useful in cases when context cannot be obtained by the middleware directly from the sensor layer.

17.3 Knowledge Modeling in KNOWME

To provide modeling language that will allow for efficient, intelligible uncertainty handling mechanisms, and provide basic mechanisms for adaptability, the HMR+ language was proposed. HMR+ extends the original HMRlanguage by introducing:

- Uncertainty modeling based on modified certainty factors algebra,
- Statistical and time-based operators, and
- Semantic annotations used by the explicit mediation mechanism.

These feature are based on the respective parts of the formal model introduced in Chap. 7.

The original HMRnotation was already discussed in Sect. 9.6, therefore in this section the primary focus will be put on the extensions. The semantic annotations were also presented in details in [5]. To give a better insight into the structure of the HMR+ file, short excerpts of the aforementioned model will be discussed here. Rules in HMR+ are defined with `xrule` keyword:

```
1  xrule 'Applications'/3:
2      [action in [travelling_home,travelling_work],
3       transportation in [driving,cycling]]
4  ==>
5      [application set [navigation]]
6  **> ['CreateNotification'] # 0.9.
```

The definition presented above ends with the certainty factors which is assigned to it, representing the overall confidence of this rule. This means, that when the rule is fired, the value of the `application` attribute will be propagated with the certainty factor not grater than 0.9. The value of the certainty factor can be assigned by a knowledge engineer, or discovered with data mining algorithms.

Another important feature of the HMR+ language in terms of building mobile context-aware systems is the time-based operators. This mechanism allows for model-based adaptability by including basic temporal modalities into the rules' conditions, which allows for richer context definitions. In HMR+ notation the time-based operators extend the standard operators by parametrising them with temporal and statistical arguments.

The example of the time-based operator was presented in the listing below:

```
1  xrule Actions/1 [
2      location eq{MIN 80% in -1h to 0} home,
3      location eq outside ]
4  ==> [action set leaving_home].
```

The time-based operators are placed after the main relational operators (in this case `eq`) in curly brackets. The first parameter of a temporal operator indicates what portion of the values in the time span defined by the last parameter needs to satisfy the relational operator in order to satisfy the entire formula. In the example above, the parametrised operator can be read as: if at least 80% of historical values of `location` attribute from an hour ago up until now equals `home`, than the entire first condition is true. Besides the `MIN` quantifier, the HMR+ notation allows also `MAX`, and `EXACTLY` parameters, which semantics is analogous to the `MIN` quantifier. The second part of the time-based operator contains a time span in which the satisfaction of relational formula is considered. The time span can be expressed in hours, minutes and seconds by placing an appropriate unit after the value, or by the number of states when no unit is given. For instance an expression `location eq{MIN 80% in -10 to 0}` `home` will analyse the first 10 historical values and the current value (indicated by 0) of the `location` attribute.

The other type of time-based operators are statistical functions. The complete list of supported statistical operators was given in [16], while a practical example is presented below:

```
1  xrule Conditions/1 : [
2          trend(temperature, -1h to 0) lte 0]
3     ==> [conditions set getting_cold].
```

The `trend` operator analyses the values of `temperature` attribute from an hour ago up until now, and calculates the trend of the values, which is defined as the slope of the trend line fitted to the attribute's values using the least–squares fit. This allows for the monitoring of the dynamics of the `temperature` attribute, as the sign and a value of the slope exposes the nature of the changes.

Uncertainty Handling

The mobile environment is highly dynamic which requires from the uncertainty handling mechanism to adjust to rapidly changing conditions. Probabilistic and machine learning approaches cope very well with the most common uncertainties types, but they need time to learn and re-learn. What is more, despite the existence of various probabilistic approaches, there is arguably no method that is able to deal with two very different sources of uncertainty: aleatoric uncertainty, and epistemic uncertainty [17] (see Sect. 7.1). The aleatoric one is caused by statistical variability and effects that are inherently random. In the area of mobile systems this can be reflected as an uncertain sensor reading which cannot be reduced due to the low quality of sensors, or external environmental conditions. Epistemic uncertainty is caused by the lack of knowledge, and can be reduced if additional information is available.

The vital source of information in mobile context-aware systems is the user, who is not only a passive observer of the system but rather its active operator. Therefore, if there is no other automatic source available, the user himself can provide additional information in order to reduce the epistemic uncertainty. However, machine learning methods use a model that is not understandable for the user, and therefore it cannot be modified by him. Fuzzy logic approaches can be used to model uncertainty in a more understandable form, but they mainly cope with uncertainty caused by the lack of human precision which is not our primary focus here.

In KNOWME we propose to handle three types of context-data uncertainty as follows: (1) an uncertainty handling mechanism based on certainty factors, (2) a pattern discovery algorithm from uncertain streaming data, and (3) implicit mediation mechanisms for ambiguity resolution in mobile context-aware systems.

Certainty factors (CF) are one of the most popular methods for handling uncertainty in rule-based expert systems. The Stanford Modified Certainty Factors Algebra [18] accommodated two types of rules with the same conclusion: cumulative rules (with an independent list of conditions) and disjunctive rules (with a dependent list of conditions). This makes the certainty factors fit ALSV(FD) logic *generalised* and *simple* attributes, which are the principle components of XTT2 rules [19]. The conditional part of every rule encoded with XTT2 can be represented as a conjunction of atomic formulae. Every atomic formula can have a certainty factor assigned, which denotes the confidence that the formula is true. Similarly, every rule can have a certainty factor assigned denoting the confidence of this rule. The calculation of the certainty factor for a single rule is a product of the minimal certainty factor of the relational formula from the conditional part and the certainty factor of the rule. The complete set of procedures to compute the certainty factors for different types of atomic formula is given in [5].

Furthermore, the uncertainty handling based on certainty factors can be extended to the table level. Rules within the same table are considered disjoint, when there is no state that is covered by more than one rule [20]. This makes all the rules within the single XTT2 table disjunctive in the understanding of certainty factors algebra. Therefore, they are considered disjunctive in the process of the evaluation of the certainty of the rules within a single XTT2 table. On the other hand, in order to model cumulative rules, it is necessary to split such rules into separate tables. Rules that are located in separate tables, but have the same attribute in their decision parts are considered cumulative with respect to certainty factors algebra, and hence cumulative interpretation applies to them. The complete description of algorithms for inference in uncertain XTT2 knowledge bases were give in [19, 21].

Pattern Discovery Algorithm from Uncertain Streaming Data

Introducing uncertainty into the learning process is an important area in knowledge discovery. One of the first methods that considered uncertainty caused by the lack of knowledge was the C4.5 decision tree algorithm, which allows for handling missing values [22]. Later more complex methods, based on probability theory were introduced like UK-means [23], Uncertain Decision Trees [24], and others were proposed [25]. For handling ambiguous data both in the feature set and in class labels, fuzzy decision trees were developed [26].

The most important drawback of the uncertain data mining methods is that they are computationally very expensive. This violates the efficiency requirement for the mobile context-aware systems defined in Sect. 17.1. Therefore, the primary objective was not to use probabilistic theory to build yet another complex and expensive solution, but to adapt VFDT [12] (or its variant for handling concept drift, called CVFDT) approach to include uncertainty statistics in the model and in the classification results. This allows for a revision of the trained model at a runtime and to discard or confirm uncertain branches of the tree by the user. The most important difference between Hoeffding trees and classical decision trees (like ID3 or C4.5 [12]) is the split criterion, which in the case of the former is approximated with use of a Hoeffding bound (or additive Chernoff bound) [27, 28]. This allows for incremental learning from data streams. The uncertainty of the training data is introduced to the VFDT algorithm by the modification of procedures for calculating entropy and information gain, which are used to choose the best split attribute. The modifications allows for including the uncertainty of the feature and class values, to strengthen the most certain values and weaken the uncertain values in the process of building the tree, a more detailed discussion is given in [5].

To fully support the learning process, and allow for offline testing of the uncertain tree generation algorithm, a custom data format was developed called Uncertain ARFF (uARFF). The uARFF format is based on the ARFF notation, but it includes the information about the uncertainty of data, which can be used to generate uncertain XTT2 models.

An example of a uARFF file was presented in Listing 17.1.

```
 1 @relation aware
 2
 3 @attribute network_traffic {None,Medium,Low,High}
 4 @attribute location {Work,Commuting,Home}
 5 @attribute activity {Still,Walking,Vehicle}
 6 @attribute time {Evening,Night,Morning,Day,
 7     Afternoon}
 8
 9 @data
10 High,Commuting[0.9],Vehicle[0.7];Walking[0.2];Still
11     [0.1],Evening
12 High,Commuting,Vehicle[0.6];Walking[0.3];Still
13     [0.1],Evening
14 High,Commuting,Vehicle[0.6];Walking[0.2];Still
15     [0.2],Evening
16 High,Commuting[0.5],Vehicle[0.5];Walking[0.3];Still
17     [0.1],Evening
18 High,Commuting,Vehicle,Evening
19 High,Commuting,Vehicle,Evening
20 Medium,Commuting[0.9],Vehicle[0.7],Evening
21 Medium,Commuting[0.8],Vehicle[0.9],Evening
22 Medium,Commuting,Walking,Evening
23 Medium,Commuting,Walking,Evening
24 None,Home,Still[0.7],Evening
25 None,Home,Still[0.8],Evening
26 None,Home,Still[0.6],Evening
27 ...
```

Listing 17.1 An example of a uARFF file used by the tool for mining uncertain XTT2 models

Every attribute value can be followed by square brackets in which the probability of this value is given. If no probability is given, the value is considered as certain, therefore a uARFF file without any certainty information is no different than the ARFF file. If the probability is assigned only to one possible value of an attribute, the remaining probability is equally distributed among the remaining values from the domain. For instance, line 9 in the fragment of an uARFF in Listing 17.1 assigns 0.9 probability to value Commuting for an attribute location. This will automatically assign 0.05 probabilities to the remaining two values from location domain. In contrary, all of the values for the activity attribute were explicitly assigned the probability.

Implicit Mediation Mechanisms for Ambiguity Resolution

The problem of mediating knowledge originates from data exchange between relational databases of different schemas. Over the years the notion of the concept of mediation changed, and nowadays it is most often used to describe methods for modeling and executing dialogue between the system and the user or other system components. Such a dialogue aims at improving the adaptability of the system by

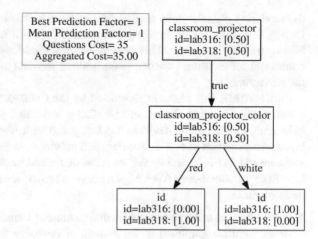

Fig. 17.3 Example of a decision tree from a question forest

indirectly obtaining a knowledge from the user, which can be used to improve the model or resolve ambiguities during the inference process.

The implicit mediation techniques assume that the user has no insight into the knowledge base, and cannot directly modify it. Instead, he is asked easy to answer questions, or is given recommendations which are easy to designate as being correct or wrong. Such information obtained from the user can be later used by the system to infer new knowledge and modify the model, or to resolve an ambiguous context and provide more accurate reasoning results. The problem of *how* to formulate a question and *what* and *when* to ask the user, is an important challenge in active mediation.

In the approach proposed in KNOWME, we use a semantic description of the environment and a mechanism for generating so-called *question forests* to choose the optimal set of questions that the user should be asked in order to resolve the ambiguous context. The question forest consists of a set of question trees that are sorted with respect to the aggregated cost of the questions they represent. The cost is calculated based on the values that are assigned to each of the question components and it differs depending on the level of focus needed by the user to answer the question. An example of a question tree was given in Fig. 17.3. The complete discussion of the question forest generation is given in [13].

17.4 KNOWME Toolset

In the KNOWME we developed several tools supporting the development of mobile rule-based systems.

Accessing and Visualizing Context Data

The context acquired by the AWARE framework in the acquisition layer has to be later modeled and processed by the inference mechanisms. However, we argue that the modeling phase should always be preceded by the initial preprocessing stage. During this stage, the system engineer should gain full insight to the data that will be an input for the system. This insight includes visualizing and initial preprocesing contextual information in order to understand the data and processes that occur in the environment.

In KNOWME, this phase is supported by the CONTEXTVIEWER application. It is a web application that allows for visualizing and initial preprocesing of contextual information from mobile devices. It is integrated with AWARE, and it allows for the basic semantization of raw sensor data and provides bindings with machine learning software like WEKA[4] and ProM[5] to allow direct and immediate analysis of context. CONTEXTVIEWER makes AWARE data easy to browse and process. It consists of two main modules:

1. *Contextual data visualization module* – aims at transforming raw AWARE data to an intuitive graphical form. Instead of browsing database tables, a user can inspect the data visually using intuitive controls.
2. *A basic semantization module* – is responsible for basic semantization of contextual data by translating raw sensor information into human readable concepts and further analyze data and process mining tools.

Although there exists number of frameworks for building context-aware systems, they all approach this issue in a standard manner, which begins with a modeling phase. However, mobile context-aware system are not yet a well established field and there is still a lot of ongoing research in the area of context modeling, automated extraction of knowledge from multiple sources, uncertainty handling, etc. Therefore, the CONTEXTVIEWER support the initial phase of the design of mobile context-aware systems which precedes the modeling phase. The output of this phase, can be later used in the modeling stage to build the model, or choose machine learning methods that best fits the characteristics of the modeled system.

CONTEXTSIMULATOR *for Simulating Mobile Environments*

CONTEXTVIEWER provides modules for data visualization and for its basic semantization. This data can be used to build models either by an expert or automatically, with an usage of machine learning methods. However, to efficiently evaluate these models, a simulation environment is needed. CONTEXTSIMULATOR provides such a functionality. It is a tool that allows for simulation of sensor events stored in relational database in a real time. It was designed to provide testing environment for context-aware applications that are based on the AWARE framework.

The architecture and features of CONTEXTSIMULATOR are presented in [29]. The input for the CONTEXTSIMULATOR application can be twofold: it can be either data fetched directly from the AWARE database, or files that were previously prepared in a ARFF, or an uARFF. Due to the potentially large volumes of data, it is not loaded entirely into the CONTEXTSIMULATOR memory, but the application automatically

---

[4]See: http://weka.waikato.ac.nz.
[5]See: http://www.processmining.org/prom/start.

reads an appropriate portion of data, to keep the simulation speed which was set by the user.

The main features of CONTEXTSIMULATOR are as follows:

1. Sensor selection, which allows for the skipping simulation of some data which are not relevant to the system.
2. Delays and speed-ups in context events arriving, which makes it possible to evaluate systems on large volume of data in a relatively short time.
3. Data *scrambling* by adding uncertainty noise to it.

CONTEXTSIMULATOR can be bound with any Java software using callback mechanisms. Every context event which appears on the CONTEXTSIMULATOR input is simulated by invoking corresponding callback. The callback is a method in Java programming language, therefore binding CONTEXTSIMULATOR with another system requires basic programming skills. It was assumed that the CONTEXTSIMULATOR will be the main simulation environment for the HEARTDROID rule inference engine, which was described in this section.

HEARTDROID *Engine*

Methods and algorithms for uncertainty handling, presented in Chap. 7 were implemented as a components of HEARTDROID. It is a rule-based inference engine dedicated for processing contextual information on mobile devices encoded with the HMR+ notation, which is a textual representation of XTT2 models. HEARTDROID is based on the HEART inference engine [30] and distributed under the GNU General Public License.[6] The HEARTDROID engine was successfully implemented in several practical use cases [4, 6, 7, 31, 32] and is part of the framework for building rule-based mobile context-aware systems [5, 33]. The architecture of the inference engine consists of three main components[7]:

1. *XTT2 Model Manager*, responsible for switching between XTT2 models.
2. *Reasoning mechanism*, that performs inference based on one of four XTT2 inference modes.
3. *Working Memory* component – a common memory for all the models, where current and historical states are stored. This repository of historical states is crucial for Bayesian interpretation of XTT2 models described in Sect. 7.6 and time-parametrised operators discussed in Sect. 7.5.

In terms of uncertainty handling, the state logging system of HEARTDROID is the most crucial component. In particular, the primary emphasis will be put on the aspect of the implementation of time-based operators in the HEARTDROID inference engine and time representation in the engine.

---

[6]See: https://bitbucket.org/sbobek/heartdroid.

[7]Besides the aforementioned components, HEARTDROID offers a prototype implementation of command-line called HAQUNA (**Heart Query Notation**).

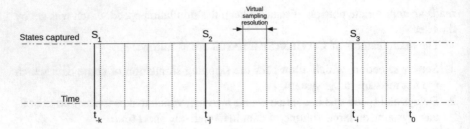

Fig. 17.4 State interpretation in the HEARTDROID inference engine

State Management Module of HEARTDROID

The state of the system in HEARTDROID is understood as a snapshot of all the attribute values registered in working memory. Each state contains a timestamp, which represents time in milliseconds indicating when the snapshot was made. Besides that, every value within a state contains a timestamp indicating the time in milliseconds when the value was assigned to the attribute (i.e. when a sensor delivered some measurement). Due to possible large growth of such state history, only a finite number of states is stored in a FILO queue. When the limit of stored states is exceeded, the older state is removed from the queue.

The state of the system is saved every time the inference is invoked. This is represented in Fig. 17.4 by bold vertical lines labeled S_1, S_2, S_3. The inference can be invoked by many different causes like new sensor reading, an expired attribute value, or on user demand. Therefore snapshots are taken in indeterministic time intervals. To allow statistical operations on attributes a virtual sampling frequency is assumed – it is denoted as vertical dotted lines in Fig. 17.4. Even though there is no actual snapshot at samples between S_1 and S_2, all of the virtual samples will refer to the last available state, in this case S_1. Such an approach allows for reliable statistical analysis of samples taken in variable intervals.

Implementation of Time-Related Operators

Section 7.5 describes parametrised operators that allows for using ALSV(FD) logic operators for time-related formulae. These operators are based on the state management system presented in the previous section. When an operator needs information about historical data it contacts the working memory module in the HEARTDROID inference engine which retrieves such data from state registry. The system stores only the number of states that is required to evaluate models. Every time a new model is registered in the working memory module, it is analyzed to retrieve point in time up to which rules in such a model refer. In order to save memory, which is of great value in mobile systems [6], only the states up to this point are stored. At the beginning the state registry is empty which may lead to errors in rules that includes referrals to the non existing data. Thus, the system does not allow for the evaluation of formulae that refer to data older than those stored in the working memory. In such a case, the inference is interrupted and the information about the cause is logged to the debug channel.

In the current version of the statistical module in the HEARTDROID inference engine, rules' conditions are evaluated every time the inference is triggered for the every state individually. In other words, the formula below will have to be evaluated at worst for 50 different states separately to test if the equality between attribute `att` and `value` was true in at least 50% percent of the cases.

```
1  IF att eq{min 50% in -50 to 0} value THEN ...
```

This can be computationally inefficient especially for large periods to which the statistical operations refers. However, this problem can be easily solved by implementing an algorithm for estimating variance over the sliding window [34].

Bayesian Learner and Reasoner Module

The module for reasoning based on the probabilistic interpretation of XTT2 models was implemented as a standalone module that embeds the HEARTDROID. The learning and querying of the Bayesian representation of the XTT2 model was implemented with the usage of the WEKA module for the Bayesian networks.[8] The XTT2 model is automatically translated to a representation of Bayesian network used by WEKA called XML BIF. The learning phase is performed by serializing the history of states that is stored in the working memory component of the HEARTDROID to ARFF file format.

The learning phase in the current version of the module is not performed online. Therefore, the system needs to relearn periodically, to follow concept drifts. Training is performed with WEKA Bayes network learner. The system performs the deterministic inference, using Algorithm 7.1. Only when reasoning fails, the probabilistic inference is triggered. The output of the reasoning engine is a set of possible values for an attribute with probabilities assigned. Listing 17.4 shows the sample output from the hybrid reasoner.

```
1  Warning: Found missing value in test set,
2  filling in values.
3  application_news {no,yes}
4      no -> 0,99909
5      yes -> 0,00091
6  application_calendar {no,yes}
7      no -> 0,99909
8      yes -> 0,00091
9  application_mail {no,yes}
10     no -> 0,99909
11     yes -> 0,00091
12 application_navigation {no,yes}
13     no -> 0,99909
14     yes -> 0,00091
15 application_clock {yes,no}
16     yes -> 0,99909
17     no -> 0,00091
```

---

[8] See: http://www.cs.waikato.ac.nz/~remco/weka_bn.

```
18  application_weather {no,yes}
19      no -> 0,99909
20      yes -> 0,00091
21  application_sport_tracker {no,yes}
22      no -> 0,99909
23      yes -> 0,00091
24  application_trip_advisor {no,yes}
25      no -> 0,99909
26      yes -> 0,00091
27  application_restaurants {no,yes}
28      no -> 0,99909
29      yes -> 0,00091
```

The most probable values are then taken for further processing, as shown in Algorithm 7.2. The next section shows the evaluation studies of the hybrid reasoning performed on the XTT2 model presented previously in Fig. 7.3.

Integration of HEARTDROID *with Context-Aware Systems*

The integration of the HEARTDROID inference engine with a context-aware application can be performed twofold:

1. API-level binding. In such a type of integration, a communication with application's logic is done using the programming interface of HEARTDROID.
2. Model-level binding. The HMR+ language offers two possibilities of connecting the XTT2 model with the application's logic: callbacks and actions.

The first binding type is purely programming. It is based on the Java API provided by the HEARTDROID. Additionally to execute models, the API allows for dynamic changes to the XTT2 models by adding, removing, or modifying rules and tables. Therefore, this type of binding is most suitable for the internal context-aware components which provide learning and adaptability features.

Components which are responsible for making use of XTT2 models rather than creating them, will use model-level binding. It is based on so called *callbacks* and *actions* mechanisms which are integral components of the HMR+ notation. Callbacks are associated with attributes, while actions with rules. The example of an callback declaration is presented below. Its definition is implemented in Java.

```
1  xattr [ name: day,
2          class: simple,
3          type: day,
4          comm: in,
5          callback: 'DayCallback'
6        ].
```

The example above declares a callback *DayCallback* that will be used to obtain a value of an attribute *day*. The callback mechanism is based on Java reflections, therefore the callback name is effectively a name for a class that needs to be called in order to execute callback action. This class should implement the *Callback* interface. The *comm* element in the attribute definition determines behavior of the callback.

The second type of model-based binding are *actions* which are components of HMR+ language associated with rules definition. The example of a declaration of an action of a name *ProfileOffline* is presented below.

```
1  xrule 'Profile'/3:
2      [action eq sleeping]
3      ==>
4      [profile set offline] **> ['ProfileOffline'].
```

Similarly to callbacks, actions are based on the Java reflection mechanism. Every class that is an action should implement the *Action* interface. The action is triggered only when the rule to which it is assigned is fired. The other difference between callbacks is that actions are used to perform tasks which does not affect the attributes values, nor the model structure explicitly. However, they can change the system state indirectly, which includes turning sensors on or off, changing sampling rates, etc.

HMR+ language supports also semantic annotations that can be assigned to different elements of XTT2 models such as tables, rules, attributes and types. This allows for binding the formal description of the model with concepts which are more familiar to inexperienced users. These concepts may explain the purpose of different system components, or reveal inference scenarios and their impact on the system state.

17.5 Feedback Loop In System Development

As the result of the project, the three-phased design approach was enriched with a feedback loop, which primary objective was to improve the adaptability of the system. This loop includes: *collecting feedback*, *adapting models* and *adjusting context providers*. The process of this looped three-phased approach was presented in Fig. 17.5, as a list of consecutive procedures. However, the real dependency between the phases is much more complex, and the process itself involves the use of a several different methods and tools listed here. It can be seen that there are no strict borders that separate methods used in particular phases. A more common situation is when the same method is used across several phases to achieve different goals.

The primary goal of the *collecting feedback* phase is to obtain information that can be later used to improve the model, or to instantly correct the result of the reasoning process. Therefore, the main methods for obtaining feedback are mediation techniques [13], and the main source of this information is the user. However, the span of this phase is much wider, and may include other sources and methods for obtaining feedback, depending on the phase for which the feedback will be used.

For the *processing phase*, when feedback is mostly used to resolve ambiguities on the runtime level, the implicit mediation techniques are used to feed the system with more information that can be used to improve the accuracy of reasoning. However, the feedback may also be interpreted as historical states of the system which will be later used in other phases, to rebuild the model, or to evaluate the time-based or

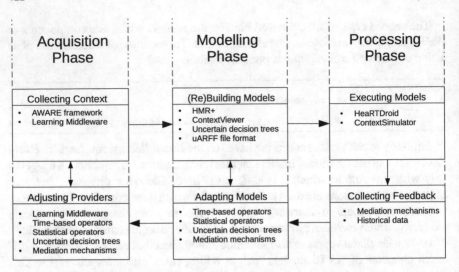

Fig. 17.5 Phases of building mobile CAS supported by KNOWME

statistical operators, improving overall system adaptability. Therefore, the working
memory component of HEARTDROID that is used to store and access historical data
can also be labeled as a special case of feedback collecting mechanism.

The phase of *adapting models* is used to modify existing models, or to build new
one, that best fits user preferences, needs and environmental conditions. The set of
tools and methods used to achieve this goal includes several elements, discussed
briefly in the following paragraphs. One of the most basic, yet powerful method of
adapting models is the use of user feedback. However, the feedback has to be obtained
in a non-intrusive way. Methods that allow for providing seamless communication
between the system and the user are called mediation techniques. Another way of
self-adapting models is the use of statistical and time-based operators. Although the
model itself does not change over time, operators are able to capture the dynamics of
the environment and appropriately react to it, thus improving the adaptability. Finally,
the use of uncertain Hoeffding trees can be used to assure constant adaptability to
changing environment conditions and user preferences.

The phase of *adjusting context-providers* aims at improving the efficiency and
responsiveness of the system by providing an intelligent middleware layer, which acts
as a proxy between the inference engine and the context acquisition layer. The idea of
the learning middleware component was already mentioned. Because the middleware
itself acts as an intelligent subsystem of the context-aware application, the number of
methods it can achieve its goals covers all the previously mentioned. In particular this
includes using implicit mediation techniques to better fit the current environmental
conditions and turn off unnecessary sensors to save energy consumption. The implicit
mediation could be used to determine if the user is inside the building to switch off the
GPS sensor. On the other hand, fully automatic algorithms, can be used do discover
patterns of sensor usage and adjust sampling rates to the history of sensor activity.

Finally, as depicted in Fig. 17.2, the acquisition layer contains a background knowledge component, which either can be discovered using automated methods, or can be given a priori by the designer. The former can be achieved with the afore-mentioned tool for mining uncertain XTT2 models, while the latter can exploit the strengths of statistical and time-based operators to analyze the activity of the sensors *on the fly* and adapt to it instantly.

17.6 Evaluation Studies

In this section, evaluation studies of selected components from the AMVCBC architecture were performed. Two separate use-cases were presented for the two AMVCBC components, that were primary objectives of KNOWME. Data used for the *adaptable model* use-cases presented in this section was prepared using data from six months, collected with the AWARE framework with a LG Nexus 5 mobile phone.[9] This data was used in the initial phase, that precedes the modeling phase. We used the CONTEXTVIEWER web application to analyze the data, and choose which sensors will fit best, the requirements of the models. For the sake of evaluation of the *context-based controller*, the data were obtained with a custom application for log-ging accelerometer. It was not obtained with AWARE, as at that time, the framework exposed limitations in choosing the fixed sampling rate of accelerometer, which was crucial for the success of the use case scenario.

Adaptable Model

There were two context-aware models under consideration. The objectives of the fist model was to minimize the energy consumption of the mobile network provider. There are four different levels of connectivity quality available on mobile devices (in descending order with respect to quality): LTE, 4G, 3G and Edge. This connectivity levels corresponds to the energy consumption levels, with LTE being most energy consuming and Edge least energy consuming. The main goal of the model was to minimize energy consumption by selecting the connection quality that best fits the predicted network usage. The second model was part of the context-aware personal assistant system for suggesting applications that the user might be interested and intelligently switching mobile phone profiles: offline/silent/loud.

For the first model, the analysis of the data with CONTEXTVIEWER and WEKA, allowed us to distinguish the sensors that are most correlated respectively with net-work traffic. It also allowed us to discretise and semantize the data, which gives it more human understandable meaning (i.e. using values like high, medium, low in terms of network traffic instead of numbers in kilobytes). For the first model, we per-formed numerous experiments with different types of sensors, and finally choose the best choice sensors based on which the network usage can be estimated. Figure 17.6

---

[9]See project website for details: http://glados.kis.agh.edu.pl.

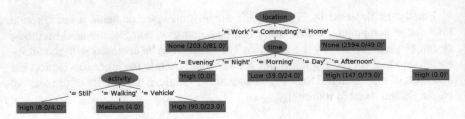

Fig. 17.6 Decision tree for the prediction of mobile network usage

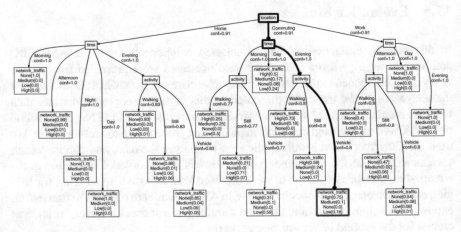

Fig. 17.7 Decision tree generated with uncertain data

presents a decision tree that was build from the data, that allows for the prediction of the possible usage of the network provider based on the selected sensors.

For the second model, using CONTEXTVIEWER we could distinguish the usual locations of the specific use. In this example, the locations from GPS were translated to concepts like *home*, *work*, etc. by a knowledge engineer. However, this information could be discovered automatically, for instance with the use of clustering algorithms like DBSCAN [35], or K-means [36].

While the first model in its objectives referred to the limitation of energy usage, and to adjusting context providers, it was localized in the learning middleware component, alongside with the intelligent GPS energy optimizer, described more extensively in [5, 7]. Because, the sensors which were selected to learn the decision tree could be delivered to the system with a large degree of uncertainty (e.g. location or activity), the uncertain Hoeffding tree generation algorithm was used for this model discovery. Figure 17.7 shows the result of the algorithm after processing data from one month for the dataset used for building a tree from Fig. 17.6. This data was translated to the XTT2 table presented in Fig. 17.8, and thus is ready to be executed by the HEARTDROID directly on the mobile device.

The intelligent personal assistant model, was more complex, and it was decided not to use automatic discovery methods for building it, but rather exploit the strengths of

(?) time	(?) location	(?) activity	(->) network_traffic
= Morning	= Home	= any	:= None(#0.91)
= Afternoon	= Home	= any	:= None(#0.9)
= Night	= Home	= any	:= None(#0.91)
= Day	= Home	= any	:= None(#0.91)
= Evening	= Home	= Walking	:= None(#0.7)
= Evening	= Home	= Still	:= None(#0.66)
= Evening	= Home	= Vehicle	:= None(#0.64)
= Morning	= Commuting	= Walking	:= Low(#0.35)
= Morning	= Commuting	= Still	:= Low(#0.5)
= Morning	= Commuting	= Vehicle	:= Low(#0.41)
= Day	= Commuting	= any	:= High(#0.46)
= Evening	= Commuting	= Walking	:= High(#0.53)
= Evening	= Commuting	= Still	:= High(#0.43)
= Evening	= Commuting	= Vehicle	:= High(#0.54)
= Afternoon	= Work	= Walking	:= None(#0.29)
= Afternoon	= Work	= Still	:= None(#0.34)
= Afternoon	= Work	= Vehicle	:= None(#0.39)
= Day	= Work	= any	:= None(#0.91)
= Evening	= Work	= any	:= None(#0.91)

Table id: id_network_traffic - network_traffic

Fig. 17.8 XTT2 table generated from the uncertain decision tree

the HMR+ features for short-term adaptability. It was also localized in the *adaptable model* component of the AMVCBC architecture. However, instead of being a part of the learning middleware, it was part of the application model, as its actions directly concerned the user (i.e. presenting an application that might be of interest to the user). The fragment of this model was presented in Fig. 17.9. Both models were executed by the HEARTDROID inference engine on the mobile device. A fragment of log from the HEARTDROID inference engine running the second model was presented in Listing 17.2.

```
1  HEART: Processing table Actions (ID: null)
2  HEART: Processing rule Actions/1 (ID: null)
3  HEART: Checking conditions of rule Actions/1 (ID: null)
4  HEART: Checking condition location EQ{min 80.0% in
5      -10.0s  to 0.0} home
6  HEART WARNING: Timestamp to obtain the state value
7    is smaller
8  than the first existing state in Working Memory.
9      The result of the
10 operation may be different than desired.
11 HEART: Condition location EQ{min 80.0% in -10.0s to
12     0.0} home satisfied
13  with certainty (0.95758194).
14 HEART: Changing the rule evaluation result to be
15    true
16    with certainty (0.95758194).
```

```
17  HEART: Checking condition location EQ home
18  HEART: Condition location EQ home satisfied with
19         certainty (0.9722968).
20  HEART: Checking condition today EQ workday
21  HEART: Condition today EQ workday satisfied with
22         certainty (1.0).
23  HEART: Checking condition daytime EQ morning
24  HEART: Condition daytime EQ morning satisfied with
25         certainty (1.0).
26  HEART: Finished evaluating rule Actions/1 (ID: null).
27      SATISFIED
28    with (0.95758194) certainty.
```

Listing 17.2 Fragment of a log from the HEART inference engine

HEARTDROID was evaluated according to the requirements presented in Sect. 17.1. Seven most popular rule engines were chosen for comparison including: Jess, ContextToolkit, Drools, EasyRules, JRuleEngine, tuHeaRT and HEARTDROID. Figure 17.10 shows the differences in reasoning times among them. Red crosses represents the last successful reasoning before the time of inference exceeded a threshold set of 60 s. It can be seen that ContextToolkit and tuHeaRT have a polynomial time of execution, and they fail to complete the inference in a 60 s time slot for more than 60 and 200 rules respectively. The remaining inference engines have

(?) location	(?) location	(?) today	(?) daytime	(->) action
= {MIN 80% in -10m to 0} home	= home	= workday	= morning	:= prepareing_to_leave
= {MIN 80% in -10m to 0} home	= outside	= workday	= morning	:= leaving_home
= {MIN 80% in -10m to 0} outside	= outside	= workday	= morning	:= travelling_work
= {MIN 80% in -60m to 0} work	= work	= workday	= dayatime	:= working
= {MIN 80% in -10m to 0} work	= work	= any	= afternoon	:= prepareing_to_leave
= {MIN 80% in -10m to 0} work	= outside	= any	= afternoon	:= leaving_work
= {MIN 80% in -20m to 0} outside	= outside	= workday	= afternoon	:= travelling_home
= {MIN 80% in -60m to 0} home	= home	= any	= evening	:= resting
= {MIN 80% in -60m to 0} home	= home	= any	= night	:= sleeping
= {MIN 80% in -60m to 0} home	= home	= weekend	= any	:= resting
= {MIN 80% in -60m to 0} outside	= any	= any	= evening	:= entertaining
= {MIN 80% in -60m to 0} outside	= outside	= any	= night	:= travelling_home
= {MIN 80% in -10m to 0} work	= outside	= workday	= dayatime	:= lunch

Table id: tab_4 - Actions

(?) action	(?) transportation	(->) application
= prepareing_to_leave	= idle	:= {news,weather}
∈ {leaving_work,leaving_home}	∈ {walking,running}	:= {clock,navigation}
∈ {travelling_home,travelling_work}	∈ {driving,cycling}	:= navigation
∈ {travelling_home,travelling_work}	∈ {bus,train}	:= {news,clock}
∈ {resting,entertaining}	∈ {running,cycling}	:= {sport_tracker,weather}
= working	= any	:= {calendar,mail}
= sleeping	= idle	:= clock
∈ {resting,entertaining}	∈ {driving,bus,train}	:= trip_advisor
∈ {lunch}	= any	:= {news,calendar,restaurants}

Table id: tab_5 - Applications

Fig. 17.9 XTT2 model for intelligent personal assistant system

Fig. 17.10 Time efficiency results for the most popular rule engines

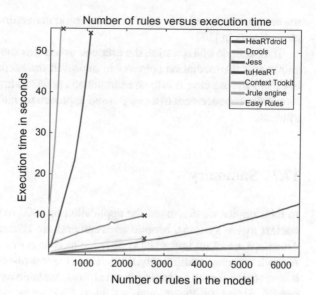

linear execution time and successfully complete reasoning for up to 2000 rules below the 10 s. Further analysis shown that the ContextToolkit and tuHeaRT are extremely sensitive to a growing number of rules that fall into different schemas. A schema is defined as a set of attributes from the conditional and decision parts of a rule. With a growing number of schemas, the number of rules that can be processed by Context-Toolkit and tuHeaRT in a 60 s time slot decreases rapidly. Furthermore, for models that contains more than 2500 rules Drools and EasyRules engines fail to compile their models, which caused inference interruption, as can be seen in Fig. 17.10. Red crosses represent the last successful reasoning before the time of inference exceeded a threshold, or a model failed to be loaded.

Although the HEARTDROID engine did not get the best score in efficiency ranking, it was shown to be the only freely available rule-based engine dedicated for mobile devices, that natively support most of the important features of mobile context-aware systems such as intelligibility and uncertainty handling (see [5]).

Context-Based Controller

The *context based controller* component was tested on the example of the indoor navigation system. The primary objective of the system was to track user location inside the building of AGH University of Science and Technology using a dead-reconning system that was solely based on the pedometer, compass and a map of the building. The tracking system was based on the particle filtering approach. However, to support the system, which very often lacked accuracy, and provided ambiguous location estimations, the mediation technique based on the question forest approach was used. Every time the tracking system provided an uncertain estimation of localization, the mediation component asked the user about the objects nearby, to narrow the possible user positions. The questions were generated based on the semantic description of the environment, and the question of the forest generation algorithm is, described in more details in [13].

The example of a question the user was asked was shown in Fig. 17.3. The results shown that the mediation component improved the accuracy of the system from 67 to 87%. This use case is also an example of a practical implementation of a feedback loop in the enhancement of a *tree-phased* approach for building mobile context-aware systems.

17.7   Summary

In this chapter we discussed the application of SKE to the development of mobile context-aware systems. Mobile environments are characterized by high dynamics. Therefore, constant update of knowledge models is one of the primary requirement for such systems. Additionally, the nature of the sensor-based systems implies that the data required for the reasoning is not always available nor certain at the time when is needed. Moreover, the amount of context data can be significant and can grow fast, constantly being processed and interpreted under soft real-time constraints. This characteristics makes these systems a challenging big data application. Furthermore, mobile platforms can impose additional constraints, e.g. related to the privacy of data, but also resource limitations, etc.

To address these challenges we discussed how the main requirements for developing such systems were addressed in the KNOWME project. The two most common causes of uncertainty in mobile environments are distinguished: uncertainty caused by the lack of knowledge (epistemic uncertainty) and by the lack of machine precision (aleatoric uncertainty). Following this classification three complementary methods were proposed for handling these types of uncertainties. We used the formal model discussed in Chap. 7. These methods are inherent parts of XTT2 rule-based knowledge modeling language. The first is based on modified certainty factors algebra, handles aleatoric uncertainty caused by the lack of machine precision. It is supported by the time-parametrised operators in XTT2 rules to handle noisy data over specified time periods These methods are applicable as long, as there are readings that come from the context providers. In the case, when the information is missing, the probabilistic interpretation of the XTT2 model can be used to reduce the epistemic uncertainty caused by the lack of knowledge.

References

1. Dey, A.K.: Providing architectural support for building context-aware applications. Ph.D. thesis, Atlanta, GA, USA, AAI9994400 (2000)
2. Dey, A.K., Abowd, G.D., Salber, D.: A conceptual framework and a toolkit for supporting the rapid prototyping of context-aware applications. Hum.-Comput. Interact. **16**(2), 97–166 (2001)
3. Pascalau, E., Nalepa, G.J., Kluza, K.: Towards a better understanding of the concept of context-aware business applications. In: Ganzha, M., Maciaszek, L.A., Paprzycki, M. (eds.) Proceedings of the Federated Conference on Computer Science and Information Systems – FedCSIS 2013, Krakow, Poland, 8–11 September 2013, pp. 959–966. IEEE (2013)
4. Nalepa, G.J., Bobek, S.: Rule-based solution for context-aware reasoning on mobile devices. Comput. Sci. Inf. Syst. **11**(1), 171–193 (2014)
5. Bobek, S.: Methods for modeling self-adaptive mobile context-aware sytems. Ph.D. thesis, AGH University of Science and Technology (2016). Supervisor: Grzegorz J. Nalepa
6. Bobek, S., Nalepa, G.J., Ślażyński, M.: Challenges for migration of rule-based reasoning engine to a mobile platform. In: Dziech, A., Czyżewski, A. (eds.) Multimedia Communications, Services and Security. Communications in Computer and Information Science, vol. 429, pp. 43–57. Springer, Berlin (2014)
7. Bobek, S., Nalepa, G.J., Ligęza, A., Adrian, W.T., Kaczor, K.: Mobile context-based framework for threat monitoring in urban environment with social threat monitor. In: Multimedia Tools and Applications (2014). https://doi.org/10.1007/s11042-014-2060-9
8. Lim, B.Y., Dey, A.K., Avrahami, D.: Why and why not explanations improve the intelligibility of context-aware intelligent systems. In: Proceedings of the SIGCHI Conference on Human Factors in Computing Systems. CHI '09, pp. 2119–2128. ACM, New York (2009)
9. Baldauf, M., Dustdar, S., Rosenberg, F.: A survey on context-aware systems. Int. J. Ad Hoc Ubiquitous Comput. **2**(4), 263–277 (2007)
10. Lim, B.Y., Dey, A.K.: Investigating intelligibility for uncertain context-aware applications. In: Proceedings of the 13th International Conference on Ubiquitous Computing. UbiComp '11, pp. 415–424. ACM, New York (2011)
11. Hunt, J.: Guide to the Unified Process featuring UML, Java and Design Patterns. Springer, Berlin (2003)
12. Domingos, P., Hulten, G.: Mining high-speed data streams. In: Proceedings of the Sixth ACM SIGKDD International Conference on Knowledge Discovery and Data Mining. KDD '00, pp. 71–80. ACM, New York (2000)
13. Köping, L., Grzegorzek, M., Deinzer, F., Bobek, S., Ślażyński, M., Nalepa, G.J.: Improving indoor localization by user feedback. In: 2015 18th International Conference on Information Fusion (Fusion), pp. 1053–1060 (2015)
14. Ferreira, D.: AWARE: a mobile context instrumentation middleware to collaboratively understand human behavior. Ph.D. thesis, University of Oulu (2013)
15. Bobek, S., Porzycki, K., Nalepa, G.J.: Learning sensors usage patterns in mobile context-aware systems. In: Ganzha, M., Maciaszek, L.A., Paprzycki, M. (eds.) Proceedings of the Federated Conference on Computer Science and Information Systems – FedCSIS 2013, Krakow, Poland, 8-11 September 2013, pp. 993–998. IEEE (2013)
16. Bobek, S., Ślażyński, M., Nalepa, G.J.: Capturing dynamics of mobile context-aware systems with rules and statistical analysis of historical data. In: Rutkowski, L., Korytkowski, M., Scherer, R., Tadeusiewicz, R., Zadeh, L.A., Zurada, J.M. (eds.) Artificial Intelligence and Soft Computing. Lecture Notes in Computer Science, vol. 9120, pp. 578–590. Springer International Publishing (2015)
17. Senge, R., Bösner, S., Dembczyński, K., Haasenritter, J., Hirsch, O., Donner-Banzhoff, N., Hüllermeier, E.: Reliable classification: learning classifiers that distinguish aleatoric and epistemic uncertainty. Inf. Sci. **255**, 16–29 (2014)
18. Parsaye, K., Chignell, M.: Expert Systems for Experts. Wiley, New York (1988)

19. Bobek, S., Nalepa, G.J.: Incomplete and uncertain data handling in context-aware rule-based systems with modified certainty factors algebra. In: Bikakis, A., Fodor, P., Roman, D. (eds.) Rules on the Web. From Theory to Applications. Lecture Notes in Computer Science, vol. 8620, pp. 157–167. Springer International Publishing (2014)
20. Nalepa, G., Bobek, S., Ligęza, A., Kaczor, K.: HalVA – rule analysis framework for XTT2 rules. In: Bassiliades, N., Governatori, G., Paschke, A. (eds.) Rule-Based Reasoning, Programming, and Applications. Lecture Notes in Computer Science, vol. 6826, pp. 337–344. Springer, Berlin (2011)
21. Bobek, S., Nalepa, G.: Compact representation of conditional probability for rule-based mobile context-aware systems. In: Bikakis, A., Fodor, P., Roman, D. (eds.) Rules on the Web. From Theory to Applications. Lecture Notes in Computer Science. Springer International Publishing (2015)
22. Quinlan, J.R.: C4.5: Programs for Machine Learning. Morgan Kaufmann Publishers Inc., San Francisco (1993)
23. Chau, M., Cheng, R., Kao, B., Ng, J.: Uncertain data mining: an example in clustering location data. In: Proceedings of the 10th Pacific-Asia Conference on Knowledge Discovery and Data Mining (PAKDD 2006). Lecture Notes in Computer Science, vol. 3918, pp. 199–204. Springer (2006)
24. Tsang, S., Kao, B., Yip, K.Y., Ho, W.S., Lee, S.D.: Decision trees for uncertain data. IEEE Trans. Knowl. Data Eng. **23**(1), 64–78 (2011)
25. Aggarwal, C.C., Yu, P.S.: A survey of uncertain data algorithms and applications. IEEE Trans. Knowl. Data Eng. **21**(5), 609–623 (2009)
26. Yuan, Y., Shaw, M.J.: Induction of fuzzy decision trees. Fuzzy Sets Syst. **69**(2), 125–139 (1995)
27. Hoeffding, W.: Probability inequalities for sums of bounded random variables. J. Am. Stat. Assoc. **58**(301), 13–30 (1963)
28. Maron, O., Moore, A.: Hoeffding races: accelerating model selection search for classification and function approximation. In: Cowan, J.D., Tesauro, G., Alspector, J. (eds.) Advances in Neural Information Processing Systems, vol. 6, pp. 59–66. Morgan Kaufmann, San Francisco (1994)
29. Bobek, S., Dziadzio, S., Jaciów, P., Ślażyński, M., Nalepa, G.J.: Understanding context with ContextViewer – tool for visualization and initial preprocessing of mobile sensors data. In: Proceedings of Modeling and Using Context: 9th International and Interdisciplinary Conference, CONTEXT 2015, Lanarca, Cyprus, November 2-6, 2015, pp. 77–90. Springer International Publishing, Cham (2015)
30. Bobek, S.: HeaRT rule inference engine in intelligent systems. PAR Pomiary Automatyka Robotyka **15**(12), 226–228 (2011). ISSN 1427-9126
31. Kiepas, P., Bobek, S., Nalepa, G.J.: Concept of rule-based configurator for Auto-WEKA using OpenML. In: Proceedings of the 2015 International Workshop on Meta-Learning and Algorithm Selection, p. 106 (2015)
32. Bobek, S., Grodzki, O., Nalepa, G.J.: Indoor microlocation with BLE beacons and incremental rule learning. In: 2015 IEEE 2nd International Conference on Cybernetics (CYBCONF), pp. 91–96 (2015)
33. Bobek, S., Nalepa, G.J.: Uncertain context data management in dynamic mobile environments. Future Gener. Comput. Syst. **66**, 110–124 (2017)
34. Zhang, L., Guan, Y.: Variance estimation over sliding windows. In: Proceedings of the Twenty-sixth ACM SIGMOD-SIGACT-SIGART Symposium on Principles of Database Systems. PODS '07, pp. 225–232. ACM, New York (2007)
35. Ester, M., peter Kriegel, H., Sander, J., Xu, X.: A density-based algorithm for discovering clusters in large spatial databases with noise. In: 2nd International Conference on Knowledge Discovery and Data Mining (KDD-96), pp. 226–231. AAAI Press (1996)
36. Bobek, S., Nalepa, G., Grodzki, O.: Automated discovery of mobile users locations with improved k-means clustering. In: Rutkowski, L., Korytkowski, M., Scherer, R., Tadeusiewicz, R., Zadeh, L.A., Zurada, J.M. (eds.) Artificial Intelligence and Soft Computing. Lecture Notes in Computer Science, vol. 9120, pp. 565–577. Springer International Publishing (2015)

Concluding Remarks

In this book, *Semantic Knowledge Engineering* is proposed as an approach for building intelligent systems based on rules. As I stated in the introduction, I wanted to
present a synthesis of the most important results of my research in the last eight years
in this area. In the book, the formal foundations are followed by practical studies.
In my opinion, these applications demonstrate the feasibility of the SKE approach. I
also believe, the rule-based technologies are still very important and useful in artificial intelligence, computer science, as well as other areas. Rules are a very basic and
intuitive, yet powerfull method for capturing our knowledge. As such, rule-based
systems are a technology that still has a lot of potential for future development.

In the book my objective was to put my research on rules in a certain context.
This is why the last part of the book is focused on the applications of rules, and the
integration of the rule-based approach with other important paradigms. I emphasize
the *semantic* aspect of my approach. By this I mean a proper interpretation of rules
and their operation in certain areas of application. I aimed at providing a generic
formal model for rules, the XTT2, which thanks to SKE can be properly interpreted
and integrated in several domains discussed in the book.

Clearly, the book has a specific perspective outlined in the introduction. It is based
on knowledge engineering and software engineering foundations. As such, there are
issues intentionally left out, and not included in the book. Today, data mining methods
based on machine learning algorithms are important and widely used techniques to
build models of knowledge from data. While I am aware of their potential, they are
out of the assumed scope of the book. They are only partially considered and included
in my work on context-aware systems.

For me this book closes a certain stage in my research. In my future work, I
am planning to shift my focus, and partially move away from classic knowledge
engineering. I plan to continue and extend my work on context-awareness, and focus
more on the applications of machine learning techniques. On the other hand, I strongly
believe, that symbolic and conceptual representation of knowledge has to be used in
many intelligent systems. Especially systems that communicate with humans need
to be understandable by them. Therefore, they should operate on a *semantic level*.

© Springer International Publishing AG 2018
G.J. Nalepa, *Modeling with Rules Using Semantic Knowledge Engineering*,
Intelligent Systems Reference Library 130,
https://doi.org/10.1007/978-3-319-66655-6

For me this book is long overdue, at least two years, or even more. I hope it serves its purpose well, that is it summarizes most of my work in the last years. Now, the book and the work can be left in peace on some shelf. Hopefully, it proves useful for someone.

In the writing of this book many people supported me. I thanked them in the introduction. During the process I also traveled a lot. Different parts of this book were written in different places around Europe, in Poland, Germany, and Spain. I also had the opportunity to discuss my results there.

Some people may notice that if the third part of the book had 11 chapters instead of 9 it would be more elegant. That way the number of chapters, and sections would be expressed only by primes. I was seriously tempted to do so. I considered to have two more chapters, summarizing my work on computer security, and cognitive science. However, I consider myself an engineer, i.e. a person that does his best to make things work well. I prefer practical quality over formal beauty. Furthermore, with time I learned that often "less is more".

Life *is* like a box of chocolates. In my case real life added an unexpected and very sad ending to my work on this book. So let me finish with a quote from Steven Wilson, whose music kept me on the surface in last months, and allowed me to finish it, instead of being finished by something else.

> *Hey brother, happy returns, it's been a while now*
> *I bet you thought that I was dead*
> *But I'm still here, nothing's changed*
>
> *Hey brother, I'd love to tell you I've been busy*
> *But that would be a lie*
> *'Cos the truth is the years just pass like trains*
> *I wave but they don't slow down*

— STEVEN WILSON

Index

© Springer International Publishing AG 2018
G.J. Nalepa, *Modeling with Rules Using Semantic Knowledge Engineering*,
Intelligent Systems Reference Library 130,
https://doi.org/10.1007/978-3-319-66655-6

Printed in the United States
By Bookmasters